NONLINEAR WAVES AND OFFSHORE STRUCTURES

ADVANCED SERIES ON OCEAN ENGINEERING

Series Editor-in-Chief
Philip L- F Liu (*Cornell University*)

Advanced Series on Ocean Engineering — Volume 27

NONLINEAR WAVES AND OFFSHORE STRUCTURES

Cheung Hun Kim

Texas A&M University, USA

World Scientific

NEW JERSEY • LONDON • SINGAPORE • BEIJING • SHANGHAI • HONG KONG • TAIPEI • CHENNAI

Published by

World Scientific Publishing Co. Pte. Ltd.

5 Toh Tuck Link, Singapore 596224

USA office: 27 Warren Street, Suite 401-402, Hackensack, NJ 07601

UK office: 57 Shelton Street, Covent Garden, London WC2H 9HE

British Library Cataloguing-in-Publication Data
A catalogue record for this book is available from the British Library.

ISBN-13 978-981-02-4884-0
ISBN-10 981-02-4884-9

ISBN-13 978-981-02-4885-7 (pbk)
ISBN-10 981-02-4885-7 (pbk)

Printed in Singapore.

Preface

This book represents a culmination of over 40 years of research in ocean wave behavior and its impact on structures. The countless hours devoted to experimentation, analysis, and reflection has yielded discoveries that have been, and always will be, rewarding. I am most blessed to have experienced this in my life. With this thought in mind, I have written this manuscript.

In short, this book discusses how the offshore structure design and analysis may employ the Volterra linear model for Gaussian seas. When the sea severity increases, the Volterra quadratic model may be employed for improved design in the weakly nonlinear seas. When the sea severity further increases from weak nonlinearity and when waves become highly nonlinear, a semi-empirical method called the universal nonlinear input-output model (UNIOM) may be employed. Interestingly, the higher order Volterra models and higher order wave and force theories are similar in their forms. Ultimately, this work provides the system concept for the advanced designs and analyses of structures.

This book may be used for the senior undergraduate, master's and Ph.D. level students.

I am grateful to my son Professor David Kim at the University of Central Arkansas, Professor Sun Hong Kwon at Pusan National University, Professor Sung Youn Boo at Korea Naval Academy, Dr. Allen Liu at FMC SOFEC in Houston and Dr. Jun Zou at Houston Offshore Engineering for their critical comments. I am indebted to Dr. Nungsoo Kim at Technip for his work in developing the figures and tables. Finally I would like to thank Mr. Seung Jae Lee for his assistance in preparing and sending the manuscript to the publisher.

Contents

Chapter 2 Fourier Transform and Wave Spectra

Chapter 3 Volterra Linear Model and Extreme Response

Division II. Linear Wave-Body Interaction

Chapter 4 Basics of Hydrodynamics and Linear Waves

Division III. Volterra Quadratic Model

Chapter 7 Volterra Quadratic Model and Cross-Bi-Spectrum

Chapter 8 Second-Order Response in Linear Wave

Chapter 9 2nd-Order Wave and 2nd-Order Force

Division IV. Universal Nonlinear Input-Output Model

Chapter 10 Volterra Cubic Model and 3rd-Order Wave and Force

Chapter 11 Highly Nonlinear Waves and UNIOM

Chapter 12   Numerical Wave Tank

Division I

Volterra Linear Model

Chapter 1

Statistics of Waves

1.1 Introduction

The theory of linear ship motion in the Gaussian seaway has been applied for the design of offshore structures in the past half century (St. Denis and Pierson, 1953). The linear theory had let us employ the popular response amplitude operator (RAO) for determining response of the structure to the random seas. Since the theory can be applied only to the Gaussian seas, the user of the theory must be able to judge if the sea considered is a Gaussian. For this purpose one needs to rely on the use of wave statistics.

Accumulated data of random seas (e.g., the North Atlantic), in terms of significant wave height H_s and zero-crossing period T_z, will be employed for the prediction of 100 year return sea.

1.1.1 Linear random wave

We measure ocean waves at a point of the ocean surface. However, no one can predict their instantaneous value at any future time. Nondeterministic data of this type are referred to as random time series. We will start with the deterministic wave that gives instantaneous wave elevation at any future time. As an example, we consider a sinusoidal progressing wave of permanent shape and small amplitude A at frequency ω with an arbitrary phase angle ε:

$$\eta(t) = A\cos(kx - \omega t + \varepsilon) \tag{1.1}$$

3

Equation (1.1) represents a time history at a fixed location x as shown in Fig. 1.1(a), and a sinusoidal wave in space at a fixed time t as illustrated in Fig. 1.1(b). The wave height H is twice of amplitude A, and the crest height H_c is identical to the trough depth H_t. T is the linear wave period and λ is the wavelength.

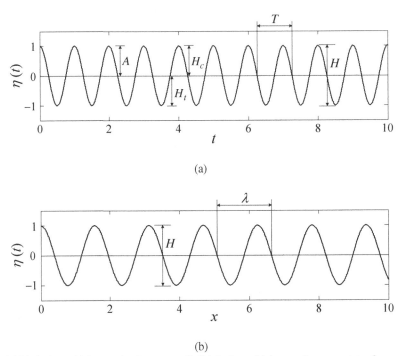

(a)

(b)

Fig. 1.1(a) A sinusoidal wave in time at $x = 0$. (b) A sinusoidal wave in space at $t = 0$.

A long-crested sea or unidirectional random wave may be represented by linear superposition of infinitely many sinusoidal wave trains with random phase angles in the form:

$$x(t) = \sum_{i=1}^{\infty} A_i \cos\left(k_i x - \omega_i t + \varepsilon_i\right) \tag{1.2}$$

where A_i is from one-sided wave energy density spectrum $U(\omega_i)$ in the form

$$A_i = \sqrt{2U(\omega_i)\Delta\omega} \qquad (1.3)$$

The two-sided energy density spectrum will be expressed by $S(\omega_i)$ hereafter. The random phase angles ε_i in Eq. (1.2) are uniformly distributed in the interval $[0, 2\pi]$, which is computed by

$$\varepsilon_i = 2\pi \times \text{rand}(n) \qquad (1.4)$$

where rand (n) provides the random number distributed in the interval $[0,1]$ with unit variance. The above procedure is called random phase method. See Sec. 2.1.4. Thus Eq. (1.2) represents a linear random wave or zero-mean Gaussian wave or long-crested wave. If we take a finite number of sinusoidal waves and assume $\Delta\omega = 2\pi/T$, the irregular wave will repeat for each period T. Thus the above numerical simulation of random wave given the wave energy spectrum is a finite data. The measured finite record is limited because one cannot predict the instantaneous wave elevation at any future time based on the finite record.

The linear random wave profile is statistically symmetric about the mean water level (MWL) of the sea surface. If the measured or theoretical wave profile is statistically asymmetric, it is a nonlinear random wave or non-Gaussian wave. Making use of Eqs. (1.2) through (1.4), one may simulate the Gaussian wave from a target energy spectrum $U(\omega)$. However, it is noted that the real waves generated in the wave tank using the same target spectrum will not necessarily produce the Gaussian (linear) wave because the real waves become nonlinear due to nonlinear wave-wave interactions during the generation stage.

1.1.2 Laboratory and field waves

Laboratory and field waves are usually non-Gaussian but one may produce Gaussian-like waves by a controlled experiment at a low sea state. At higher seas, it is certain that the waves become non-Gaussian. One needs to define the Gaussian wave using statistical tests as given in Sec. 1.2. The following is the introductory description of variety of waves.

1.1.2.1 Gaussian laboratory wave

Figure 1.2 shows a laboratory wave generated at KRISO (the Korea Research Institute of Ship and Ocean Engineering, 2000) that is approximately Gaussian. The ITTC (International Towing Tank Conference) spectrum with significant wave height of H_s = 4 m and modal period T_m of 11.26 s (where the peak energy density occurs) was used as the target. The wave shape appears to be smooth, all of the crests and troughs are almost symmetric about the vertical axes through the crest and trough points, and the elevations appear on average symmetric about MWL.

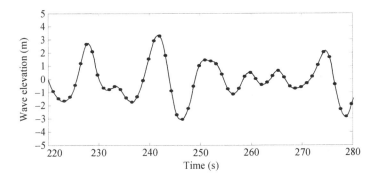

Fig. 1.2. Approximately linear random wave (courtesy of KRISO).

1.1.2.2 Non-Gaussian laboratory wave

A laboratory non-Gaussian wave is illustrated in Fig. 1.3. The target spectrum was a one hundred year return storm sea specified by JONSWAP of H_s = 15.4 m, T_m = 17.8 s, γ = 1.7, which is a replica of the wave used for the test of ringing of Heidrun TLP (tension leg platform) tendons (Stansberg et al., 1995; Statoil, 1996). This kind of wave, with the H_s being greater than 14 m, belongs to phenomenal sea (Tupper, 1996). Most crest forms are asymmetric about the vertical axes through the crest points and the elevations are also asymmetric on average about MWL axis, which will give non-zero skewness as defined in Sec. 1.2. This wave will be frequently used for illustration in the examples.

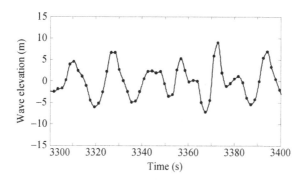

Fig. 1.3. Nonlinear laboratory random wave (courtesy of Statoil).

1.1.2.3 Hurricane Camille

The wave elevation time history of 1 hr (1500 hrs – 1600 hrs) measured during the Hurricane Camille was an analogue data. Later it was digitized at sample interval Δt of 1.0 s, while keeping the mean as shown in Fig. 1.4 (Earle, 1976). It seems that the large time step of 1.0 s might have reduced the clarity of the true nonlinear wave shape. If the sample interval were taken at 0.25 s, it might have given a truer nonlinear wave shape. The significant wave height H_s of the Camille wave was estimated 12.6 m. This wave will be utilized for illustrations of statistical parameters in the basics of statistics.

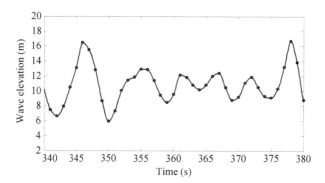

Fig. 1.4. Hurricane Camille wave (1500 hrs – 1600 hrs) (Courtesy of G. Z. Forristall).

1.1.2.4 Draupner freak wave

The freak (rogue) wave was measured at the Draupner oil rig in the North Sea off Norway on New Years Day 1995. The freak wave occurred in a zone without currents and in otherwise normal sea state. It has to be noted that the frequency of occurrence of such waves is extremely small. The maximum height of the freak wave is the highest one. If this maximum height is much greater than two times the significant wave height of the wave record, and the ratio of the crest height of the maximum height wave to significant wave height is greater than 1.25, it may be called a freak wave (Olagnon and Athanassoulis, 2000). In Fig. 1.5, we observe the crest of the freak wave being high and sharp, while the two neighboring troughs are shallow and flat. Even though the frequency of occurrence may be extremely rare, the freak wave may generate tremendous impact load on the structures as well as huge overtopping of the wave on the deck of the platform. Thus, offshore designers are highly concerned about the freak wave impacting on columns and green waters on deck. Since this kind of impact data is hardly obtainable in the field measurements, we may reconstruct freak waves in the wave tank and measure impacting loads on the model as alternative solutions (Kim and Kim, 2003).

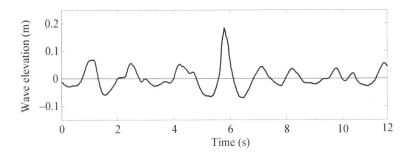

Fig. 1.5. Draupner freak wave, in 1/100 model scale. (Courtesy of S. Haver, Statoil).

1.2 Basics of Statistics

The wave elevation time history is usually measured at a point on the ocean surface or in the wave tank near the model structure, using electronic or other type of wave gauges. The foregoing continuous analogue data is digitized with an appropriate sample interval Δt and stored in the computer. The total measured time length T in the unit of second (s) is called sample length. The data is used for determination of basic statistical averages and probabilities as well as for judging the data as being a Gaussian event.

1.2.1 Mean, expected value, variance, skewness and kurtosis

The mean of a sample over T or the area under the sample divided by T may be represented by:

$$\overline{x(t)} = \langle x(t) \rangle = \lim_{T \to \infty} \frac{1}{T} \int_0^T x(t)dt = \mu_x \tag{1.5}$$

The above number is also equal to the expected value of $x(t)$ which is written as:

$$E[x] = \frac{1}{T} \int_0^T x(t)dt \tag{1.6}$$

The expected value is the average or mean value of a quantity sampled over a long time.

In the case of discrete variable:

$$E[x] = \frac{1}{N-1} \sum_{j=1}^{N} x_j = \mu_x \tag{1.7}$$

where $T = (N\text{-}1)\, \Delta t$, $x_j = x\,(j\Delta t)$.

These averaging operations can be applied to any variable such as $x^2(t)$, $x(t) \cdot y(t)$. The mean square value is defined by:

$$E[x^2] = \frac{1}{T} \int_0^T x^2(t)dt = \frac{1}{N-1} \sum_{j=1}^{N} x_j^2 \tag{1.8}$$

The mean square value about the mean μ_x is called the variance which is given by:

$$\sigma_x^2 = E[(x-\mu_x)^2] = \frac{1}{T} \int_0^T (x-\mu_x)^2\, dt = \frac{1}{N-1} \sum_{j=1}^{N} (x_j - \mu_x)^2 \tag{1.9}$$

or

$$\sigma_x^2 = E[x^2] - \mu_x^2 \tag{1.10}$$

where, σ_x is called standard deviation (std) or root mean square value (rms) of the variable or process $x(t)$ as a function of time. Thus rms value about zero-mean of the process is identical to the std of the process.

Skewness is the average of $(x - \mu_x)^3$ normalized by σ_x^3 and expressed in the form:

$$E[\frac{1}{\sigma_x^3}(x - \mu_x)^3] = \frac{1}{T}\int_0^T \frac{(x - \mu_x)^3}{\sigma_x^3} dt = \frac{1}{N-1}\sum_{j=1}^N \frac{(x_j - \mu_x)^3}{\sigma_x^3} \tag{1.11}$$

The kurtosis is the average of $(x - \mu_x)^4$ normalized by σ_x^4 and expressed in the form:

$$E[\frac{1}{\sigma_x^4}(x - \mu_x)^4] = \frac{1}{T}\int_0^T \frac{(x - \mu_x)^4}{\sigma_x^4} dt = \frac{1}{N-1}\sum_{j=1}^N \frac{(x_j - \mu_x)^4}{\sigma_x^4} \tag{1.12}$$

Example 1.1 Estimate of the averages
A zero-crossing analysis code (e.g., Krafft and Kim, 1987), developed with reference to (Tucker, 1963), is utilized to estimate a variety of averages of Hurricane Camille data (1500 hrs – 1600 hrs) as shown in Fig. 1.4, including the mean, std, variance, skewness and kurtosis. The analysis employing the foregoing formulas results in; mean = 10.834 m, std = 3.162 m, variance = 9.999 m^2, skewness = 0.289, and kurtosis = 3.187.

1.2.2 Probability of one variable

Probability of one random variable $x(t)$ is called 1st-order probability. In reference to Fig. 1.6, we define the probability of $x(t)$ lying in the band between $x(t)$ and $x(t) + dx$ during T is identical to the probability of the total sum of time dt_i lying in the band over the total sample length T:

$$dP = p(x)dx = \frac{(dt_1 + dt_2 + dt_3 + \cdots)}{T} = \frac{\sum dt}{T} \tag{1.13}$$

where $p(x)$ is the probability density. Consider that we have substituted the right hand side (RHS) of Eq. (1.13) into Eq. (1.6), and that $x(t)$ is assumed to be independent of t in this transform of variable. Then the mean value of a random variable x or random process $x(t)$ is:

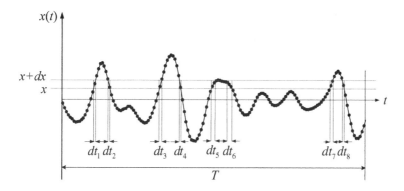

Fig. 1.6. Sketch of data points lying in band dx.

$$E[x] = \frac{1}{T}\int_0^T x(t)dt = \int_{-\infty}^{\infty} xp(x)dx \qquad (1.14)$$

Similarly the mean of x^2 and variance of the process are given by the integral of x:

$$E[x^2] = \int_{-\infty}^{\infty} x^2 p(x)dx \qquad (1.15)$$

$$\sigma_x^2 = \int_{-\infty}^{\infty} (x - \mu_x)^2 p(x)dx \qquad (1.16)$$

These formulas will be used for determining the averages of the process if the probability density distribution of the process is known as function of x instead of the time series.

It is common to measure the data electronically, hence the probability dP Eq. (1.13) is equal to the ratio of the sum of the measured data points dn_j lying in the same band dx to the total N data points recorded during T as shown in Fig. 1.6:

$$p(x)dx = \frac{\left(\sum_{dx} dn_j\right)}{N} \qquad (1.17)$$

Example 1.2 Histogram

A zero-crossing analysis of a wave time series gives random distribution of wave heights with a total of 198. To make a histogram of the data we create 5 classes with difference of unit between heights. Sorting out the data in an ascending rank, we find the number of waves belonging to each class. The number is also called occurrence in each class or range. Dividing the occurrence by the total number 198, we obtain the frequency of occurrence in percentile. Thus the total of frequency of occurrence is unity. These data are expressed in a histogram in Fig. 1.7.

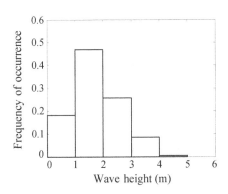

Fig. 1.7. Example of histogram of wave height.

Table 1.1. Example of histogram, frequency of occurrence of wave height distribution.

Class	Range (m)	Occurrence	Frequency of occurrence
1	0.0 ~ 1.0	36	0.18
2	1.0 ~ 2.0	93	0.47
3	2.0 ~ 3.0	51	0.26
4	3.0 ~ 4.0	17	0.09
5	4.0 ~ 5.0	1	0.01
		$\Sigma = 198$	$\Sigma = 1.00$

1.2.2.1 Gaussian distribution

The probability density function of a process is well represented by a bell shaped Gaussian (normal) equation:

$$p(x) = \frac{1}{\sqrt{2\pi}\sigma_x} \exp\left(-\frac{(x-\mu_x)^2}{2\sigma_x^2}\right), \quad -\infty \le x \le \infty \qquad (1.18)$$

The probability density in Eq. (1.18) is normalized by std σ_x:

$$p(z) = \frac{1}{\sqrt{2\pi}} \exp\left(-\frac{z^2}{2}\right), \quad z = \frac{(x-\mu_x)}{\sigma_x}, \quad -\infty \le z \le \infty \qquad (1.19)$$

The dimensional expression in Eq. (1.18) depends on the std σ_x while the non-dimensional presentation Eq. (1.19) is identical to the case of unit std of the dimensional expression Eq. (1.18).

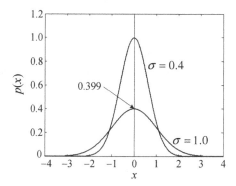

Fig. 1.8. Gaussian probability density distribution abou $\mu_x = 0$.

It is known that for the zero-mean Gaussian process, the skewness and kurtosis are zero and 3.0, respectively. If the skewness and kurtosis of the measured data deviate from the above values, the process may be called non-Gaussian, which means the wave elevation is statistically asymmetric about the zero-mean, i.e., the average of crest height is statistically higher than that of trough depth. The above criteria are frequently used for defining the Gaussianity of the process. Example 1.1 shows that the Camille wave

(1500 hrs–1600 hrs) has the skewness of 0.289 and kurtosis of 3.187. Thus the Camille wave is a non-Gaussian process.

1.2.2.2 Uniform distribution

The uniform probability density function is defined:

$$p(x) = \frac{1}{b-a}, \quad a < x < b \tag{1.20}$$

The mean and variance of the process are estimated employing Eqs. (1.15) and (1.16):

$$\mu_x = \int_a^b xp(x)dx = \frac{a+b}{2} \tag{1.21}$$

$$\sigma_x^2 = \int_a^b (x-\mu_x)^2 p(x)dx = \frac{(b-a)^2}{12} \tag{1.22}$$

The uniform distribution is used for producing the Gaussian random phase in the interval $[0, 2\pi]$. The most frequently used probability density functions are Rayleigh, lognormal and Weibull distributions which will be addressed later.

1.2.2.3 Cumulative distribution

The probability of the process x lying between $-\infty$ and x is called cumulative probability distribution function or cumulative distribution of x,

$$P(x) = \Pr\{-\infty \le x\} = \int_{-\infty}^x p(x)dx \tag{1.23}$$

Thus the probability of x between $-\infty$ and ∞

$$P(\infty) = \Pr\{-\infty \le x \le \infty\} = \int_{-\infty}^\infty p(x)dx = 1.0 \tag{1.24}$$

1.2.3 Joint probability of two variables

The 1st-order probability density function $p(x)$ of one random variable $x(t)$ specifies the probability $p(x)dx$ that a random variable $x(t)$ lies in the band dx. The 2nd-order probability density function $p(x, y)$ is defined in the same way by extending to the case of two variables x, y. The probability of the x lying in the range between $[x, x + dx]$ and the y lying in the range $[y, y + dy]$ is given by $p(x, y)dxdy$.

If both variables are sampled at an arbitrary time t_0, then the joint probability of variables of x and y lying in the differential area $dxdy$ is:

$$\Pr\{x \le x(t_0) \le x + dx \text{ and } y \le y(t_0) \le y + dy\} = p(x, y)dxdy \quad (1.25)$$

The probability of the random variables x and y lying in the finite domain is expressed by,

$$\Pr\{-x_1 \le x(t_0) \le x_2 \text{ and } - y_1 \le y(t_0) \le y_2\} = \int_{-x_1}^{x_2} \int_{-y_1}^{y_2} p(x, y)dxdy \quad (1.26)$$

The probability of the random variables x and y lying in the infinite domain:

$$\Pr\{-\infty \le x(t_0) \le \infty \text{ and } -\infty \le y(t_0) \le \infty\} = \int_{-\infty}^{\infty} \int_{-\infty}^{\infty} p(x, y)dxdy = 1 \quad (1.27)$$

Equation (1.27) indicates the volume underneath the joint probability density surface $p(x, y)$. The 2nd-order average of a function $f(x, y)$ of two random variables is similarly defined as the average of one variable x with the probability density $p(x)$ as defined in Eq. (1.14):

$$E[f(x, y)] = \int_{-\infty}^{\infty} \int_{-\infty}^{\infty} f(x, y)p(x, y)dxdy \quad (1.28)$$

which is a general formula for calculating the average of an arbitrary function $f(x, y)$. We consider finding the probability of the random variable $y(t_0)$ within the band of y to $y + dy$ independently of the value of the random variable $x(t_0)$:

$$\Pr\{y \le y(t_0) \le y + dy\} = p(y)dy \quad (1.29)$$

The same result must be obtained by using the 2nd-order probability density function $p(x, y)$ and allowing the band of values for $x(t_0)$ to extend from $-\infty$ to ∞ keeping the band dy as constant:

$$\Pr\{-\infty \le x(t_0) \le \infty, y \le y(t_0) \le y + dy\} = dy \int_{-\infty}^{\infty} p(x, y)dx \quad (1.30)$$

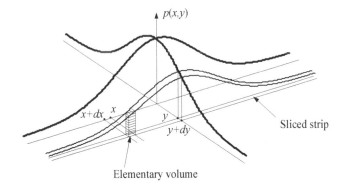

Fig. 1.9. Sketch of joint probability distributed in the band [y, y+dy] and [x, x+dx].

Equating Eqs. (1.29) and (1.30), one has:

$$p(y) = \int_{-\infty}^{\infty} p(x, y)dx \qquad (1.31)$$

which is called marginal probability density function of y.

1.2.3.1 Conditional probability density

We define a conditional probability density of x varying from $-\infty$ to ∞ for a specific y lying in the band [y, y+dy], referring to Fig. 1.9:

$$p(x \mid y)dx = \frac{p(x, y)dxdy}{dy \int_{-\infty}^{\infty} p(x, y)dx} = \frac{\text{elementary volume}}{\text{sliced strip volume}} \qquad (1.32)$$

From Eqs. (1.31) and (1.32), we have the conditional probability density of x for the given y:

$$p(x \mid y) = \frac{p(x, y)}{p(y)} \qquad (1.33)$$

If the conditional distribution of variable x is independent of variable y then it becomes function of only x:

$$p(x \mid y) = p(x) \qquad (1.34)$$

Then

$$p(x, y) = p(x)p(y) \qquad (1.35)$$

Equation (1.35) is the condition of statistical independence of variables x and y.

If x and y are statistically independent and Gaussian, the joint probability is:

$$p(x, y) = \frac{1}{(2\pi)^{1/2}\sigma_x}\exp\left[-\frac{(x-\mu_x)^2}{2\sigma_x^2}\right]\frac{1}{(2\pi)^{1/2}\sigma_y}\exp\left[-\frac{(y-\mu_y)^2}{2\sigma_y^2}\right]$$

(1.36)

If x and y are statistically dependent to each other and Gaussian:

$$p(x, y) = \frac{1}{2\pi\sigma_x\sigma_y(1-\rho_{xy}^2)^{1/2}}$$

$$\exp\left[-\frac{1}{2(1-\rho_{xy}^2)}\left\{\frac{(x-\mu_x)^2}{\sigma_x^2}+\frac{(y-\mu_y)^2}{\sigma_y^2}-\frac{2\rho_{xy}(x-\mu_x)(y-\mu_y)}{\sigma_x\sigma_y}\right\}\right]$$

(1.37)

where

$$\rho_{xy} = \frac{E[(x-\mu_x)(y-\mu_y)]}{\sigma_x\sigma_y}$$

(1.38)

ρ_{xy} is called correlation coefficient of x and y. If it is zero, x and y are statistically independent and the probability density function reduces to Eq. (1.35) or (1.36).

1.2.3.2 Joint cumulative probability distribution

The cumulative joint probability is defined in the form:

$$P(x, y) = \Pr\{-\infty \leq x, \ -\infty \leq y\} = \int_{-\infty}^{x}\int_{-\infty}^{y} p(x, y)dxdy \qquad (1.39)$$

The joint conditional cumulative probability distribution function x against a specific y is defined by:

$$P(x \mid y) = \Pr\{-\infty \leq x \mid y\} = \int_{-\infty}^{x} p(x \mid y)dx \qquad (1.40)$$

1.2.4 Ensemble averages, stationary, and ergodic

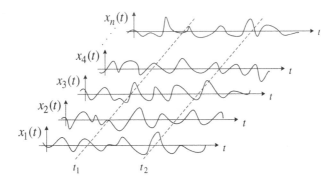

Fig. 1.10. Sketch of samples in t-direction and x-direction $[x_1(t), x_2(t), \cdots, x_n(t)]$.

Because the measured sample lengths are finite we use the ensemble averages in determining the representative population of sample values. The concept is based on the collection of sample functions: $x(t) = x_1(t) + x_2(t) + x_3(t) + \cdots$ which together make up the random process $x(t)$ consisting of a theoretically infinite number of sample functions, each of which can be thought of as resulting from a separate experiment. The collection of finite samples constitutes an approximation for the infinite ensemble of a random process.

The random process is said to be stationary if the probability distributions obtained for the ensemble $x(t)$ do not depend on absolute time. This implies that all the averages are independent of absolute time and specifically, that the mean, mean square, variance and std are independent of time altogether.

If the averages computed from each sample (data in t-direction) are the same as those of any other sample and equal to the averages of ensemble $x(t)$ (collection of data in the normal direction to t), then the random process is said to be ergodic as shown in Fig. 1.10.

We assume all the processes we deal with here are stationary and ergodic. The assumption allows us to use the averages of a sample (data in *t*-direction) as the statistical representative population of sample values, as far as the statistical averages are independent of the sample length.

1.2.4.1 Autocorrelation and spectral density

We assume the process $x(t)$ is stationary and Gaussian. Hence the statistical averages are independent of the real time length. From the sample $x(t)$ with zero-mean we may extract another sample $x(t + \tau)$ with a finite separation τ. (see Sec. 1.2.4). Then, the autocorrelation function (or auto-covariance) is formed as a time average of the product of two processes:

$$R_{xx}(\tau) = E[x(t)x(t+\tau)], \quad -\infty \le \tau \le \infty \tag{1.41}$$

where t is absolute time while τ is a separation or lag time. The above average depends on the lag time since the process is stationary and independent of t. Thus we may write:

$$\begin{aligned} R_{xx}(\tau) &= E[x(t)x(t+\tau)] \\ &= E[x(t)(x(t-\tau)] \\ &= R_{xx}(-\tau), \quad -\infty \le \tau \le \infty \end{aligned} \tag{1.42}$$

which implies that the autocorrelation is an even function of τ. When $\tau = 0$ the maximum auto correlation occurs, while it diminishes as the separation time increases. A numerical example is found in Fig. 2.4(a).

Fourier transform (FT) is valid, if $\int_{-\infty}^{\infty} |R_{xx}(t)| \, dt < \infty$ is satisfied. The energy spectral density is defined as the FT of autocorrelation function:

$$S_{xx}(\omega) = \frac{1}{2\pi} \int_{-\infty}^{\infty} R_{xx}(\tau) e^{-i\omega\tau} d\tau \tag{1.43}$$

By definition, the inverse FT of the auto energy spectral density function gives the autocorrelation function:

$$R_{xx}(\tau) = \int_{-\infty}^{\infty} S_{xx}(\omega) e^{i\omega\tau} d\omega \tag{1.44}$$

Since the auto correlation is an even function, the auto energy density spectrum is also an even function of frequency. When $\tau = 0$, $R_{xx}(\tau)$ is the maximum correlation, which is identical to the area under the two-sided auto-spectral density $S_{xx}(\omega)$ or the variance of the zero-mean process $x(t)$:

$$E\left[x^2\right] = R_{xx}(0) = \int_{-\infty}^{\infty} S_{xx}(\omega)d\omega = 2\int_{0}^{\infty} S_{xx}(\omega)d\omega \qquad (1.45)$$

It is noted that Eq. (1.45) is the result of the use of FT given in Eq. (1.43), which is FT of convention B as defined in Chap. 2. We may also use FT of convention A in other cases.

1.2.4.2 Cross-correlation and cross-spectrum

Cross-correlation functions between two different stationary random processes x (t) and y (t) are defined:

$$R_{xy}(\tau) = E[x(t)y(t+\tau)]$$
$$R_{yx}(\tau) = E[y(t)x(t+\tau)] \qquad (1.46)$$

Because the processes are stationary, it follows that:

$$R_{xy}(\tau) = E[x(t-\tau)y(t)] = R_{yx}(-\tau)$$
$$R_{yx}(\tau) = E[y(t-\tau)x(t)] = R_{xy}(-\tau) \qquad (1.47)$$

Equations (1.46) and (1.47) indicate that cross-correlations $R_{xy}(\tau)$ and $R_{yx}(\tau)$ are not the same and, unlike the autocorrelation function, they are not an even or odd function of τ.

Similarly the cross-spectral density is defined as the Fourier transform of the corresponding cross-correlation functions:

$$S_{xy}(\omega) = \frac{1}{2\pi} \int_{-\infty}^{\infty} R_{xy}(\tau)e^{-i\omega\tau}d\tau$$
$$S_{yx}(\omega) = \frac{1}{2\pi} \int_{-\infty}^{\infty} R_{yx}(\tau)e^{-i\omega\tau}d\tau \qquad (1.48)$$

And the inverse transforms of Eq. (1.48) give the cross-correlations:

$$R_{xy}(\tau) = \int_{-\infty}^{\infty} S_{xy}(\omega)e^{i\omega\tau}d\tau$$
$$R_{yx}(\tau) = \int_{-\infty}^{\infty} S_{yx}(\omega)e^{i\omega\tau}d\tau \qquad (1.49)$$

Substituting Eq. (1.47) into Eq. (1.49), we have:

$$S_{xy}(\omega) = \frac{1}{2\pi} \int_{-\infty}^{\infty} R_{yx}(-\tau)e^{-i\omega\tau}d\tau \qquad (1.50)$$

After some manipulation in Eq. (1.50), we have:

$$S_{xy}(\omega) = \frac{1}{2\pi} \int_{-\infty}^{\infty} R_{yx}(\tau) e^{i\omega\tau} d\tau \tag{1.51}$$

Comparing the second equation of Eq. (1.48) to Eq. (1.51), one has:

$$S_{xy}(\omega) = S_{yx}^{*}(\omega) \tag{1.52}$$

Equation (1.52) indicates that the cross-spectrum of x to y is identical to the conjugate of the cross-spectrum y to x.

1.2.5 Energy spectrum of velocity and acceleration

Let us derive the correlation functions and spectral densities of the velocity $\dot{x}(t)$ and acceleration $\ddot{x}(t)$ of the stationary and zero-mean Gaussian process $x(t)$. The derivative of autocorrelation function Eq. (1.42):

$$\frac{d}{d\tau} R_{xx}(\tau) = E[x(t)\dot{x}(t+\tau)] \tag{1.53}$$

Since the processes x and $\dot{x}$ are stationary, the above may be written:

$$\frac{d}{d\tau} R_{xx}(\tau) = E[x(t-\tau)\dot{x}(t)] \tag{1.54}$$

The derivative of Eq. (1.54) is given by:

$$\frac{d^2}{d\tau^2} R_{xx}(\tau) = -E[\dot{x}(t-\tau)\dot{x}(t)] = -R_{\dot{x}\dot{x}}(\tau) \tag{1.55}$$

On the other hand the 1st and 2nd derivatives of Eq. (1.44):

$$\frac{d}{d\tau} R_{xx}(\tau) = \frac{1}{2\pi} \int_{-\infty}^{\infty} i\omega S_{xx}(\omega) e^{i\omega\tau} d\omega \tag{1.56}$$

and

$$\frac{d^2}{d\tau^2} R_{xx}(\tau) = -\frac{1}{2\pi} \int_{-\infty}^{\infty} \omega^2 S_{xx}(\omega) e^{i\omega\tau} d\omega \tag{1.57}$$

Equating Eqs. (1.55) and (1.57), one has:

$$R_{\dot{x}\dot{x}}(\tau) = \frac{1}{2\pi} \int_{-\infty}^{\infty} \omega^2 S_{xx}(\omega) e^{i\omega\tau} d\omega \tag{1.58}$$

Since the velocity is assumed stationary, the auto-energy density of the velocity is similar to Eq. (1.44):

$$R_{\dot{x}\dot{x}}(\tau) = \frac{1}{2\pi} \int_{-\infty}^{\infty} S_{\dot{x}\dot{x}}(\omega) e^{i\omega\tau} d\omega \qquad (1.59)$$

From Eqs. (1.58) and (1.59):

$$S_{\dot{x}\dot{x}}(\omega) = \omega^2 S_{xx}(\omega) \qquad (1.60)$$

thus the variance of velocity is given in terms of the spectral density of the displacement x. When x is the wave elevation, the velocity is the vertical elevation velocity. Using Eqs. (1.59) and (1.60), one obtains the variance of velocity:

$$E[\dot{x}^2] = \frac{1}{2\pi} \int_{-\infty}^{\infty} S_{\dot{x}\dot{x}}(\omega) d\omega = \frac{1}{2\pi} \int_{-\infty}^{\infty} \omega^2 S_{xx}(\omega) d\omega \qquad (1.61)$$

Similarly the variance of acceleration is given in terms of the spectral density of the displacement $x(t)$:

$$E[\ddot{x}^2] = \frac{1}{2\pi} \int_{-\infty}^{\infty} \omega^4 S_{xx}(\omega) d\omega \qquad (1.62)$$

If displacement energy spectral density is known, one can determine the variances of velocity and acceleration by Eqs. (1.61) and (1.62).

1.2.5.1 The spectral moment

We define nth spectral moment of one-sided spectrum $U(\omega)$:

$$m_n = \int_0^{\infty} \omega^n U_{xx}(\omega) d\omega \qquad (1.63)$$

where m_0 stands for the area of the spectrum or the variance of zero-mean Gaussian process. The standard deviations of the displacement, velocity and acceleration, are given by, respectively:

$$\sigma_x = \sqrt{m_0}, \ \sigma_{\dot{x}} = \sqrt{m_2}, \ \sigma_{\ddot{x}} = \sqrt{m_4} \qquad (1.64)$$

1.2.5.2 Statistical independence between x and $\dot{x}$

The correlation coefficient or the normalized covariance between the zero-mean process x and velocity $\dot{x}$:

$$\rho_{x\dot{x}} = \frac{E[x\dot{x}]}{\sigma_x \sigma_{\dot{x}}} \qquad (1.65)$$

Comparing Eq. (1.54) and Eq. (1.56), one has:

$$E[x\dot{x}] = \frac{d}{d\tau} R_{xx}(\tau)\big|_{\tau=0} = \frac{1}{2\pi} \int_{-\infty}^{\infty} i\omega S_{xx}(\omega)d\omega \qquad (1.66)$$

and since the integrand in Eq. (1.66) is an odd function of ω, Eq. (1.65) vanishes:

$$\rho_{x\dot{x}} = 0 \qquad (1.67)$$

Hence if the processes of the displacement and velocity are both stationary, they are statistically independent with each other. Since the zero-mean stationary processes x and $\dot{x}$ are statistically independent and zero-mean Gaussian, the joint probability distribution of these is according to Eq. (1.36):

$$p(x,\dot{x}) = \frac{1}{2\pi\sigma_{xx}\sigma_{\dot{x}\dot{x}}} \exp\left[-\frac{1}{2}\left\{ \frac{x^2}{\sigma_{xx}^2} + \frac{\dot{x}^2}{\sigma_{\dot{x}\dot{x}}^2} \right\} \right] \qquad (1.68)$$

1.2.6 Gaussian narrow-band process

The probability of positive peaks of a process that exceeds a reference positive peak is important for statistical comparison of peak values of the process. The probability of exceedence can easily be found by counting the peaks of the time series that exceed a certain reference peak. We will digitally simulate a narrow-band zero-mean Gaussian process in Example 1.3 and determine the probability of exceedence. Rayleigh model of probability of exceedence will be derived in Sec. 1.2.6.1 and compared with experimental data.

Example 1.3 Simulation of narrow-band spectrum
Consider a narrow-band energy spectrum that is distributed in a narrow-band $\omega = 0.4\sim0.6$ rad/s with constant density 5 m$^2\cdot$s so as to ensure the process to have unit variance as shown in Fig. 1.11(a).

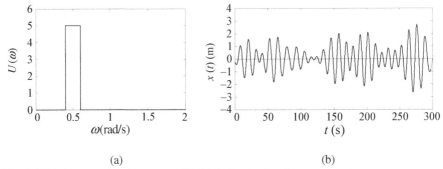

| (a) | (b) |

Fig. 1.11(a) A narrow-band spectrum. (b) Digitally simulated narrow-band process.

The narrow-band process is digitally simulated employing the random phase method Eqs. (1.2), (1.3) and (1.4). The sample length was taken $T = 1800$ s. The narrow-band process $x(t)$ in Fig. 1.11(b) has a feature similar to sinusoidal waves of slowly varying amplitudes and of nearly equal periods. The histogram of the narrow-band process agrees well with the Gaussian distribution with unit variance as shown in Fig. 1.12.

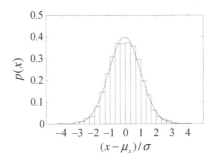

Fig. 1.12. Probability density of the narrow-band process defined in Fig. 1.11(b) is compared to the normal density distribution in Fig. 1.8.

If the sample length were much longer than the above 1800 s, it would be much closer to the theory according to the central theorem by Rice (1956). Since the distribution is Gaussian, the elevation has to be statistically symmetric about the mean level.

1.2.6.1 Probability of exceedence

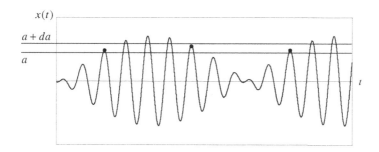

Fig. 1.13. Sketch of a distribution of peaks in the band da in crossing analysis.

Let the probability of peaks lying in the range a to $a + da$, in Fig. 1.13, be denoted by $p_p(a)\,da$, then the probability that exceeds a reference peak a may be given in the form:

$$\Pr\{\text{peak value} > a\} = \int_a^\infty p_p(a)da \qquad (1.69)$$

Assume T is the sample length, then the average number of cycles crossing $x = a$ is $N_a^+ = f_a^+ T$, where f_a^+ is the average crossing frequency of positive slope at $x = a$. Similarly the average number of cycles of zero-upcrossing at $x = 0$ is $N_0^+ = f_0^+ T$, where f_0^+ is the average zero-upcrossing frequency. Since the process is assumed to be narrow-banded (see Fig. 1.12 or 1.13) the average number of cycles of zero-upcrossing is identical to that of the total positive peaks. Hence the ratio of the number of cycles exceeding $x = a$ to that of zero-upcrossing is the probability of peaks exceeding the reference peak a:

$$\int_a^\infty p_p(a)da = \frac{N_a^+}{N_0^+} = \frac{f_a^+}{f_0^+} \qquad (1.70)$$

The average frequency of the positive slope crossing reference peak a is according to (Newland, 1984):

$$f_a^+ = \int_0^\infty p(a,\dot{x})\dot{x}d\dot{x} \tag{1.71}$$

In the above $p(a,\dot{x})$ is the joint probability density function between the random reference peak a and random positive slope $\dot{x}$. Since these are both Gaussian and represent the displacement and velocity, we may write, according to Eq. (1.68):

$$p(a,\dot{x}) = \frac{1}{(2\pi)^{1/2}\sigma_x}\exp\left[-\frac{a^2}{2\sigma_x^2}\right]\frac{1}{(2\pi)^{1/2}\sigma_{\dot{x}}}\exp\left[-\frac{\dot{x}^2}{2\sigma_{\dot{x}}^2}\right] \tag{1.72}$$

Substituting Eq. (1.72) into Eq. (1.71), we have:

$$f_a^+ = \frac{1}{(2\pi)^{1/2}\sigma_x}\exp\left[-\frac{a^2}{2\sigma_x^2}\right]\frac{1}{(2\pi)^{1/2}\sigma_{\dot{x}}}\int_0^\infty \dot{x}\exp\left[-\frac{\dot{x}^2}{2\sigma_{\dot{x}}^2}\right]d\dot{x} \tag{1.73}$$

Since $\int_0^\infty \dot{x}\exp\left[-\dfrac{\dot{x}^2}{2\sigma_{\dot{x}}^2}\right]d\dot{x} = \sigma_{\dot{x}}^2$ in the above equation, we have

$$f_a^+ = \frac{1}{2\pi}\frac{\sigma_{\dot{x}}}{\sigma_x}\exp\left[-\frac{a^2}{2\sigma_x^2}\right] \tag{1.74}$$

When $a = 0$ in Eq. (1.74), we have the average frequency of zero-upcrossing of Gaussian narrow-band process. Thus employing Eq. (1.64), one writes the average zero-upcrossing frequency in the form:

$$f_0^+ = \frac{1}{2\pi}\frac{\sigma_{\dot{x}}}{\sigma_x} = \frac{1}{2\pi}\sqrt{\frac{m_2}{m_0}} \tag{1.75}$$

Given the wave energy spectrum, we can determine the spectral moments and the average zero up-crossing period or simply, zero-crossing period of narrow-band Gaussian process:

$$T_z = \frac{1}{f_0^+} = 2\pi\sqrt{\frac{m_0}{m_2}} \tag{1.76}$$

Substituting Eqs. (1.74) and (1.75) into Eq. (1.70), we have the probability of the positive peaks exceeding reference peak a in the form:

$$\Pr\{\text{peaks} > a\} = \int_a^\infty p_p(a)da = \exp\left[-\frac{a^2}{2m_0}\right] \qquad (1.77)$$

In non-dimensional form:

$$\Pr\{\text{peaks} > z\} = \exp\left[-\frac{z^2}{2}\right], \quad z = \frac{a}{\sqrt{m_0}} \qquad (1.78)$$

Equations (1.77) and (1.78) are called the Rayleigh probability distribution exceeding reference peak a. It is valid for the narrow-band zero-mean Gaussian or linear random process. Given the time series or energy density spectrum of a process, one may estimate the variance and consequently the probability exceeding the reference peak a. The Rayleigh distribution is frequently compared with the experimental data as shown in Example 1.4.

Example 1.4 Probability of exceedence of laboratory wave
It is a common practice to count the probability exceeding the reference (measured) peak by using a computer code and compare the result with theoretical Rayleigh formula Eq. (1.77). The probability of exceedence of an experimental data is compared with Rayleigh as shown in Fig. 1.14(a). The nonlinear laboratory wave data in Fig. 1.3 exceeds Rayleigh (linear) peaks. The comparison demonstrates the nonlinearity (non-Gaussianity) of the measured wave. The Rayleigh curves are plotted as shown in Fig. 1.14(b). It shows that the probability of exceedence increases with std. When the std is unity, it represents the non-dimensional expression Eq. (1.78).

1.2.6.2 Rayleigh probability density

Differentiating the probability of exceedence Eq. (1.77) with respect to a one obtains the probability density of peak values a:

$$p(a) = \frac{a}{m_0}\exp\left[-\frac{a^2}{2m_0}\right] \qquad 0 \le a \le \infty \qquad (1.79)$$

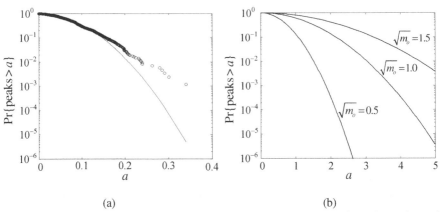

(a)   (b)

Fig. 1.14(a) Comparison of the probability exceeding reference crest height a of experimental data (model scale) given in Fig. 1.3, where $m_0 = 0.0047$ m^2, with Rayleigh probability of exceedence (solid line). (b) Rayleigh probability of exceedence with a variation of standard deviation.

Rayleigh probability density function of positive peak a of a Gaussian narrow-band process is shown in Fig. 1.15(a). It should be noted that in Eq. (1.79) one has dropped the subscript in p_p for simplicity. We consider only the positive peak values such as wave amplitudes, crest heights and wave height.

Rayleigh probability density in Fig. 1.15(a) is determined given the variance of the process. It can be shown that the maximum value of the probability density occurs when a is equal to the unit standard deviation.

Example 1.5. Probability density of crest height of narrow-band process
In Fig. 1.15(b) we compare the histogram of crest height (positive peak) of the narrow-band process as illustrated in Fig. 1.11(b), with Rayleigh density function. These are in excellent agreement as expected because the simulated data is narrow-band Gaussian. However, as will be shown later, for the broad-band process, the agreement will become poor.

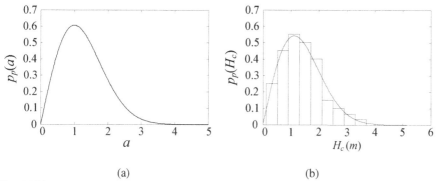

(a)   (b)

Fig. 1.15(a) Rayleigh probability density of positive peak. (b) Comparison of histogram of crest height of the simulated narrow-band process given in Fig. 1.11(b) with Rayleigh density distribution (solid line).

1.2.6.3 Rayleigh cumulative probability

Referring to Eq. (1.77), one obtains the Rayleigh cumulative probability up-to positive peak a:

$$P(a) = \Pr\{\text{peaks} \leq a\} = 1 - \exp\left[-\frac{a^2}{2m_0}\right] \qquad (1.80)$$

1.2.6.4 Average of $1/n$th highest peaks

Assume 300 peaks (amplitudes) of the wave are displayed according to the rank in the order of ascendance. Then we have the lowest peak at the beginning and the highest peak at the end. The average peak of the 1/3rd highest can be determined by dividing the sum of the 100 peak values counted from the 201st peak to the 300th peak (e.g., wave amplitude, or crest height) by 100. The formula of average peak of the $1/n$th highest is identical to the moment arm of the area distributed in the $1/n$th highest peaks of the probability density curve:

$$\bar{a}_{1/n} = \frac{\int_{a_{1/n}}^{\infty} a p(a)\,da}{\int_{a_{1/n}}^{\infty} p(a)\,da} \qquad (1.81)$$

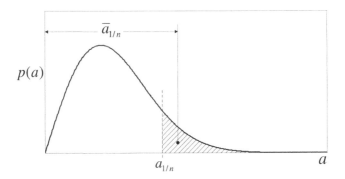

Fig. 1.16. Definition of average of the 1/nth highest peaks.

Since the probability of peaks contained in the shaded area is given by:

$$\frac{1}{n} = \int_{a_{1/n}}^{\infty} p(a)da \qquad (1.82)$$

we may write Eq. (1.81):

$$\overline{a}_{1/n} = n \int_{a_{1/n}}^{\infty} ap(a)da \qquad (1.83)$$

Average amplitudes of 1/nth highest of the Gaussian narrow-band process:

$\overline{a}_1$ = Average amplitude = $1.25\sigma_x = 1.25\sqrt{m_0}$

$\overline{a}_{1/3}$ = Average amplitude of 1/3rd highest amplitudes

$\quad$ = (significant amplitude $\overline{a}_s$) = $2.0\sqrt{m_0}$

$\overline{a}_{1/10}$ = Average amplitude of 1/10th highest amplitudes = $2.55\sqrt{m_0}$

$$(1.84)$$

1.2.6.5 The probable extreme amplitude

Referring to Eq. (1.80), we may consider the probable peak response among N observations in the following expression:

$$\frac{1}{N} = 1 - P(a) \qquad (1.85)$$

where $P(a)$ is Rayleigh cumulative probability up to positive peak a as expressed in Eq. (1.80). Thus the most probable peak amplitude in N observations is given by:

$$\hat{a}_N = \sqrt{2 \ln N} \sqrt{m_0} \qquad (1.86)$$

If we consider the risk parameter α (Ochi, 1978):

$$\hat{a}_N = \sqrt{2 \ln(N/\alpha)} \sqrt{m_0} \qquad (1.87)$$

where $\alpha = 0.01$.

The N amplitudes measured in T (hours) can be evaluated using zero-crossing period of Gaussian narrow-band process T_z in Eq. (1.76) :

$$N = \frac{3600T}{T_z} \qquad (1.88)$$

Hence:

$$\hat{a}_N = \sqrt{2 \ln \left[\frac{3600T}{2\pi\alpha} \sqrt{\frac{m_2}{m_0}} \right]} \sqrt{m_0} \qquad (1.89)$$

If the wave energy spectrum is given, one can estimate the probable extreme and extreme wave amplitude. It should be noted that we have assumed narrow-band Gaussian process.

1.2.6.6 Distribution of wave height

The transformation of probability density function of wave amplitude to wave height gives (Price and Bishop, 1974):

$$p(H) = \frac{p(a)}{\left| \dfrac{dH}{da} \right|} \qquad (1.90)$$

Since the Gaussian narrow-band wave is statistically symmetric on average about MWL, $H = 2a$, $p(H) = 1/2[\, p(a)]$. Substituting these in Eq. (1.79), we have the probability density of the wave height:

$$p(H) = \frac{H}{4m_0} \exp \left[-\frac{H^2}{8m_0} \right] \qquad (1.91)$$

In the same way, the probability of wave height exceeding a reference wave height $\hat{H}$ is derived from Eq. (1.77):

$$\Pr\{H \geq \hat{H}\} = \exp\left[-\frac{\hat{H}^2}{8m_0}\right] \tag{1.92}$$

Hence the cumulative distribution of wave height of Gaussian narrow-band process:

$$P(\hat{H}) = \int_0^{\hat{H}} p(H)\, dH = 1 - \exp\left[-\frac{\hat{H}^2}{8m_0}\right] \tag{1.93}$$

The relation between the variance or total energy and significant wave height of narrow-band Gaussian process is in accordance to Eq. (1.84):

$$m_0 = H_s^2/16 \tag{1.94}$$

which will be used in the derivation of formulas of the wave energy spectral density in Sec. 2.3.

Since the variance of a component sinusoidal wave is $a_i^2/2$, the relation between the total variance is given in terms of wave height rms of the process:

$$m_0 = \frac{1}{N}\sum_{i=1}^{N}\frac{H_i^2}{8} = \frac{H_{rms}^2}{8} \tag{1.95}$$

Substituting Eq. (1.94) into Eq. (1.91) we obtain the probability density of wave height and the probability of H exceeding a reference height $\hat{H}$ in terms of significant wave height H_s:

$$p(H) = \frac{4H}{H_s^2}\exp\left[-\frac{2H^2}{H_s^2}\right] \tag{1.96}$$

$$\Pr\{H > \hat{H}\} = \exp\left[-\frac{2\hat{H}^2}{H_s^2}\right] \tag{1.97}$$

Similarly the probability density of the wave height in Eq. (1.93) may be expressed in terms of rms wave height in accordance to Eq. (1.95):

$$p(H) = \frac{2H}{H_{rms}^2}\exp\left[-\frac{H^2}{H_{rms}^2}\right] \tag{1.98}$$

1.2.6.7 Average of 1/nth highest wave heights

The average of 1/nth highest wave heights may be derived as the average of 1/nth highest amplitudes in Eq. (1.81):

$$\bar{H}_{1/n} = \frac{\int_{H_{1/n}}^{\infty} H \cdot p(H)\, dH}{\int_{H_{1/n}}^{\infty} p(H)\, dH} \tag{1.99}$$

However, we may obtain it by simply doubling the average of 1/nth highest amplitudes $\bar{a}_{1/n}$ in Eq. (1.84), because the process is narrow-band and Gaussian:

$$\begin{aligned} \bar{H}_1 &= 2.5\sqrt{m_0} \\ \bar{H}_{1/3} &= H_s \,(\text{significant wave height}) = 4.0\sqrt{m_0} \\ \bar{H}_{1/10} &= 5.1\sqrt{m_0} \end{aligned} \tag{1.100}$$

1.2.7 Gaussian broad-band process

The relative bandwidth of the spectrum is obtained given the energy density spectrum (Price and Bishop, 1974):

$$\varepsilon = \left(1 - \frac{m_2^2}{m_0 m_4}\right)^{1/2} \qquad 0 \le \varepsilon \le 1 \tag{1.101}$$

Small ε indicates narrow banded whereas large one broad-banded spectrum.

We will investigate the broad-band process by simulation of the broad-band spectrum as shown in Example 1.6.

Example 1.6 Digital simulation of a broad-band process

We consider a broad-band spectrum of unit variance distributed in the band $\omega_{\min} = 1 \times 10^{-9}$ rad/s and $\omega_{\max} = 15.0$ rad/s as shown in Fig. 1.17(a).

Using the above broad-band spectrum Fig. 1.17(a), we can simulate the broad-band random wave time series applying the random phase method in Eqs. (1.2), (1.3), and (1.4) as shown in Fig. 1.17(b). The broad-band process is very different from the narrow-band as shown in Fig. 1.11. However, the

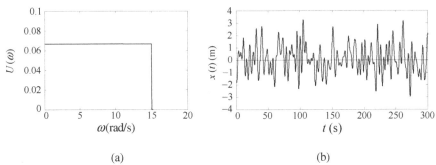

(a)   (b)

Fig. 1.17(a) A broad-band spectrum with unit variance. (b) A simulated broad-band process derived from the spectrum in Fig. 1.17(a).

histogram of the wave elevation is in excellent agreement with the Gaussian distribution as shown in Fig. 1.18(a). This fact demonstrates the broad-band process is Gaussian. Hence the broad-band wave is Gaussian and the elevation is on average statistically symmetric about zero mean level.

We also compare the histogram of the crest height H_c (positive peak) of the foregoing broad-band process with Rayleigh distribution Eq. (1.79) as shown in Fig. 1.18(b), which demonstrates that the distribution of the crest height H_c of the broad-band process does not agree with narrow-band Rayleigh distribution as expected.

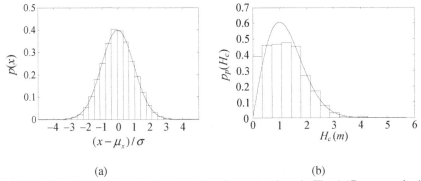

(a)   (b)

Fig. 1.18(a) Probability density of the broad-band process given in Fig. 1.17 compared with the normal probability density distribution. (b) Crest height distribution of the broad-band process in Fig. 1.17 compared with Rayleigh distribution (solid line).

1.2.7.1 Alternative estimation of spectral bandwidth

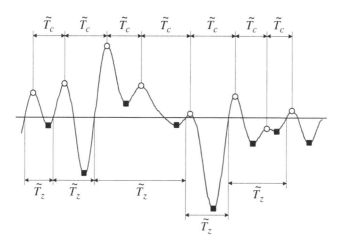

Fig. 1.19. Sketch of all maxima and the local zero up-crossing and crest-to-crest periods maxima ($\circ$), minima ($\blacksquare$).

The spectral bandwidth ε can be estimated from the time series by paying attention on the maxima. All maxima are all the crests above and below the mean water level. Thus all maxima have positive and negative values. Similarly all minima (all troughs) have also positive or negative values. See Fig. 1.19. From the above data, we may estimate the average crest-to-crest period and average zero-crossing period. The average period of all maxima or the average of crest-to-crest period T_c (Cartwright and Longuet-Higgins, 1956):

$$T_c = 2\pi\sqrt{\frac{m_2}{m_4}} \tag{1.102}$$

Employing zero-crossing period Eq. (1.76) of the narrow-band process and average crest-to-crest period of broad-band process, one may represent the relative bandwidth in terms of T_c and T_z :

$$\varepsilon = \left\{1 - \left(\frac{T_c}{T_z}\right)^2\right\}^{1/2} \tag{1.103}$$

If the process is extremely narrow-band, $\varepsilon \to 0$, then $T_c \to T_z$ indicating all waves have nearly equal periods. In other words, there are no multiple peaks in any excursions of the process. If $T_c \ll T_z$, then $\varepsilon \to 1$. The spectrum is in this case very broad, where many high frequency short waves appear. Using zero-crossing analysis, we frequently determine T_c, T_z and ε. However, the estimated value of ε is usually not identical to that given in Eq. (1.103).

1.2.7.2 Probability density of all maxima

The probability density ξ of all maxima of the Gaussian broad-band and zero-mean process $x(t)$ (Cartwright and Longuet-Higgins, 1956):

$$p(\xi) = \frac{1}{(2\pi m_0)^{1/2}} [\varepsilon \exp(-\frac{1}{2}\frac{\xi^2}{m_0}\frac{1}{\varepsilon^2}) +$$

$$(1-\varepsilon^2)^{1/2} \frac{\xi}{m_0^{1/2}} \exp(-\frac{1}{2}\frac{\xi^2}{m_0}) \int_{-\infty}^{\frac{\xi}{m_0^{1/2}}\frac{(1-\varepsilon^2)^{1/2}}{\varepsilon}} \exp(-\frac{1}{2}x^2)dx] \tag{1.104}$$

If one considers the condition of extremely narrow-band process $\varepsilon \to 0$ in Eq. (1.104), the probability density of all maxima becomes Rayleigh density distribution of positive peaks as given in Eq. (1.79):

$$p(\xi) = \frac{\xi}{m_0} \exp(-\frac{\xi^2}{2m_0}) \tag{1.105}$$

For extremely broad-band process $\varepsilon \to 1$, the probability density of all maxima is reduced to Gaussian distribution of wave elevation Eq. (1.18):

$$p(\xi) = \frac{1}{(2\pi m_0)^{1/2}} \exp(-\frac{\xi^2}{2m_0}) \tag{1.106}$$

Non-dimensionalizing all maxima ξ by std $m_0^{1/2}$; $\eta = \xi / m_0^{1/2}$:

$$p(\eta) = m_0^{1/2} p(\xi) = \frac{1}{(2\pi)^{1/2}} [\varepsilon \exp(-\frac{1}{2}\eta^2\frac{1}{\varepsilon^2})$$

$$+ (1-\varepsilon^2)^{1/2}\eta\exp(-\frac{1}{2}\eta^2) \int_{-\infty}^{\frac{(1-\varepsilon^2)^{1/2}}{\varepsilon}\eta} \exp(-\frac{1}{2}x^2)dx] \tag{1.107}$$

The probability density of non-dimensional of all maxima Eq. (1.107) as a function of the relative bandwidth ε is shown in Fig. 1.20(a).

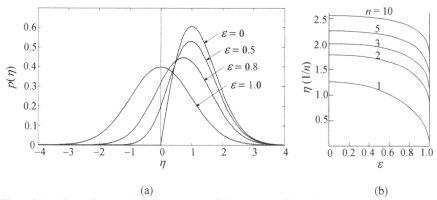

(a)   (b)

Fig. 1.20(a) Normalized probability density of all maxima affected by the spectral bandwidth. (b) Average of 1/nth highest maxima versus ε (Cartwright and Longuet-Higgins, 1956).

From the probability density function of all maxima in Eq. (1.107), we obtain the average of 1/nth highest maxima that depends on ε, for instance, $\eta^{(1)}$ = average of all maxima, $\eta^{(1/3)}$ = average of 1/3rd highest maxima and $\eta^{(1/10)}$ average of 1/10th highest maxima as shown in Fig. 1.20(b). The bandwidth of the process in ship responses is usually less than 0.6. The error due to the use of Rayleigh ($\varepsilon = 0$) against 1/nth highest maxima is about 10 % as presented in Fig. 1.20(b). Thus one may use the Rayleigh formula for practical approximation.

Example 1.7 Probability density distribution of the simulated broad-band process

We will investigate the probability density of the simulated broad-band process as shown in Example 1.6. The histogram of all maxima of the simulated broad-band process of $\varepsilon = 0.669$ in Fig. 1.17(b) is compared with the theoretical probability density of all maxima Eq. (1.107) as shown in Fig. 1.21(a). It is noted that the density distribution of all maxima is asymmetric about ξ/σ = zero as the theory indicates in Fig. 1.20(a).

The probability density of all maxima is asymmetric about $\xi/\sigma = 0$ as presented in Fig. 1.21(a). It should be noted that the probability density of the wave elevation time series as presented in Fig. 1.18(a) is different from the probability density of all maxima.

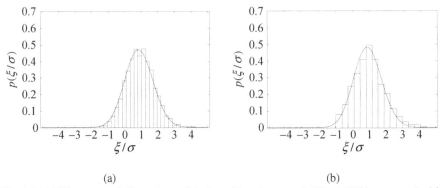

(a)   (b)

Fig. 1.21(a) Histogram of all maxima of the broad-band process in Fig. 1.17(b) compared with theoretical probability density of all maxima of broad-band process of ε=0.669, in Eq. (1.107). (b) Histogram of all maxima of Hurricane Camille wave (1500 hrs–1600 hrs) compared with the theoretical probability density distribution of broad-band process in Eq. (1.107) ($\varepsilon = 0.64$).

Example 1.8 Distribution of all maxima of Camille wave

Figure 1.21(b) illustrates a comparison of the histogram of all maxima of the Camille wave (1500 hrs – 1600 hrs) with the theoretical distribution Eq. (1.107). They are in excellent agreement. Hence the Camille wave is regarded as a broad band process.

1.3 Joint Distribution of Wave Height and Period

The wave height and period of a given sea are jointly distributed. Longuet-Higgins (1983) derived the theoretical formula of joint distribution of wave height and period given the sea spectrum.

The joint distribution of the significant wave height H_s and zero-crossing period T_z of the seas in a specified region of the North Sea accumulated over past years will be used for determining the significant wave height of 100 year return sea. Using these two parameters we will determine the Bretschneider spectrum (B-spectrum) of 100 year return sea.

1.3.1 Joint distribution of wave height and period

For the design and operation of offshore structures, consideration of wave height alone is not sufficient; instead, probabilistic information on wave height together with associated period is extremely important. To elaborate, let us consider the heaving motion of a floating offshore structure in a seaway. The heaving motion is induced by waves, and hence a wave with large height results in a large heaving motion, in general. However, the response characteristic of the structure's motion in a seaway is also dependent on the wave frequency. If the wave frequency (and thereby wavelength) is either sufficiently large or small in comparison with the natural heaving frequency of the structure, the heaving motion may not be serious even though the wave height is large. On the other hand, violent heaving motion may take place under the resonance condition in which the wave frequency is very close to the natural heaving frequency of the structure. Thus the probabilistic knowledge of wave height and associated period will provide information vital to the safe operation and design of the offshore structure.

For the foregoing problem we have to consider the joint probability distribution of wave height and period (Longuet-Higgins, 1983). The theory is created assuming Gaussian narrow-band process but introducing a broadness factor v to take into account of slightly broad-band process. Here H and T are random variables:

$$p(H,T) = \frac{(2\pi)^{1/2}}{8} \frac{2}{m_0^{1/2} m_1} \frac{1}{v} \left(1 + \frac{v^2}{4}\right) \frac{H^2}{T^2}$$

$$\cdot \exp\{(-\frac{H^2}{8m_0})[1 + \left(1 - \frac{2\pi m_0}{m_1 T}\right)^2 \frac{1}{v^2}]\}$$

(1.108)

$$v = \left(\frac{m_2 m_0}{m_1^2} - 1\right)^{1/2}$$

(1.109)

The maximum joint probability density:

$$p_{\max}(H,T) = \frac{1}{e(2\pi)^{3/2}} \left(\frac{4 + v^2}{4v}\right) \frac{m_2}{m_0^{1/2} m_1}$$

(1.110)

occurs at

$$H = 2(2)^{1/2} \left(\frac{m_1}{m_2^{1/2}}\right)$$

(1.111)

and

$$T = 2\pi \sqrt{\frac{m_1}{m_2}} \qquad (1.112)$$

The marginal probability density function of period T:

$$p(T) = \frac{\pi m_0}{m_1} \frac{1}{v} \left(1 + \frac{v^2}{4}\right) \frac{1}{T^2} \left[1 + \left(1 - \frac{2\pi m_0}{m_1 T}\right)^2 \frac{1}{v^2}\right]^{-3/2} \qquad (1.113)$$

The marginal probability density function of H agrees with Rayleigh distribution when the broadness factor v is zero.

Example 1.9 Joint distribution of the wave height and period of Camille

The wave energy spectrum of the Camille (1500 hrs – 1600 hrs) is estimated as shown in Fig. 2.3 and Fig. 2.4(b) in Chap. 2, from which we have the spectral moments $m_0 = 9.99$ m^2, $m_1 = 6.16$ m^2/s, and $m_2 = 4.46$ (m/s)^2, and the broadness factor $v = 0.42$. Substituting these in Eqs. (1.111) and (1.112), one obtains the wave height and period $H = 8.25$ m and $T = 7.4$ s at which the maximum probability density occurs as shown in Fig. 1.22(a).

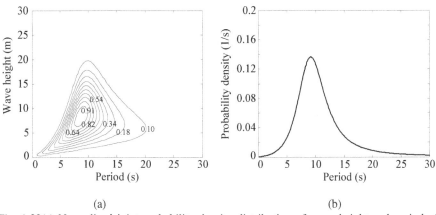

(a)   (b)

Fig. 1.22(a) Normalized joint probability density distribution of wave height and period of Camille wave (1500 – 1600). (b) Probability density of wave period irrespective of the wave height of Camille wave.

The period irrespective of the wave height or marginal wave period given in Eq. (1.113), at which the maximum probability density of period T occurs is 9.0 s, as shown in Fig. 1.22(b). It is to be noticed that these two periods 9 s and 7.4 s are quite different from each other.

Suppose we design a floating structure in the Camille wave of a certain height for instance 7 m. Then we have to consider to detune the heaving resonance period from that of the peak density of the joint distribution as shown in Fig. 1.22(a).

Joint distributions of height and period of ocean waves were observed by Chakrabarti and Cooley (1977) and Goda (1978).

1.3.2 Joint distribution of H_s and T_z in the North Atlantic

Analysis of an adequately long wave time series gives the wave statistics such as the significant wave height and zero-crossing period of the seas. An accumulated data of such parameters H_s and T_z are listed in Table 1.2, which may be used for estimating the extreme significant wave height and associated zero-crossing period of 100 year return sea in the same specified region. For this purpose we need a model for the probability density distribution. For instance the Weibull or lognormal distribution may be used for such extreme wave statistics. A reference is made to Ochi (1978).

1.3.2.1 Lognormal distribution of H_s and T_z

The joint distribution of significant wave height H_s and zero-crossing period T_z measured at weather station I located at 19° longitude west and 59° latitude north in the North Atlantic is given Table 1.2. The data will be used for determining histograms of H_s irrespective of T_z, and vice versa. For convenience sake, we may use a generalized variable x for H_s or T_z and consider the lognormal probability density function of x:

$$p(x) = \frac{1}{\sqrt{2\pi}\sigma_{\ln x}} \frac{1}{x} \exp\left[-\frac{1}{2}\left(\frac{\ln x - \mu_{\ln x}}{\sigma_{\ln x}} \right)^2 \right], \quad 0 < x < \infty \qquad (1.114)$$

Table 1.2. H_s (ft) and T_z (s) of Weather Station I (Draper and Squire, 1967) (Ochi, 1978).

		Zero crossing wave period															
		6.0	6.5	7.0	7.5	8.0	8.5	9.0	9.5	10.0	10.5	11.0	11.5	12.0	12.5	13.0	13.5	
		ι	ι	ι	ι	ι	ι	ι	ι	ι	ι	ι	ι	ι	ι	ι	ι	sec
		6.5	7.0	7.5	8.0	8.5	9.0	9.5	10.0	10.5	11.0	11.5	12.0	12.5	13.0	13.5	14.0	
	ft*																	
	0~2	0	1	0	0	2	0	1	2	0	1	0	1	1	0	0	0	9
	2~4	3	4	3	12	17	16	12	5	1	0	0	0	0	0	0	0	73
	4~6	0	4	7	24	39	24	24	12	6	1	1	0	0	0	0	0	142
	6~8	0	3	5	15	39	23	42	20	13	5	2	0	0	0	0	0	167
	8~10	1	1	2	12	21	26	39	24	12	8	3	1	1	0	0	0	151
	10~12	0	2	2	4	13	18	24	22	16	8	5	3	1	0	0	0	118
	12~14	0	0	0	2	8	8	19	14	15	9	5	4	2	1	0	0	87
Significant wave height	14~16	0	0	0	1	4	4	13	9	10	10	8	4	3	1	0	1	68
	16~18	0	0	1	0	2	4	14	7	10	6	4	1	3	1	1	1	55
	18~20	0	0	0	0	1	2	6	5	9	3	6	1	2	1	0	0	36
	20~22	0	0	0	0	1	0	1	1	5	5	4	1	1	0	0	0	19
	22~24	0	0	0	1	0	0	1	3	4	3	3	1	0	2	0	0	18
	24~26	0	0	0	0	0	0	0	1	2	4	7	1	1	0	0	1	17
	26~28	0	0	0	0	0	0	0	0	1	0	4	1	1	1	0	0	8
	28~30	0	0	0	0	0	0	0	1	1	1	2	2	2	1	0	1	11
	30~32	0	0	0	0	0	0	0	0	1	2	0	1	0	0	1	0	5
	32~34	0	0	0	0	0	0	0	0	0	0	2	1	1	0	0	0	4
	34~36	0	0	0	0	0	0	0	0	0	1	1	1	1	1	0	1	6
	36~38	0	0	0	0	0	0	0	0	0	1	0	0	1	0	0	0	2
	38~40	0	0	0	0	0	0	0	0	0	0	0	1	1	0	0	0	2
	40~42	0	0	0	0	0	0	0	0	0	0	0	0	0	0	1	0	1
	42~44	0	0	0	0	0	0	0	0	0	0	0	0	0	0	0	0	0
	44~46	0	0	0	0	0	0	0	0	0	0	0	0	0	0	0	0	0
	46~48	0	0	0	0	0	0	0	0	0	0	0	0	0	0	1	1	1
		4	15	20	71	147	125	196	126	106	68	57	25	22	9	3	6	1000

*BG unit for length is used due to the limited space of the paper.

where $\mu_{\ln x}$ and $\sigma_{\ln x}$ are the mean and std of $\ln x$ respectively. Substitution of $p(x)$ in Eq. (1.114) into Eqs. (1.14) and (1.16) results in the mean and variance of the process:

$$\mu_x = \int_0^\infty x p(x)\, dx = \exp\left[\mu_{\ln x} + \frac{\sigma_{\ln x}^2}{2}\right] \qquad (1.115)$$

$$\sigma_x^2 = \int_0^\infty (x-\mu_x)^2 p(x)dx = \left[\exp(\sigma_{\ln x}^2)-1\right]\exp\left[2\mu_{\ln x}+\sigma_{\ln x}^2\right] \quad (1.116)$$

On the other hand, from the data in Table 1.2, we can construct the histograms of x $(= H_s)$ irrespective of T_z and vice versa as given in Fig. 1.23(a) and Fig. 1.23(b). Then we can estimate the mean and variance of x $(= H_s)$ from the histogram. Substituting the mean and variance into left hand sides of Eqs. (1.115) and (1.116), we have simultaneous equations of unknowns $\mu_{\ln x}$ and $\sigma_{\ln x}$. These unknowns are determined by solving the equations simultaneously. Then we substitute the knowns $\mu_{\ln x}$ and $\sigma_{\ln x}$ into Eq. (1.114) that results in the lognormal probability density distribution as shown in Figs. 1.23(a) and 1.23(b). The lognormal distributions are in excellent agreement with the histograms. Similar work is carried out with the Weibull distribution. However, the model is in poor agreement compared to the lognormal as shown in Fig. 1.23(a).

1.3.2.2 Weibull distribution

The Weibull probability density function may be given in the form:

$$p(x) = c\lambda^c x^{c-1} e^{-(\lambda x)^c} \quad 0 < x < \infty \quad (1.117)$$

The mean μ_x and variance σ_x^2 of the variable x are given in the form:

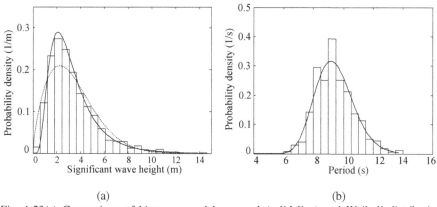

(a)   (b)

Fig. 1.23(a) Comparison of histogram and lognormal (solid line) and Weibull distribution (dotted line) of H_s irrespective of T_z. (b) Comparison of histogram and lognormal distribution of T_z irrespective of H_s.

$$\mu_x = \frac{1}{\lambda}\Gamma\left(1+\frac{1}{c}\right) \tag{1.118}$$

$$\sigma_x^2 = \frac{1}{\lambda^2}\left\{\Gamma\left(1+\frac{2}{c}\right)-\Gamma^2\left(1+\frac{1}{c}\right)\right\} \tag{1.119}$$

where Γ represents Gamma function.

Because μ_x and σ_x^2 are available from the histogram of the sample as shown in Fig. 1.23, the two unknown parameters λ and c in Eq. (1.117) are determined by solving Eqs. (1.118) and (1.119) simultaneously. Substituting the above two parameters in Eq. (1.117), one determines the Weibull probability density distribution.

1.3.2.3 Conditional lognormal distribution

The joint lognormal distribution of two variables x (= H_s) and y (= T_z) is similar to the joint distribution of statistically dependent Gaussian processes x and y as given in Eq. (1.37):

$$p(x,y) = \frac{1}{2\pi\sigma_{\ln x}\sigma_{\ln y}(1-\rho_{xy}^2)^{1/2}}\frac{1}{xy}$$
$$\cdot \exp\left[-\frac{1}{2(1-\rho_{xy}^2)}\left(\begin{array}{c}\dfrac{(\ln x-\mu_{\ln x})^2}{\sigma_{\ln x}^2}+\dfrac{(\ln y-\mu_{\ln y})^2}{\sigma_{\ln y}^2}\\[2mm]-\dfrac{2\rho_{xy}(\ln x-\mu_{\ln x})(\ln y-\mu_{\ln y})}{\sigma_{\ln x}\sigma_{\ln y}}\end{array}\right)\right] \tag{1.120}$$

The joint distribution of x and y is represented in terms of five parameters, $\mu_{\ln x}$, $\mu_{\ln y}$, $\sigma_{\ln x}$, $\sigma_{\ln y}$, ρ_{xy}. The conditional probability lognormal density distribution of the zero-crossing period y against significant wave height x:

$$p(y\,|\,x) = \frac{1}{\sqrt{2\pi}\,y(1-\rho_{xy}^2)^{1/2}\sigma_{\ln y}}\exp\left[-\frac{1}{2}\frac{(\ln y-\mu_{\ln y|\ln x})^2}{(1-\rho_{xy}^2)\sigma_{\ln y}^2}\right] \tag{1.121}$$

where

$$\mu_{\ln y|\ln x} = \mu_{\ln y}+\rho_{xy}\frac{\sigma_{\ln y}}{\sigma_{\ln x}}(\ln x-\mu_{\ln x}) \tag{1.122}$$

$$\rho_{xy} = \frac{E[(x - \mu_x)(y - \mu_y)]}{\sigma_x \sigma_y} \qquad (1.123)$$

It is evident, from Eqs. (1.121) through (1.123), that the conditional lognormal distribution of zero-crossing period y for a specific significant wave height x is given in terms of the five parameters μ_{lnx}, μ_{lny}, σ_{lnx}, σ_{lny}, ρ_{xy}. Since we have already determined the two sets of four parameters μ_{lnx}, μ_{lny}, σ_{lnx}, σ_{lny} and μ_x, μ_y, σ_x, σ_y in Sec. 1.3.2.1, the remaining term that needs to be determined is the numerator of the cross correlation coefficient ρ_{xy} in Eq. (1.123):

$$E[(x - \mu_x)(y - \mu_y)] = \sum_{j=1}^{L}\sum_{i=1}^{K}(x_i - \mu_x)(y_j - \mu_y)p(x_i, y_j)\Delta x_i \Delta y_j \qquad (1.124)$$

Equation (1.124) is computed using the matrix of joint distribution of x and y in Table 1.2. Since we computed all the five parameters including the cross correlation coefficient, we can now determine the conditional lognormal distribution of zero crossing period y $(= T_z)$ for the two specific significant wave heights x $(= H_s)$ of 3.35 m and 10.7 m using Eq. (1.121), which are indicated in solid lines, in Fig. 1.24(a) and Fig. 1.24(b).

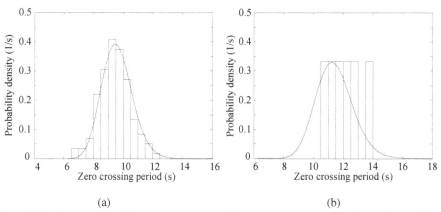

(a)   (b)

Fig. 1.24(a) Comparison of histogram and conditional lognormal distribution of T_z for H_s of 3.35 m (Table 1.2). (b) Comparison of histogram and conditional lognormal distribution of T_z for H_s 10.7 m (Table 1.2).

Figure 1.24(a) illustrates a comparison of the histogram of T_z for H_s of 3.35 m, to the conditional lognormal distribution of T_z for the same H_s, which is in good agreement. However, if H_s is 10.7 m, as shown in Fig. 1.24(b), the comparison is poor. This is due to the sparse data as shown in the Table 1.2.

1.3.3 100 year return sea

1.3.3.1 Significant wave height

For the design of offshore structure, it is essential to estimate the severest sea condition in terms of significant wave height that is expected in the lifetime of the structure. It is usual to take 20 – 25 years for ocean-going ships and 50 – 100 years for offshore production structures. This kind of work is carried out from the accumulated measured wave data in certain region. For example, the data in Table 1.3 shows the frequency of occurrence of significant wave height H_s and zero-crossing period T_z obtained over 3 years period in the North Sea near 2° longitude east and between 53° and 54° north latitude (Bouws, 1978). From the data, we construct the histogram, which agrees with the lognormal distribution (solid line) as illustrated in Fig. 1.25. The number of total H_s in 3 years is 5302 (see Table 1.3).

Table 1.3. H_s (m) and T_z (s) (Bowus, 1978) (Ochi and Whalen, 1980).

		Wave height (m)																				
		0.0	0.5	1.0	1.5	2.0	2.5	3.0	3.5	4.0	4.5	5.0	5.5	6.0	6.5	7.0	7.5	8.0	8.5	9.0	> 9.0	
	0	0	0	0	0	0	0	0	0	0	0	0	0	0	0	0	0	0	0	0	0	0
	1	0	0	0	0	0	0	0	0	0	0	0	0	0	0	0	0	0	0	0	0	0
	2	0	0	0	0	0	0	0	0	0	0	0	0	0	0	0	0	0	0	0	0	0
	3	2	389	36	0	0	0	0	0	0	0	0	0	0	0	0	0	0	0	0	0	427
	4	16	699	946	305	12	1	0	0	0	0	0	0	0	0	0	0	0	0	0	0	1979
	5	0	126	474	635	456	185	10	1	0	0	0	0	0	0	0	0	0	0	0	0	1887
(s)	6	0	21	50	101	138	202	153	65	16	0	0	0	0	0	0	0	0	0	0	0	746
period	7	0	5	3	7	9	11	29	44	47	33	12	2	1	0	0	0	0	0	0	0	203
Wave	8	0	1	1	2	1	0	0	5	0	6	6	19	6	4	1	0	0	0	0	0	52
	9	0	1	0	1	0	0	0	0	0	0	0	0	0	4	1	1	0	0	0	0	8
	10	0	0	0	0	0	0	0	0	0	0	0	0	0	0	0	0	0	0	0	0	0
	11	0	0	0	0	0	0	0	0	0	0	0	0	0	0	0	0	0	0	0	0	0
	12	0	0	0	0	0	0	0	0	0	0	0	0	0	0	0	0	0	0	0	0	0
	13	0	0	0	0	0	0	0	0	0	0	0	0	0	0	0	0	0	0	0	0	0
	14	0	0	0	0	0	0	0	0	0	0	0	0	0	0	0	0	0	0	0	0	0
	15	0	0	0	0	0	0	0	0	0	0	0	0	0	0	0	0	0	0	0	0	0
	> 15	0	0	0	0	0	0	0	0	0	0	0	0	0	0	0	0	0	0	0	0	0
		18	1242	1510	1051	616	399	192	115	63	39	18	21	7	8	2	1	0	0	0	0	5302

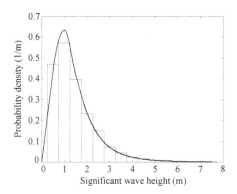

Fig. 1.25. Comparison of histogram and lognormal distribution (solid line) of H_s irrespective of T_z.

Estimation of H_s (x) of 100 year return sea can be made only by extending its cumulative distribution function such that the probable extreme significant wave height can be estimated as the value of x (H_s) which satisfies the following relationship (Sec. 1.2.6.5):

$$\frac{1}{N} = 1 - P(x) \qquad (1.125)$$

or

$$N = \frac{1}{\left[1 - P(x)\right]} \qquad (1.126)$$

where, $P(x)$ is the cumulative distribution of significant wave height H_s and N is the unknown number of significant wave height expected in, for instance, 100 years. N is frequently called return period. The cumulative distribution function $P(x)$ of significant wave height H_s can be obtained by integrating the histogram or the lognormal density distribution of H_s in Fig. 1.25. The cumulative distribution of significant wave height plotted on lognormal probability paper is shown in Fig. 1.26. However, the cumulative distribution from histogram is not satisfactorily represented over the entire range of significant wave height H_s by the lognormal probability distribution (straight line). In fact the theory and data agree up to 0.99.

For a precise estimation, it may be desirable to extend the mathematical cumulative distribution function representing the data over the entire range of the values. One method of achieving a precise representation of the data is to express the cumulative distribution in the following form (Ochi and Whalen, 1980):

$$P(x) = 1 - \exp[-q(x)] \qquad (1.127)$$

where

$$q(x) = ax^m \exp\left[-px^k\right] \qquad (1.128)$$

The parameters, a, m, p, and k may be evaluated numerically from the field data by applying nonlinear least square or curve-fitting method. Taking logarithms of Eq. (1.127):

$$G(x) = \ln[q(x)] = \ln[-\ln(1 - P(x))] = \ln a + m \ln x - px^k \qquad (1.129)$$

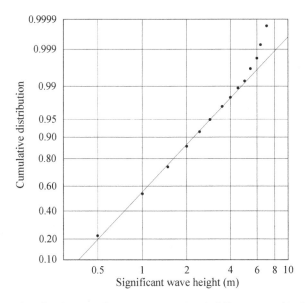

Fig. 1.26. Cumulative distribution of H_s on lognormal probability paper (dot: from data; solid line from lognormal distribution curve in Fig. 1.25).

The four unknown parameters are determined given the data of x and $G(x)$ by a curve-fitting method, i.e., $a = 0.8388$, $k = -1.664$, $m = 1.146$, $p = 0.1104$. These values are slightly different, from Ochi and Whalen (1980), though negligibly small. In the above we have determined the cumulative function $P(x)$, which will be used for determining the logarithm of the inverse of $[1-P(x)]$ as shown in Fig. 1.27:

$$\ln(N) = \ln\left[\frac{1}{1 - P(x)}\right] \tag{1.130}$$

which is necessary for estimation of the extreme value of $x = H_s$ for given years of life time.

The return period gives the number of significant wave height N in 50 and or 100 years. The number of total H_s in 3 years was 5302 (see Table 1.3). But since we have counted 18 less due to the choice of class interval, the total number of H_s is estimated as 5284 during three years. Therefore, the total number of H_s during 50 years will be $N = 5284 \times 50/3 = 88067$. Thus we have ln (88067) = 11.39. We enter the line with the foregoing value, then we find H_s of 9.8 m. Similarly the extreme significant wave height during 100 years is 10.30 m.

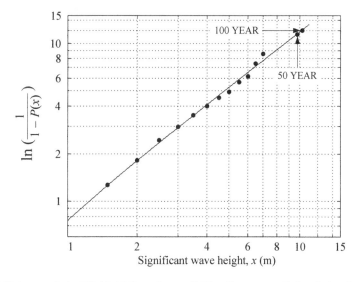

Fig. 1.27. Return period of H_s (dots) from data, solid line from curve-fitting, in log-log scale.

1.3.3.2 Zero-crossing period

We search for zero-crossing period associated with the extreme significant wave height of 50 or 100 year return sea obtained in Sec. 1.3.3.1. This may be done in the same way as in Sec. 1.3.2 by computing the mean and variance $\mu_{\ln x}$ and $\sigma^2_{\ln x}$ of logarithm of T_z against given H_s in Table 1.3. For each $H_s(m) = 0.5, 1.5, 2.5$ and 3.5, we have computed the parameters $\mu_{\ln x}$ and $\sigma^2_{\ln x}$, which are plotted for four significant wave heights on lognormal probability papers as shown in Fig. 1.28(a) and Fig. 1.28(b). Extending the straight line we can determine the parameters corresponding to the previously determined extreme significant wave heights 9.8 m and 10.3 m, i.e., $\mu_{\ln x} = 2.27$ and 2.32 and $\sigma^2_{\ln x} = 0.0034$ and 0.0032. Substituting these into Eqs. (1.115) and (1.116), we obtain the mean of the zero-crossing period as 9.69 s and 10.19 s, respectively. The modal period (period of peak energy spectral density) is related to zero-crossing period:

$$T_z = 0.710 T_m \qquad (1.131)$$

Hence the modal frequencies are 0.46 rad/s and 0.44 rad/s, respectively.

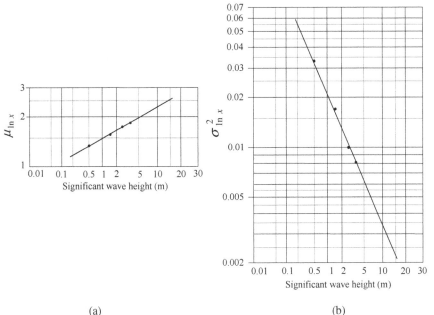

(a)   (b)

Fig. 1.28(a) The mean of $\ln x$ where $(x = H_s)$. (b) The variance of $\ln x$ where $(x = H_s)$, plotted on lognormal probability papers.

1.3.3.3 B-spectra for 100 year return sea

Using the H_s and ω_m of 50 and 100 year return seas obtained in the above
section, we have determined B-spectra, from which we have the variances
$m_0 = 6.0$ m^2 and 6.6 m^2, respectively. B-spectra in Sec. 2.3.2 for 50 and 100
year return seas are plotted as shown in Fig. 1.29. Since we have energy
spectra of 50 and 100 year return seas, we may use them for design of
offshore structures. For instance we may apply the joint distribution of wave
height and period given in Eq. (1.108). It is a common practice to estimate
the resonance period of motion of floating structure. The resonance period
will be plotted on the joint distribution and the largest wave height may be
estimated.

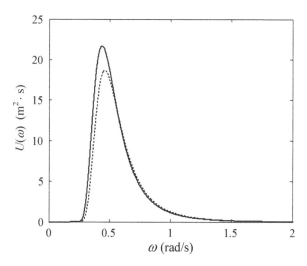

Fig. 1.29. Bretschneider spectra for 50 and 100 year return sea. 50 year (dotted line) and 100
year (solid line): $\omega_m = 0.46$ rad/s and 0.44 rad/s for 50 year and 100 year respectively.

References

Journal references:

Cartwright, D.E. and Longuet-Higgins, M.S. (1956) The Statistical Distribution of the Maxima of a Random Function, Proc. Roy. Soc. London, Ser. A, Vol. 237, pp. 212–232.

Chakrabarti, S.K. and Cooley, R.P. (1977) Statistical distribution of the periods and heights of ocean waves, J. Geophys. Res., 82, 1363–1368.

Earle, M.D. (1976) Extreme Wave Conditions During Hurricane Camile, J. Geophys. Res., Vol. 80, No. 3, pp. 377–379.

Longuet-Higgins, M.S. (1983) On the Joint Distribution of Wave Periods and Amplitudes in a Random Wave Field, Proc. Roy. Soc. London, Ser. A, Vol. 389, pp. 241–258.

Michel, W.H. (1999) Sea Spectra Revisited, Maine Technology, Vol. 36, No. 4.

Ochi, M. K. (1978) Wave Statistics for the Design of Ships and Ocean Structures, SNAME Transactions, Vol. 86, pp. 47–76.

St. Denis, M. and Pierson, W.J. (1953) On the Motions of Ships in Confused Seas, Transactions, SNAME, Vol. 61, 280-357.

Book references:

Brebbia, C.A. and Walker, S. (1979) Dynamic Analysis of Offshore Structures, Newnes-Butterworths, London, Boston.

Chakrabarti, S.K. (1987) Hydrodynamics of Offshore Structures Computational Mechanics, Publications, Southhamption, Boston.

Chakrabarti, S.K. (1900) Nonlinear Methods in Offshore Engineering, Elsevier, Amsterdam-Oxford-New York-Tokyo.

Goda,Y (2000) Random Seas and Design of Marine Structures, Advanced Series on Ocean Engineering–Vol. 15, World Scientific Publishing Company, Singapore.

Lewis, E.V. (1989) Motions in Waves and Controllability, Principles of Naval Architecture, Vol. 3, The Society of Naval Architects and Marine Engineers, Jersey City, NJ.

Newland, D. E. (1986) An Introduction to Random Vibrations and Spectral Analysis, Longman Inc., New York.

Price, W.G. and Bishop, R.E.D. (1974) Probabilistic Theory of Ship Dynamics, Chapman and Hall Ltd.

Rice, S.O. (1954) Selected Papers on Noise and Stochastic Processes, Dover, New York.

Sarpkaya T. and Isaacson, M. (1981) Mechanics of Wave Forces on Offshore Structures, Van Nostrand Reinhold Co., New York.

Thomson, W.T. (1981) Theory of Vibration with Applications, Second Edition, Prentice-Hall, Inc., Englewood Cliffs, New Jersey.

Tupper, E. (1998) Introduction to Naval Architecture, Soc. of Naval Architects and Marine Engineers, Jersey City, New Jersey.

Proceedings references:

Bouws, E. (1978). Wind and Wave Climate in the Netherlands Sector of the North Sea Between 53 degree and 56.4 degree North Latitude, Koninklijk Netherlands Meteorologisch Instituut Report 78-9.

Goda, Y. (1978). The observed joint distribution of periods and heights of sea waves, Proc. 16th Int. Conf. Coastal Eng., Sydney, Australia, pp. 227–246.

Krafft, M.J. and Kim, C.H. (1987). Zero-Crossing Analysis, Ocean Engineering Program, Civil Engineering Department, Texas A & M University.

Ochi, M.K. and Whalen, J.E. (1980). Prediction of the Severest Significant Wave Height, Proc. 7th Coastal Engineering Conference, Sydney, Australia, ASCE, Vol. 1, pp. 587–599.

Olagnon, M. and Athanasssoulis, G.A. (2000). ROGUE WAVES 2000, Proc. Workshop, Ifremer, France.

Stansberg, C.T., Huse, E, Krogstad, J.R. and Lehn, E. (1995). Experimental Study of Non-Linear Loads on Vertical Cylinders in Steep Random Waves, Proc. 5th Int. Offshore Polar Eng. Conf., ISOPE, Vol. 1, pp. 75–82.

Statoil (1996). Single Column Test Data Produced at MARINTEK, Stavanger, Norway.

Tucker, M.J. (1963). Simple Measurement of Wave Records-Wave Recording for Civil Engineers, Proc. Conf. The National Institute of Oceanography, pp. 2–3.

Chapter 2

Fourier Transform and Wave Spectra

2.1 Introduction

The measured wave elevation time series are usually transformed to the variety of spectra, including discrete complex amplitude spectrum, discrete complex amplitude density spectrum, discrete energy (power) spectrum, and discrete energy density spectrum (or wave spectrum). The wave spectrum is to store the information of the measured wave time series. Wave spectrum loses the phase spectrum, which contradicts to the feature of the complex amplitude spectrum (amplitude and phase spectrum) but can store large amount of data.

The Fourier transforms are usually carried out by the fast Fourier transform (FFT). Parseval's theorem is useful to express cross- and auto-spectra using FFT, which is called direct method. We will review Blackman-Tukey's indirect method in-depth because it has been widely employed in naval architecture and ocean engineering with some advantages.

The wave spectra used in designs and analyses of response of offshore structures were formulated based on many field data, most of which are presented as a function of wave frequency with two parameters: the significant wave height and modal frequency at which the peak energy density is measured. These two parameters are statistically associated with each other as discussed in Chap. 1. We will review various wave spectra and family of wave spectra that are mostly fitted to the wave data in the North Sea region, which may be utilized for the short- and long-term prediction of responses of offshore structures in Chap. 3.

2.1.1 Fourier transform

Let a sample time series $x(t)$ of length T be expressed in terms of infinite trigonometric function series:

$$x(t) = A_0 + \sum_{n=1}^{\infty} \left(A_n \cos \omega_n t + B_n \sin \omega_n t \right) \qquad (2.1)$$

where ω_n is the circular frequency of nth harmonic term. Let T be the fundamental wave period, then the frequency of fundamental harmonic is $\omega_1 = 2\pi/T = \Delta\omega$, $(\Delta f = 1/T)$. Consequently the frequency of the nth harmonic term becomes $\omega_n = n\omega_1$. The coefficients of the trigonometric function series in Eq. (2.1) are obtained applying the orthogonal property of trigonometric functions:

$$A_0 = \frac{1}{T} \int_0^T x(t)dt = \frac{1}{T} \int_{-T/2}^{T/2} x(t)dt$$

$$A_n = \frac{2}{T} \int_0^T x(t)\cos\omega_n t\, dt = \frac{2}{T} \int_{-T/2}^{T/2} x(t)\cos\omega_n t\, dt \qquad (2.2)$$

$$B_n = \frac{2}{T} \int_0^T x(t)\sin\omega_n t\, dt = \frac{2}{T} \int_{-T/2}^{T/2} x(t)\sin\omega_n t\, dt$$

Employing FFT, one determines the above unknown coefficients numerically. The basic approach of the algorithm is briefly reviewed. First one determines the mean value from $x(t)$ and subtracts it from the original time series to make it zero-mean time series, which makes $A_0 = 0$. Then there remains unknowns (A_n, B_n) to be determined. The sample data points are usually determined using the computer at sample intervals; Δt, $2\Delta t$, $3\Delta t$, $\cdots$ $N\Delta t$, giving $T = N\Delta t$. In the second place, the trigonometric series is expressed as a function of frequency; $\omega_1 = \Delta\omega, \cdots, \omega_n = n\Delta\omega, \cdots, \omega_{max} = (N/2)\Delta\omega$, so that the total number of unknown coefficients (A_n, B_n) may become N. In this manner one constructs N number of real algebraic equations for the N unknowns in the form:

$$x(t_m) = \text{Re}\left[\sum_{n=1}^{N/2} (A_n - iB_n)e^{i\omega_n t} \right] = \text{Re}\left[\sum_{n=1}^{N/2} C_n e^{i\omega_n t} \right], m = 1,2,3,\cdots,N \qquad (2.3)$$

where C_n is the one-sided discrete complex amplitude spectrum in dimension of length, which is illustrated in Fig. 2.1(a). Solving the above algebraic equation system one may determine the coefficients $(A_n, B_n) = C_{n..}$

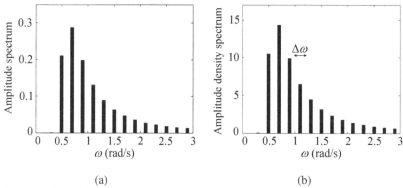

(a)   (b)

Fig. 2.1(a) Moduli of discrete amplitude spectrum C_n. (b) Discrete amplitude density spectrum c_n.

However, instead of determining C_n directly in the foregoing, one determines amplitude density spectrum c_n as defined below. One denotes the two integrals in Eq. (2.2) by half of a_n and b_n and designate them the amplitude densities:

$$\frac{1}{2}\begin{pmatrix} a_n \\ b_n \end{pmatrix} = \begin{Bmatrix} \int_{-T/2}^{T/2} x(t)\cos\omega_n t\, dt \\ \int_{-T/2}^{T/2} x(t)\sin\omega_n t\, dt \end{Bmatrix} = \frac{1}{2\Delta f}\begin{pmatrix} A_n \\ B_n \end{pmatrix} \tag{2.4}$$

Then the complex amplitude density spectrum can be expressed as follows:

$$c_n = \frac{a_n - ib_n}{2} = \frac{A_n - iB_n}{2\Delta f} = \frac{C_n}{2\Delta f} \tag{2.5}$$

Thus, $x(t)$ in Eq. (2.3) referring to Eq. (2.5) is given in terms of amplitude density spectrum c_n in the form:

$$x(t) = \text{Re}\left[\Delta f \sum_{n=1}^{N/2} 2c_n e^{i\omega_n t} \right] \tag{2.6}$$

Equation (2.6) may be expressed using the sum of complex and complex conjugate amplitude density in the form:

$$x(t) = \Delta f \sum_{n=1}^{N/2} \left[c_n e^{i\omega_n t} + c_n^* e^{-i\omega_n t} \right] \tag{2.7}$$

or

$$x(t) = \Delta f \left[\sum_{n=1}^{N/2} c_n e^{i\omega_n t} + \sum_{n=-1}^{-N/2} c_n^* e^{i\omega_n t} \right] \tag{2.8}$$

with

$$c_{-n} = c_n^*, \quad \omega_{-n} = -\omega_n \tag{2.9}$$

where the c_n is located at the positive frequency ω_n, while its conjugate c_n^* is at the negative frequency, $-\omega_n$. Thus, Eq. (2.8) is given in compact form:

$$x(t) = \Delta f \sum_{n=-N/2}^{N/2} c_n e^{i\omega_n t} \tag{2.10}$$

where c_n is also expressed referring to Eq. (2.4) in the form:

$$c_n = \int_{-T/2}^{T/2} x(t) e^{-i\omega_n t} dt \tag{2.11}$$

Discretizing Eqs. (2.10) and (2.11), one has:

$$x(t_m) = \Delta f \sum_{n=-N/2}^{N/2} c_n e^{i\frac{2\pi nm}{N}} \tag{2.12}$$

$$c_n = \Delta t \sum_{m=-N/2}^{N/2} x(t_m) e^{-i\frac{2\pi nm}{N}} \tag{2.13}$$

In Eqs. (2.10) and (2.13), one has used $T = N\Delta t$, $t_m = m\Delta t$, $\omega_n = 2\pi n/T = 2\pi n/(N\Delta t)$ and m, n and N (positive integers). Equation (2.12) is a linear algebraic equation for the N complex amplitude density c_n, while the number of the given values of $x(t)$ are N. Since c_n is complex constituting of two components, and RHS of Eq. (2.12) is real, the total number of the complex amplitude density c_n is $N/2$.

Now one renames the complex amplitude density c_n in Eqs. (2.12) and (2.13) $X(\omega_n)$ the Fourier transform of $x(t)$:

$$X(\omega_n) = \Delta t \sum_{m=-N/2}^{N/2} x(t_m) e^{-i\frac{2\pi nm}{N}} \tag{2.14}$$

$$x(t_m) = \Delta f \sum_{n=-N/2}^{N/2} X(\omega_n) e^{i\frac{2\pi nm}{N}} \tag{2.15}$$

Equations (2.14) and (2.15) are the discrete Fourier transform (DFT) pair. The same holds when f_n is used instead of ω_n in Eq. (2.14).

2.1.2 Note for using fft and ifft in MATLAB

Given the sample time series $x(t_m)$, one may determine the Fourier transform X_n by FFT code [fft(x)] in MATLAB. However, it should be noted that the code does not contain Δt. Likewise the inverse fast Fourier transform (IFFT) code [ifft(X)] does not contain Δf. $X_n = fft(x)$ provides N pairs of complex Fourier transform without Δt. The first $N/2$ complex pairs are for the positive frequency while the rest of $N / 2$ pairs are for negative frequency components which are complex conjugate of the first half. These positive and conjugate components are for the two-sided spectrum. The negative frequency part has to be discarded when one considers one-sided spectrum.

2.1.3 One-sided complex amplitude and amplitude density spectrum

Given the time series $x(t)$, we can determine Fourier transform X_n which is identical to complex amplitude density c_n and thus referring to Eq. (2.5), we determine the one-sided complex amplitude spectrum in the form:

$$C_n = 2X_n \Delta f \tag{2.16}$$

which is calculated using the [fft] in MATLAB:

$$C_n = \frac{2X_n}{T} = \frac{2 fft(x)\Delta t}{N\Delta t} = \frac{2 fft(x)}{N} \tag{2.17}$$

Figures 2.1(a) and 2.1(b) illustrate discrete complex amplitude spectrum C_n and discrete complex amplitude density spectrum c_n as function of frequency ω. Use of Eq. (2.17) in Eq. (2.3) will reproduce precisely the same time series over the same original time interval in the form:

$$x(t_m) = \text{Re} \sum_{n=1}^{N/2} C_n e^{i\omega_n t_m} \quad m=1, 2, \cdots, N. \tag{2.18}$$

The above reproduction procedure may be used to check if the FFT was performed properly.

The two-sided amplitude spectrum may simply be constructed by plotting $X_n \Delta f$ or $C_n / 2$ on the positive frequency axis and the conjugate components on the negative frequency axis.

2.1.4 Nyquist frequency and aliasing

In Sec. 2.1.1, one found that the DFT automatically gives the maximum frequency contained in the time series $x(t)$.

$$\omega_{max} = \omega_c = \frac{N}{2}\Delta\omega = \frac{\pi}{\Delta t} \qquad (2.19)$$

The maximum frequency is called Nyquist frequency or folding frequency. If there are frequencies above $\pi/\Delta t$ these introduce a distorted spectrum of the Fourier transform X_n called aliasing. The high frequency components contribute to the time series and falsely distort the Fourier spectrum X_n calculated by the DFT for frequencies below $\pi/\Delta t$. If ω_0 is the maximum frequency component in the time series, then aliasing can be avoided by taking Δt small enough such that:

$$\frac{\pi}{\Delta t} > \omega_0,$$

or if $f_0 = \omega_0/2\pi$, $\qquad\qquad (2.20)$

$$\frac{1}{2\Delta t} > f_0$$

Different from the foregoing subject, there is a problem of random wave simulation for a desired record length T given the wave energy spectrum. In this case uncertainty arises on the determination of cut-off frequency in the wave spectrum. The cut-off frequency is usually assumed as the frequency where the density takes 1% of the peak density. Since T ($= 2\pi/\Delta\omega$) has been given, assigning Δt one can determine N and vice versa referring to Eq. (2.19). In this manner the frequency interval $\Delta\omega$ and maximum number of frequency interval $N/2$ are determined according to Eq. (2.19). In the random wave simulation we usually employ the energy spectrum and random phase angles as shown in Sec. 1.1.1.

Tucker et al. (1984) note that there arises error in the random phase method if the amplitudes are assumed deterministic when they are in fact random.

2.1.5 Various definitions of continuous Fourier transform

The DFT pair Eqs. (2.14) and (2.15) may be brought to integral forms by increasing the number of data points to infinity:

$$X(f) = \int_{-\infty}^{\infty} x(t)e^{-i2\pi ft} dt \tag{2.21}$$

$$x(t) = \int_{-\infty}^{\infty} X(f)e^{i2\pi ft} df \tag{2.22}$$

There are three conventions of Fourier transform (FT) in ω-domain:
FT convention A:

$$X(\omega) = \int_{-\infty}^{\infty} x(t)e^{-i\omega t} dt \tag{2.23}$$

$$x(t) = \frac{1}{2\pi} \int_{-\infty}^{\infty} X(\omega)e^{i\omega t} d\omega \tag{2.24}$$

FT convention A is identical to FT given in Eqs. (2.21) and (2.22).
FT convention B:

$$X(\omega) = \frac{1}{2\pi} \int_{-\infty}^{\infty} x(t)e^{-i\omega t} dt \tag{2.25}$$

$$x(t) = \int_{-\infty}^{\infty} X(\omega)e^{i\omega t} d\omega \tag{2.26}$$

FT convention C:

$$X(\omega) = \frac{1}{\sqrt{2\pi}} \int_{-\infty}^{\infty} x(t)e^{-i\omega t} dt \tag{2.27}$$

$$x(t) = \frac{1}{\sqrt{2\pi}} \int_{-\infty}^{\infty} X(\omega)e^{i\omega t} d\omega \tag{2.28}$$

Regardless of the type of the conventions the FT formulas will give the same energy density spectrum, and will reproduce the original time series. All of the above FTs have to satisfy the condition:

$$\int_{-\infty}^{\infty} |x(t)| \, dt < \infty \qquad (2.29)$$

2.1.6 Properties of Dirac delta function

Dirac delta function is a useful tool in treating the random data involved with Fourier transform. It has the following basic properties:

$$\delta(x) = 0 \ \text{ if } x \neq 0$$

$$\int_{-\infty}^{\infty} \delta(x) dx = 1 \qquad (2.30)$$

$$\int_{-\infty}^{\infty} \delta(x - x_0) s(x) dx = s(x_0)$$

where x may be either time or frequency in our application. We will define Dirac delta function both in the time and frequency domain as shown below.

2.1.6.1 Dirac delta function in time domain

Consider a rectangular impulse (force $A \times$ time t or wave elevation $A \times$ time t) (Price and Bishop, 1974):

$$g(t) = \begin{cases} A & |t| < T/2 \\ 0 & |t| > T/2 \end{cases} \qquad (2.31)$$

The Fourier transform of impulse $g(t)$ with FT convention A gives,

$$G(\omega) = \int_{-\infty}^{\infty} g(t) e^{-i\omega t} dt = AT \frac{\sin\left(\omega \dfrac{T}{2}\right)}{\omega \dfrac{T}{2}} \qquad (2.32)$$

If the area AT of the rectangular impulse is kept equal to unity as the breadth of the impulse approaches zero and the height becomes infinite, we have:

$$G(\omega) = \lim_{T/2 \to 0} AT \frac{\sin\left(\dfrac{\omega T}{2}\right)}{\dfrac{\omega T}{2}} = 1 \tag{2.33}$$

Referring to Eqs. (2.32) and (2.33), one will obtain Dirac delta function $\delta(t)$ as will be shown below:

$$G(\omega) = \int_{-\infty}^{\infty} \delta(t)\, e^{-i\omega t}\, dt = 1.0 \tag{2.34}$$

This satisfies the property of Dirac delta function in Eq. (2.30). Hence the inverse Fourier transform of $G(\omega)$ gives the Dirac delta function in the form:

$$\delta(t) = \frac{1}{2\pi} \int_{-\infty}^{\infty} e^{i\omega t}\, d\omega \tag{2.35}$$

If FT of convention B is applied, $G(\omega) = 1\,/\,2\pi$, which gives the same definition of Dirac delta function Eq. (2.35).

2.1.6.2 Dirac delta function in frequency domain

Consider a sinusoidal wave of amplitude A and frequency ω_n:

$$x(t) = A e^{i\omega_n t} \tag{2.36}$$

Applying the FT of convention B:

$$A e^{i\omega_n t} = \int_{-\infty}^{\infty} X(\omega) e^{i\omega t}\, d\omega \tag{2.37}$$

Recognizing the properties of δ function in Eq. (2.30), the above equation holds if:

$$X(\omega) = A\delta(\omega - \omega_n) \tag{2.38}$$

Substituting Eqs. (2.36) and (2.38) into Eq. (2.25), one has:

$$\delta(\omega - \omega_n) = \frac{1}{2\pi} \int_{-\infty}^{\infty} e^{-i(\omega - \omega_n)t}\, dt \tag{2.39}$$

If FT of convention A is applied, one will have the same dirac delta function as given in Eq. (2.39).

2.2 Estimation of Energy Density Spectra

2.2.1 Discrete energy and energy density spectrum

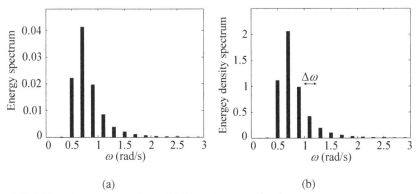

(a)   (b)

Fig. 2.2(a) Discrete energy spectrum. (b) Discrete energy density spectrum.

The one-sided discrete energy spectrum or power spectrum is defined as the variance of each component wave, in f_n domain, referring to Sec. 2.1.3:

$$G(f_n) = \frac{C_n C_n^*}{2} = 2X_n X_n^* (\Delta f)^2 , \quad 0 \le f_n \le \infty \tag{2.40}$$

An example of discrete one-sided energy spectrum as a function of circular frequency is shown in Fig. 2.2(a).

The one-sided discrete energy density spectrum is equal to the discrete energy spectrum Eq. (2.40) distributed on the band Δf:

$$U_{xx}(f_n) = 2X_n X_n^* \Delta f, \quad 0 \le f_n \le \infty \tag{2.41}$$

A sketch of discrete one-sided energy density spectrum in circular frequency domain is shown in Fig. 2.2(b). Increasing n to infinity; one may have the continuous one-sided energy density spectrum in the form:

$$U_{xx}(f) = \lim_{T \to \infty} \frac{2X(f)X^*(f)}{T}, \quad 0 \le f \le \infty \tag{2.42}$$

2.2.2 Parseval's theorem for energy density spectra

Parseval's theorem is written in the form:

$$\int_{-\infty}^{\infty} x_1(t)x_2(t)dt = \int_{-\infty}^{\infty} X_1(f)X_2^*(f)df$$

$$= \int_{-\infty}^{\infty} X_1^*(f)X_2(f)df \tag{2.43}$$

where * denotes the complex conjugate and $X_1(f)$ and $X_2(f)$ are Fourier transforms of real time functions $x_1(t)$ and $x_2(t)$, respectively. Equation (2.43) may be proved using the FT as follows:

$$x_1(t)x_2(t) = x_2(t)\int_{-\infty}^{\infty} X_1(f)e^{i2\pi ft}df$$

$$\int_{-\infty}^{\infty} x_1(t)x_2(t)dt = \int_{-\infty}^{\infty} x_2(t)dt \int_{-\infty}^{\infty} X_1(f)e^{i2\pi ft}df$$

$$= \int_{-\infty}^{\infty} X_1(f)\int_{-\infty}^{\infty} x_2(t)e^{i2\pi ft}dtdf \tag{2.44}$$

$$= \int_{-\infty}^{\infty} X_1(f)X_2^*(f)df$$

The mean square value, autocorrelation, and cross-correlation can now be expressed in terms of Fourier transform by Parseval's theorem. The mean square value is given in terms of the Fourier transform:

$$E[x^2(t)] = \lim_{T\to\infty}\frac{1}{T}\int_{-T/2}^{T/2} x^2(t)dt = \int_{-\infty}^{\infty}\lim_{T\to\infty}\frac{1}{T}X(f)X^*(f)df \tag{2.45}$$

while the same can be expressed by the area under the curve of energy density spectra:

$$E[x^2(t)] = \int_{-\infty}^{\infty} S_{xx}(f)df \tag{2.46}$$

Thus comparing Eqs. (2.45) and (2.46), one has:

$$S_{xx}(f) = \lim_{T\to\infty}\frac{1}{T}X(f)X^*(f) \tag{2.47}$$

Since two-sided auto spectral density S_{xx} is an even function of f, we can fold them to make one-sided auto-energy density spectrum:

$$U_{xx}(f) = \lim_{T\to\infty}\frac{2}{T}X(f)X^*(f) \tag{2.48}$$

which is identical to Eq. (2.42) that was derived from the discrete energy density spectrum.

The cross-correlation of the stationary processes $x(t)$ and $y(t)$ is formed as:

$$R_{xy}(\tau) = E[x(t)y(t+\tau)] = \lim_{T \to \infty} \frac{1}{T} \int_{-\infty}^{\infty} x(t)y(t+\tau)dt \qquad (2.49)$$

Applying Parseval's theorem Eq. (2.43) to cross-correlation Eq. (2.49), one has:

$$
\begin{aligned}
R_{xy}(\tau) &= \lim_{T \to \infty} \frac{1}{T} \int_{-\infty}^{\infty} x(t)y(t+\tau)dt \\
&= \lim_{T \to \infty} \frac{1}{T} \int_{-\infty}^{\infty} x(t) \int_{-\infty}^{\infty} Y(f)e^{i2\pi f(t+\tau)}dfdt \\
&= \int_{-\infty}^{\infty} \lim_{T \to \infty} \frac{1}{T} \left\{ \int_{-\infty}^{\infty} x(t)e^{i2\pi ft}dt \right\} Y(f)e^{i2\pi f\tau}df \\
&= \int_{-\infty}^{\infty} \lim_{T \to \infty} \frac{1}{T} X^*(f)Y(f)e^{i2\pi f\tau}df
\end{aligned}
\qquad (2.50)
$$

which is also given by definition in the form:

$$R_{xy}(\tau) = \int_{-\infty}^{\infty} S_{xy}(f)e^{i2\pi f\tau}df \qquad (2.51)$$

Thus equating Eqs. (2.50) and (2.51), one has:

$$
\begin{aligned}
S_{xy}(f) &= \lim_{T \to \infty} \frac{1}{T} X^*(f)Y(f) \\
&= \lim_{T \to \infty} \frac{1}{T} X(f)Y^*(f) \\
&= S_{xy}^*(f) = S_{xy}(-f); \quad -\infty \le f \le \infty
\end{aligned}
\qquad (2.52)
$$

As shown in Eq. (2.52), the two-sided spectral density S_{xy} is an even functions of f. Thus, the two-sided spectrum Eq. (2.52) can be folded to make one-sided spectrum:

$$U_{xy}(f) = \lim_{T \to \infty} \frac{2X^*(f)Y(f)}{T} \qquad (2.53)$$

which is in parallel to auto spectral density Eq. (2.48).

The cross-correlation is, in general, not an even function of lag τ. The cross-correlation can be transformed to even and odd function of τ:

$$R_{xy}(\tau) = \frac{R_{xy}(\tau) + R_{xy}(-\tau)}{2} + \frac{R_{xy}(\tau) - R_{xy}(-\tau)}{2} \tag{2.54}$$

Applying IFT to cross-correlation in Eq. (2.51), one obtains:

$$
\begin{aligned}
S_{xy}(f) &= \int_{-\infty}^{\infty} R_{xy}(\tau) e^{-i2\pi f\tau} d\tau \\
&= \int_{-\infty}^{\infty} \frac{R_{xy}(\tau) + R_{xy}(-\tau)}{2} \cos 2\pi f\tau d\tau \\
&\quad - i \int_{-\infty}^{\infty} \frac{R_{xy}(\tau) - R_{xy}(-\tau)}{2} \sin 2\pi f\tau d\tau \\
&= c(f) - iq(f)
\end{aligned} \tag{2.55}
$$

where, $c(f)$ and $q(f)$ are called cross- and quadrature-spectra, respectively. It is noted that the cross-spectrum derived from cross-correlation of an uneven function of τ consists of the co- and quadrature-spectra.

Since the variance is the same $S_{xx}(\omega)d\omega = S_{xx}(f)df$ in the f - and ω - domain and $X(\omega) = X(f)$, one has the relations between the spectral densities as functions of f and ω in the forms:

$$
\begin{aligned}
S_{xx}(f) &= 2\pi S_{xx}(\omega) \\
U_{xx}(f) &= 2\pi U_{xx}(\omega)
\end{aligned} \tag{2.56}
$$

$$
\begin{aligned}
S_{xy}(f) &= 2\pi S_{xy}(\omega) \\
U_{xy}(f) &= 2\pi U_{xy}(\omega)
\end{aligned} \tag{2.57}
$$

2.2.3 Summary of discrete auto- and cross-energy density spectra

The discrete energy density spectra are computed using the following formulas:

$$S_{xx}(f_n) = \frac{1}{N} X(f_n) X^*(f_n) \Delta t \tag{2.58}$$

$$U_{xx}(f_n) = \frac{2}{N} X(f_n) X^*(f_n) \Delta t \tag{2.59}$$

$$S_{xy}(f_n) = \frac{1}{N} X^*(f_n)Y(f_n)\Delta t \tag{2.60}$$

$$U_{xy}(f_n) = \frac{2}{N} X^*(f_n)Y(f_n)\Delta t \tag{2.61}$$

$$S_{xx}(\omega_n) = \frac{X(\omega_n)X^*(\omega_n)\Delta t}{2\pi N} \tag{2.62}$$

$$U_{xx}(\omega_n) = \frac{X(\omega_n)X^*(\omega_n)\Delta t}{\pi N} \tag{2.63}$$

$$S_{xy}(\omega_n) = \frac{X^*(\omega_n)Y(\omega_n)\Delta t}{2\pi N} \tag{2.64}$$

$$U_{xy}(\omega_n) = \frac{X^*(\omega_n)Y(\omega_n)\Delta t}{\pi N} \tag{2.65}$$

Example 2.1 Estimate of energy density spectrum of Camille wave

Hurricane Camille wave (1500 hrs – 1600 hrs) is used for the direct estimation of energy density spectrum as shown in Fig. 2.3. The dark solid line represents the smoothed result by applying moving average from the fluctuating raw data.

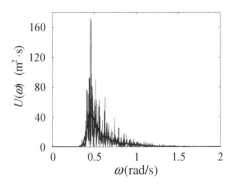

Fig. 2.3. Spectrum of Camille wave (1500 hrs – 1600 hrs) by direct method.

2.2.4 Ensemble average method for energy density spectra

For stationary random process, ensemble average method for one-sided spectra (Bendat, 1990):

$$U_{xx}(f) = \frac{2}{T} E[|X(f)|^2] \tag{2.66}$$

$$U_{yy}(f) = \frac{2}{T} E[|Y(f)|^2] \tag{2.67}$$

$$U_{xy}(f) = \frac{2}{T} E[X^*(f)Y(f)] \tag{2.68}$$

$E[\cdot]$ denotes an expected value ensemble average over the quantities inside the brackets. Here the terms $X(f)$ and $Y(f)$ are Fourier transforms of finite length T for distinct digitized data, f is restricted to discrete values spaced $(1/T)$ apart in a bounded frequency range.

2.2.5 Blackman-Tukey's indirect estimates

The estimation of the energy density spectra in the previous sections is called direct method because it estimates the spectra directly from FFT. The indirect estimate of energy density spectra has three steps, i.e., first one estimates the correlation functions, second conducts the Fourier transform of the estimated correlation functions and third smoothes the raw spectral densities by Hamming method (Blackman-Tukey, 1956; Korvin-Kroukovski, 1961; Kinsman, 1965).

2.2.5.1 Estimate of one-sided auto-energy spectrum

For the stationary process, one forms the autocorrelation function:

$$R(\tau) = E[x(t)x(t+\tau)] = \frac{1}{T} \int_{-T/2}^{T/2} x(t)x(t+\tau)dt, \quad \tau \geq 0 \tag{2.69}$$

The auto-energy density spectrum is obtained through Fourier transform of convention B as given in Eq. (2.25):

$$S(\omega) = \frac{1}{2\pi} \int_{-\infty}^{\infty} R(\tau)e^{-i\omega\tau}d\tau, -\infty \leq \omega \leq \infty \tag{2.70}$$

The inverse Fourier transform of Eq. (2.70):

$$R(\tau) = \int_{-\infty}^{\infty} S(\omega)e^{i\omega\tau}d\omega \qquad (2.71)$$

Since the above autocorrelation function and auto-energy density spectrum are even functions of time and frequency, respectively, the one-sided auto-energy spectrum and autocorrelation function are given by:

$$U(\omega) = \frac{1}{\pi} \int_{0}^{\infty} R(\tau)\cos\omega\tau d\tau \qquad (2.72)$$

$$R(\tau) = \int_{0}^{\infty} U(\omega)\cos\omega\tau d\omega \qquad (2.73)$$

Thus, the autocorrelation function Eq. (2.69) is written in a discrete form:

$$R_p = \frac{2}{N-p} \sum_{i=1}^{N-p} x(t_i) \cdot x(t_i + p\Delta\tau) \qquad (2.74)$$

where $\Delta\tau = \Delta t$; $T = N\Delta\tau$ and $\tau = p\Delta\tau$, $p = 0, 1, 2, \cdots, m$, and m denotes the maximum lag number which will be determined according to the Guideline as shown below. The number 2 in the numerator takes into account the symmetry of the autocorrelation. The above equation provides $m + 1$ discrete, equally spaced values of the sample autocorrelations at 0, $\Delta\tau$, $2\Delta\tau, \cdots, m\Delta\tau$. Since the integration with respect to time t in Eq. (2.69) is over the sample length $T - \tau = (N - p)\Delta t$, one has the denominator $N - p$ in Eq. (2.74). When $p = 0$, the autocorrelation becomes the variance of $x(t)$. For the estimate of one-sided spectral density, one discretizes Eq. (2.72) using $m + 1$ equally spaced values of the autocorrelations at frequencies 0, $\Delta\omega$, $2\Delta\omega, \cdots$, $m\Delta\omega$.

$$\Delta\omega = \frac{\omega_c}{m} = \frac{\pi}{m\Delta\tau} \qquad (2.75)$$

where ω_c is the Nyquist frequency. Using $\omega = h\Delta\omega$ and Eq. (2.75), one has:

$$\omega\tau = (h\Delta\omega)(p\Delta\tau) = \frac{\pi ph}{m} \qquad (2.76)$$

The trapezoidal rule is used for integration Eq. (2.72). Hence the raw spectral density is given in the form:

$$L_h = \frac{1}{m\Delta\omega}\left[R_0 + 2\sum_{p=1}^{m-1} R_p \cos\frac{\pi ph}{m} + R_m \cos\pi h \right], \quad h = 0,1,2,\cdots,m \quad (2.77)$$

In the above, we have employed Eqs. (2.72), (2.74), and (2.76) with the multiplier $\Delta\tau/2$ for trapezoidal integration. In addition, $\Delta\tau/\pi$ in Eq. (2.72) is identical to $1/(m\Delta\omega)$ according to Eq. (2.75). This is implemented into Eq. (2.77) as a cofactor. The above raw spectral density is smoothed by the moving, weighted average (Hamming process) in the form:

$$U(h\Delta\omega) = U_h \equiv \sum_{i=0}^{m} a_{h,i} L_h, \quad h = 1,2,\cdots,m-1 \quad (2.78)$$

$$\begin{cases} a_{h,h-1} = 0.23 \\ a_{h,h} = 0.54 \\ a_{h,h+1} = 0.23 \\ a_{h,i} = 0.00, \quad \text{for } i \neq h-1, h \text{ or } h+1 \end{cases} \quad (2.79)$$

And those for the end points, $h = 0$ and m are given by:

$$U_0 = 0.54L_0 + 0.46L_1$$
$$U_m = 0.46L_{m-1} + 0.54L_m \quad (2.80)$$

Thus the entire smoothed estimates are given in the forms:

$$U_0 = 0.54L_0 + 0.46L_1$$
$$U_h = 0.23L_{h-1} + 0.54L_h + 0.23L_{h+1}, \quad h = 1,2,\cdots,m-1 \quad (2.81)$$
$$U_m = 0.46L_{m-1} + 0.54L_m$$

U_h in Eq. (2.81) gives the spectral density distribution.

2.2.5.2 Estimate of one-sided cross-spectrum

The cross-correlation of two stationary processes $x(t)$ and $y(t)$ is formed:

$$R_{xy}(\tau) = E[x(t)y(t+\tau)] \quad (2.82)$$

where the cross-correlation is in general not even function of τ. Applying the FT of convention B to Eq. (2.82), one obtains the same form as Eq. (2.55):

$$S_{xy}(\omega) = \frac{1}{2\pi}\int_{-\infty}^{\infty} R_{xy}(\tau)e^{-i\omega\tau}d\tau = c(\omega) - iq(\omega) \quad (2.83)$$

Temporarily dropping suffix xy, we consider cross-correlation $R(\tau)$ that is not even function in τ $(-\infty, \infty)$. Rearranging $R(\tau)$ to give even and odd function:

$$R(\tau) = \frac{R(\tau) + R(-\tau)}{2} + \frac{R(\tau) - R(-\tau)}{2} \tag{2.84}$$

and using $e^{-i\omega t} = \cos \omega t - i \sin \omega t$ in Eq. (2.83), we obtain the one-sided cross-spectral density in the form:

$$U_{xy}(\omega) = C(\omega) - iQ(\omega) \tag{2.85}$$

$$C(\omega) = \frac{1}{\pi} \int_0^\infty \left(\frac{R(\tau) + R(-\tau)}{2} \right) \cos \omega \tau d\tau \tag{2.86}$$

$$Q(\omega) = \frac{1}{\pi} \int_0^\infty \left(\frac{R(\tau) - R(-\tau)}{2} \right) \sin \omega \tau d\tau \tag{2.87}$$

The integrands of cross-correlations associated with the co- and quadrature-spectra Eqs. (2.86) and (2.87) may be given with different notations as:

$$F_p = \frac{R_p + R_{-p}}{2} \quad \text{and} \quad G_p = \frac{R_p - R_{-p}}{2} \tag{2.88}$$

In the above the variables $x(t)$ and $y(t)$ are stationary, thus we use cross-equations in Sec. 1.2.4.2 for one-sided cross-correlations in the same manner as in the autocorrelation Eq. (2.74):

$$R_p = \frac{2}{N-p} \sum_{i=1}^{N-p} x(t_i) y(t_i + p\Delta t), \qquad p = 0, 1, 2, \cdots, m$$

$$R_{-p} = \frac{2}{N-p} \sum_{i=1}^{N-p} x(t_i + p\Delta \tau) y(t_i), \quad p = 0, 1, 2, \cdots, m \tag{2.89}$$

Equation (2.89) provides $m + 1$ discrete, equally spaced values of the sample cross-correlations at 0, $\Delta\tau$, $2\Delta\tau$, $\cdots$, $m\Delta\tau$. Thus the raw cross-spectral densities in the two components are obtained precisely in the same way as the auto-energy density spectrum Eq. (2.77):

$$L_{C_h} = \frac{1}{m\Delta\omega}\left[F_0 + 2\sum_{p=1}^{m-1}F_p\cos\frac{\pi ph}{m} + F_m\cos\pi h\right]$$

$$L_{Q_h} = \frac{1}{m\Delta\omega}\left[G_0 + 2\sum_{p=1}^{m-1}G_p\sin\frac{\pi ph}{m} + G_m\sin\pi h\right]$$

(2.90)

where

$$\omega = \frac{\pi h}{m\Delta t} = h\Delta\omega, \ \ h=0, 1, 2, \cdots, m$$ (2.91)

The raw estimates of the co- and quadrature-spectrum are similarly smoothed as the auto-energy density spectrum:

$$C_0 = 0.54L_{C_0} + 0.46L_{C_1}$$
$$C_h = 0.23L_{C_{h-1}} + 0.54L_{C_h} + 0.23L_{C_{h+1}}$$
$$C_m = 0.46L_{C_{m-1}} + 0.54L_{C_m}$$

(2.92)

$$Q_0 = 0.54L_{Q_0} + 0.46L_{Q_1}$$
$$Q_h = 0.23L_{Q_{h-1}} + 0.54L_{Q_h} + 0.23L_{Q_{h+1}}$$
$$Q_m = 0.46L_{Q_{m-1}} + 0.54L_{Q_m}$$

(2.93)

Thus the magnitude and phase of the smoothed cross-spectrum are estimated by:

$$U_h = \sqrt{C_h^2 + Q_h^2}\ , \quad \phi = \tan^{-1}\left(-\frac{Q_h}{C_h}\right), \quad h=0,1,2$$ (2.94)

2.2.5.3 Guideline in determining maximum lag number

How many lags should one take to obtain the best spectral estimate? In fact the spectral estimate varies depending on the choice of number of maximum lag numbers. And do we have confidence in the chosen maximum lag number m? If so, to what degree does it have confidence? The answer to these questions is given in Kinsman (1965, pp. 449): "As a practical matter it is seldom worth planning for more than 100 degrees of freedom df in the field measurement of waves, and 50 to 60 is a very reasonable number":

$$df = 2\left(\frac{N}{m} - \frac{1}{4}\right) \tag{2.95}$$

Example 2.2 Indirect estimate of energy density spectrum of Camille wave

Consider Hurricane Camille wave (1500 hrs – 1600 hrs), $\Delta t = 1$ s, $N = 3600$. If we assume degrees of freedom $df = 60$, we have the maximum lag $m = 119$. The estimated autocorrelation and the spectral estimate are shown in Figs. 2.4(a) and (b). The variance derived from the energy spectrum is 9.99 m², which is identical to the autocorrelation value at $\tau = 0$ in Fig. 2.4(a). It was also confirmed that the variance from the time series is identical to the above.

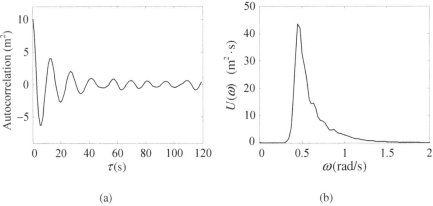

(a)   (b)

Fig. 2.4(a) Autocorrelation function). (b) One-sided wave spectrum of Hurricane Camille wave (1500 hrs – 1600 hrs).

A comparison of the direct method for estimation of the auto-energy density spectra as shown in Fig. 2.3 to indirect method shown in Fig. 2.4(b) is in excellent agreement. The indirect method seems to be less favorable for the user because of its lengthy procedure. However, since we use the computer code it is not lengthy at all. The indirect method automatically provides the smoothed result directly.

2.3 Wave Spectra

2.3.1 Introduction

Formulas for wave density spectra are developed based on the accumulated data in a region and usually expressed in terms of significant wave height and modal frequency.

Table 2.1. Sea state code (Price and Bishop,1974) and in Tupper (1996).

Code	Description of sea	H_s (m)	Frequency of occurrence		
			World wide	North Atlantic	Northern North Atlantic
0	Calm (glassy)	0.0			
1	Calm (rippled)	0.00 ~ 0.10	11.2486	8.3103	6.0616
2	Smooth (wavelets)	0.10 ~ 0.50			
3	Slight	0.50 ~ 1.25	31.6851	28.1996	21.5683
4	Moderate	1.25 ~ 2.50	40.1944	42.0273	40.9915
5	Rough	2.50 ~ 4.00	12.8005	15.4435	21.2383
6	Very rough	4.00 ~ 6.00	3.0253	4.2938	7.0101
7	High	6.00 ~ 9.00	0.9263	1.4968	2.6931
8	Very high	9.00 ~ 14.00	0.1190	0.2263	0.4346
9	Phenomenal	over 14.00	0.0009	0.0016	0.0035

The wind-generated sea state is usually classified by code # as listed in Table 2.1, which represents the sea state or sea severity in terms of the significant wave height. It also shows the frequency of occurrence depending on the regions. The Northern North Atlantic region has the highest frequencies in the occurrence of very high sea severities.

The empirical formulas of the wave energy density spectra (wave spectra) have been evolutionarily developed: Though there are quite a few more formulas, we review here the works of Bretschneider (1959), Pierson and Moskowitz (1964), Hasselmann et al. (1973), Houmb and Overik (1976), Ochi (1978), Lee and Bales (1980), Isherwood (1986). More detailed information about the wave spectra may be referred to Michel (1999).

The empirical wave spectra represent Gaussian waves regardless of the sea severity. Because the wave spectra determined from the sea records have lost the phase angles, the waves simulated from the spectra Sec. 2.1.4 will become Gaussian. The seakeeping analysis depends totally on the empirical wave spectra and linear RAOs which will be discussed in Chap. 3.

2.3.2 Bretschneider two-parameter spectrum

The Bretschneider frequency spectrum is the result of evolution of a long-term research. Originally, Bretschneider (1959) derived period energy spectrum in terms of two parameters, the total energy E and true average period T_a:

$$U(T) = 2.7E\frac{T^3}{T_a^4}\exp\left[-0.675(\frac{T}{T_a})^4\right]$$ (2.96)

where

$$T_a = \frac{\int_0^\infty TU(T)dT}{\int_0^\infty U(T)dT}$$ (2.97)

Use of $dU(T) / dT = 0$ gives the period at which the spectral density becomes peak:

$$T_p = 1.027T_a$$ (2.98)

Thus, the above period spectrum can be transformed to a function of total energy E and peak period T_p:

$$U(T) = \frac{3E}{T_p^4}T^3\exp\frac{-3}{4}\left(\frac{T}{T_p}\right)^4$$ (2.99)

The period spectrum above was transformed to the frequency energy spectrum employing the relations for the identity of total energy:

$$\int_0^\infty U(T)dT = -\int_0^\infty U(\omega)d\omega = -\int_0^\infty U(f)df$$ (2.100)

with,

$$T = \frac{2\pi}{\omega} = \frac{1}{f} \tag{2.101}$$

The frequency-period relation Eq. (2.101) is used to create a relation between the period and frequency spectrum:

$$U(\omega) = U(T)\frac{T^2}{2\pi} = 3E\left(\frac{2\pi}{T_p}\right)^4 \omega^{-5} \exp\left[-\frac{3}{4}\left(\frac{2\pi}{T_p}\right)^4 \omega^{-4}\right] \tag{2.102}$$

$$U(f) = T^2 U(T) = 3E\frac{1}{T_p^4} f^{-5} \exp\left[-\frac{3}{4}\frac{1}{T_p^4} f^{-4}\right] \tag{2.103}$$

From the above spectra, we obtain the modal frequency (frequency of maximum energy spectral density), by taking $dU(\omega)/d\omega = 0$:

$$\omega_m = \left(\frac{3}{5}\right)^{1/4}\frac{2\pi}{T_p} \quad \text{and} \quad f_m = \left(\frac{3}{5}\right)^{1/4}\frac{1}{T_p} \tag{2.104}$$

Substituting Eq. (2.104) into Eqs. (2.102) and (2.103), one obtains the frequency spectrum that is widely used today:

$$U(\omega) = 5E\,\omega_m^4\,\omega^{-5} \exp\left[-1.25\left(\frac{\omega_m}{\omega}\right)^4\right] \tag{2.105}$$

$$U(f) = 5E\,f_m^4\,f^{-5} \exp\left[-1.25\left(\frac{f_m}{f}\right)^4\right] \tag{2.106}$$

Since here one assumes Gaussian and narrow-band process, we have $E = m_0 = H_s^2/16$ as given in Sec. 1.2.6.6. Thus, the spectra Eqs. (2.105) and (2.106) are written in terms of modal frequency and significant wave height in the forms:

$$U(\omega) = \frac{5}{16}H_s^2\omega_m^4\omega^{-5} \exp\left[-1.25\left(\frac{\omega_m}{\omega}\right)^4\right] \tag{2.107}$$

$$U(f) = \frac{5}{16}H_s^2 f_m^4 f^{-5} \exp\left[-1.25\left(\frac{f_m}{f}\right)^4\right] \tag{2.108}$$

If the process is extremely broad-banded $\varepsilon = 1$, it is suggested to use $E = m_0 = H_s^2 / 9$ (St. Denis, 1980). B-spectrum represents the usual partially developed seas as shown in the family spectra in Fig. 2.8.

2.3.3 One-parameter Pierson-Moskowitz spectrum

Pierson-Moskowitz (P-M) spectrum (1964) was originally developed based on measured wind speed and wave elevations for fully developed seas in the North Atlantic:

$$U(\omega) = \alpha g^2 \omega^{-5} \exp\left[-\beta \left(\frac{g}{V} \right)^4 \omega^{-4} \right] \qquad (2.109)$$

where V is the wind speed measured at 19.5 m above the mean sea surface, g gravitational acceleration, $\alpha = 0.0081$, $\beta = 0.74$ for the North Sea.

2.3.4 Modified one-parameter P-M spectrum

Sometime later the data of P-M spectrum Eq. (2.109) were analyzed to develop the relationship:

$$0.74 \frac{g^4}{V^4} = \frac{5}{4} \omega_m^4 \qquad (2.110)$$

Use of Eq. (2.109) makes P-M spectrum expressed in terms of modal frequency:

$$U(\omega) = \alpha g^2 \omega^{-5} \exp\left[-\frac{5}{4} \left(\frac{\omega}{\omega_m} \right)^{-4} \right] \qquad (2.111)$$

Noting that the same variance both in the f- and ω-domain, i.e. $U(\omega) = U(f)/2\pi$ in accordance to Eq. (2.56), one has:

$$U(f) = \alpha g^2 (2\pi)^{-4} f^{-5} \exp\left[-\frac{5}{4} \left(\frac{f}{f_m} \right)^{-4} \right] \qquad (2.112)$$

Equation (2.111) or (2.112) is the modified one parameter P-M spectrum for the fully developed sea, which is used in the formulation of JONSWAP spectrum as will be shown later.

2.3.5 Modified two-parameter P-M spectrum

The one parameter frequency spectrum Eq. (2.111) may be expressed in the same form as two parameter B-spectrum in Eq. (2.107) by simply assuming:

$$\alpha g^2 = \frac{5}{16} H_s^2 \omega_m^4 \tag{2.113}$$

Then, the so-called modified two parameter P-M spectrum becomes the two-parameter B-spectrum Eq. (2.107), which preceded the basic P-M spectrum Eq. (2.109). However, the modified two-parameter P-M spectrum through usage has gained some undeserved acceptance (Michel, 1999).

2.3.6 Definitions of wave periods

Modal period T_m and the modal frequency ω_m (frequency of peak spectral density) are derived from the B-spectrum in Eq. (2.102). It is frequently found in literature that the modal period is represented by T_p, indicating the period for peak energy density. However, T_p was long before defined as the peak period of the period energy spectrum by Bretschneider (1959) as given in Eq. (2.98). The relationships between modal period T_m and other periods in practical applications are given below.

2.3.6.1 Average period

The period T_1 is the average period derived from the average frequency of component waves of B-spectrum Eq. (2.107):

$$T_1 = 2\pi \frac{m_0}{m_1} = \frac{2\pi \int_0^\infty U(\omega)d\omega}{\int_0^\infty \omega U(\omega)d\omega} = \frac{T_m}{\left(\frac{5}{4}\right)^{1/4} \Gamma\left(\frac{3}{4}\right)} \tag{2.114}$$

T_1 was named characteristic period of the ITTC spectrum and also adopted as the average period of ISSC spectrum with a different notation $T_v = T_1$, which will be shown shortly.

2.3.6.2 Zero-crossing period

The average of zero-crossing period is derived from the average of zero-crossing frequency of narrow-band Gaussian process in Sec. 1.2.6.1, which is calculated by substituting B-spectrum Eq. (2.107):

$$T_z = 2\pi \left(\frac{m_0}{m_2} \right)^{1/2} = 2\pi \left(\frac{\int_0^\infty U(\omega)d\omega}{\int_0^\infty \omega^2 U(\omega)d\omega} \right)^{1/2} = \left(\frac{4}{5\pi} \right)^{1/4} T_m \qquad (2.115)$$

2.3.6.3 Average crest-to-crest period

The average of crest-to-crest period was derived from the average frequency of all maxima of broad-band process in Sec. 1.2.7.1, which may be expressed using B-spectrum Eq. (2.107):

$$T_c = 2\pi \left(\frac{m_2}{m_4} \right)^{1/2} = 2\pi \left(\frac{\int_0^\infty \omega^2 U(\omega)d\omega}{\int_0^\infty \omega^4 U(\omega)d\omega} \right)^{1/2} \qquad (2.116)$$

However, as the denominator approaches $\to \infty$, it diverges for the B-spectrum in Eq. (2.107). Thus this period should not be used as an index with the theoretical spectrum.

2.3.6.4 Significant period

Significant period is defined as the average period of one third of the highest waves on the record. It is determined directly by measuring the time between significant wave crests of the recorded data. But there is no way to derive this from theory. In various studies, Bretschneider found that the significant period falls between values of T_p (peak period of period spectrum) and T_m and proposed the value:

$$T_s = \left(\frac{4}{3}\right)^{1/4} T_p = \left(\frac{4}{5}\right)^{1/4} T_m \qquad (2.117)$$

2.3.6.5 Summary of the periods in relation to modal period

$$T_s = 0.946T_m, \; T_p = 0.880T_m, \; T_a = 0.857T_m$$
$$T_z = 0.710T_m, \; T_1 = 0.772T_m \qquad (2.118)$$

It is noted that T_p peak density period of period energy spectrum is frequently used in literature today to indicate modal period.

2.3.7 ISSC spectrum (1964)

$$S(f) = 0.11H_s^2 T_v (T_v f)^{-5} \exp\left[-0.44(T_v f)^{-4}\right] \qquad (2.119)$$

where $T_v = T_1$ and Eq. (2.119) is identical to the B-spectrum Eq. (2.106).

2.3.8 ITTC spectrum (1978)

The 15th International Towing Tank Conference (ITTC 1978) recommended the use of a form of the B-spectrum for average conditions and not for fully developed seas, given the significant wave height and characteristic period:

$$U(\omega) = \frac{A}{\omega^5} \exp\left(-\frac{B}{\omega^4}\right), \quad A = 173H_s^2 T_1^{-4}, \; B = 691T_1^{-4} \qquad (2.120)$$

If characteristic period T_1 is not known, one may use approximately $A = 8.10 \times 10^{-3} g^2$ and $B = 3.11 / H_s^2$.

2.3.9 JONSWAP spectrum

The Joint North Sea Wave Project (JONSWAP) was carried out in 1968 along a line extending over 160 km into the North Sea from the Sylt Island, Germany. From an analysis of the about 2000 wave records, Hasselmann et al. (1973) derived JONSWAP spectrum being representative of wind-generated seas with a fetch limitation. It uses five parameters of $f_m, \alpha, \gamma, \sigma_a, \sigma_b$:

$$U(f) = \alpha g^2 (2\pi)^{-4} f^{-5} \exp\left[-\frac{5}{4}\left(\frac{f}{f_m}\right)^{-4}\right] \gamma^{\exp\left[-\frac{(f-f_m)^2}{2\sigma^2 f_m^2}\right]} \tag{2.121}$$

where

$\sigma = \sigma_a = 0.07$ for $f \leq f_m$ is the absolute constant indicating average width to the left of the spectrum,

$\sigma = \sigma_b = 0.09$ for $f > f_m$ is the absolute constant indicating average width to the right of the spectrum, and,

$\alpha = 0.076 \bar{x}^{-0.22}$ is allowed to vary due to non-dimensional fetch $\bar{x}$, which is different from 0.0081 of P-M spectrum of fully developed sea.

$f_m = 3.5 \dfrac{g}{U} \bar{x}^{-0.33}$ is the modal frequency depending on fetches and wind velocity.

$\bar{x} = \dfrac{\sqrt{gx}}{U}$ is a dimensionless fetch.

U is wind speed at 10 m above sea level, where x is a fetch.

γ is a peakedness parameter.

Since σ_a and σ_b are fixed as constants, JONSWAP spectrum is a function of three parameters α, f_m, and γ.

(i) $\alpha g^2 (2\pi)^{-4}$: scale factor independent of frequency f

(ii) $f^{-5} \exp\left[-\dfrac{5}{4}\left(\dfrac{f}{f_m}\right)^{-4}\right]$: spectrum shape term

(iii) $\gamma^{\exp\left[-\frac{(f-f_m)^2}{2\sigma^2 f_m^2}\right]}$: enhancement term

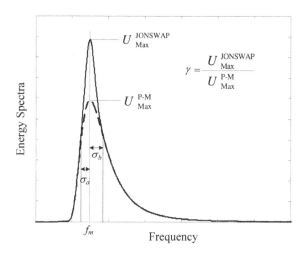

Fig. 2.5. Comparison of P-M and JONSWAP spectrum.

The product of (i) and (ii) represents the fully developed one parameter P-M spectrum Eq. (2.112). (iii) denotes the enhancement factor. Thus the spectrum ranges from the fully developed one parameter P-M spectrum to the sharply peaked spectrum shape. And γ is the ratio of the maximum of JONSWAP spectral density to the maximum of P-M spectral density. If $\gamma = 1$, JONSWAP is identical to the P-M spectrum. γ is a random number varying typically from 1 to 7, the mean of which is 3.3. This spectrum can represent the spectra of steep waves in the North Sea that cannot be represented by P-M spectrum. Regardless of the peakedness γ of the spectrum, it represents the linear random wave if simulated digitally in the time domain. It may be possible to make higher nonlinear random waves in the wave tank if it is used as the target wave spectrum to generate the random waves. Another effect of the high peakedness of the spectral density is that it may give larger response at the modal frequency than P-M spectrum when employed as input to determine the response spectrum of the offshore structure in the frequency domain.

2.3.10 JONSWAP spectrum in circular frequency

Corresponding to Eq. (2.121), we have the spectrum as function of angular frequency:

$$U(\omega) = \alpha g^2 (\omega)^{-5} \exp\left[-\frac{5}{4}\left(\frac{\omega}{\omega_m}\right)^{-4} \right] \gamma^{\exp\left[-\frac{(\omega-\omega_m)^2}{2\sigma^2 \omega_m^2} \right]} \qquad (2.122)$$

where

$$\sigma = (0.07 \text{ for } \omega \leq \omega_m, \;\; 0.09 \text{ for } \omega > \omega_m) \qquad (2.123)$$

2.3.11 Norwegian west-coast sea spectra

Houmb and Overik (1976) noted the need for data of many different sea states near the Norwegian west coast for a systematic parameterization. They obtained 115 spectra and fitted them to the JONSWAP spectrum, which has five parameters α, f_m, γ, σ_a, σ_b, among which σ_a, σ_b are regarded absolute constants for expressing the width of the spectrum. But the rest three parameters are not well known for different sea states. These authors provided the result in a table to determine three parameters α, f_m, and γ given the two parameters H_s and T_z for engineering application.

2.3.12 Ochi-Hubble six parameter spectrum

Waves generated by a local wind or storms are called wind-generated seas. On the other hand, if waves are generated by a travel of distant storm for a great distance, then the short steep waves decay during the travel and only the relatively long waves remain. This situation is called swell. Indeed, we occasionally observe very long waves in the ocean even though there are no local winds. Also we often observe the wind-generated seas and swell simultaneously as shown in Fig. 2.6. In this case, the wave spectrum has double peaks, one of which is a very sharp peak at low frequency representing the swell. The above two-peak spectrum can be fitted by the six-parameter spectrum:

$$U(\omega) = \frac{1}{4} \sum_{j=1}^{2} \frac{\left[\left(\lambda_j + \frac{1}{4}\right)\omega_m^4\right]^{\lambda_j}}{\Gamma(\lambda_j)} \frac{H_{Sj}^2}{\omega^{4\lambda_j+1}} \exp\left[-\left(\lambda_j + \frac{1}{4}\right)\left(\frac{\omega_{mj}}{\omega}\right)^4\right] \quad (2.124)$$

$\Gamma(\lambda_j)$ = Gamma function

Ochi (1978) combined two of his three-parameter spectra. The three-parameter spectrum is expressed by a formula of three parameters, namely, significant wave height H_s, modal frequency ω_m, and shape parameter λ.

When $\lambda = 1$, $U(\omega)$ reduces to the two-parameter B-spectrum in Eq. (2.107). Application of this formula to "Hurricane Georges" showed good agreement (Michel, 1999). Combination of this formula for low and high frequency can create the six parameters H_{sj}, ω_{mj}, λ_j ($j = 1, 2$), which are determined numerically and the difference between the theory and observed spectra is minimal. An example of the sum of these three parameter spectra is taken from data (Ochi, 1978) as shown in Fig. 2.6.

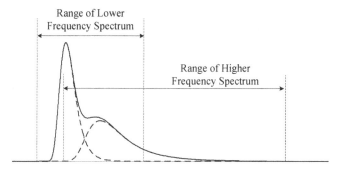

Fig. 2.6. Sketch of Ochi-Hubble six parameter spectrum.

2.3.13 Modified two-parameter JONSWAP spectrum

For simplicity as well as consistency with the seakeeping performance assessment throughout the world, it is desirable to have a modified version of JONSWAP spectrum that depends only on the significant wave height and modal frequency for a given γ (Lee and Bales, 1980). The foregoing approach might have been developed by others to take the form as shown below. It is noted that no specific author's name is found in literature:

$$U(\omega) = \frac{5}{16} H_s^2 \omega_m^4 \omega^{-5} \exp\left[-1.25\left(\frac{\omega_m}{\omega}\right)^4\right](1-0.287\ln\gamma)\gamma^{\exp\left[-\frac{(\omega-\omega_m)^2}{2\sigma^2\omega_m^2}\right]} \qquad (2.125)$$

where

$$\sigma = \begin{cases} 0.07 & \text{for } \omega \le \omega_m \\ 0.09 & \text{for } \omega > \omega_m \end{cases} \qquad (2.126)$$

Equation (2.125) represents B-spectrum Eq. (2.107) multiplied by the product of the term $(1-0.287\ln\gamma)$ and enhancement term. The modified spectrum is the most widely used JONSWAP spectrum. For instance it is used for the 100 year return sea, in the Gulf of Mexico, with $H_s = 12.5$ m, $T_m = 13.5-16.5$ s and $\gamma = 2$ (Demirbilek, 1989).

2.3.14 Isherwood formula

To make the data of Houmb and Overik (1976) in Sec. 2.3.10 more convenient for engineering work, Isherwood (1986) presented the above data in the form of algebraic equations with figures. Given the wave energy spectrum one can estimate average zero-crossing period:

$$T_z = 2\pi\sqrt{\frac{m_0}{m_2}} \qquad (2.127)$$

and the slope s:

$$s = \frac{2\pi H_s}{g T_z^2} \qquad (2.128)$$

With fixed absolute constants $\sigma = \sigma_a = 0.07$ for $f \le f_m$, and $\sigma = \sigma_b = 0.09$ for $f > f_m$, one can evaluate three parameters α, f_m and γ from the following three equations.

The 1st equation:

$$f_m T_z = 0.6063 + 0.1164\gamma^{1/2} - 0.01224\gamma \qquad (2.129)$$

which is valid for $0.6 < \gamma < 8.0$ with standard error $= 0.0001$.
The 2nd equation:

$$\frac{\alpha}{s^2} = 2.964 + 0.4788\gamma^{1/2} - 0.3430\gamma + 0.04225\gamma^{3/2} \qquad (2.130)$$

which is valid for $0.6 < \gamma < 8.0$ with standard error $= 0.0002$.
The 3rd equation:

$$\gamma = 10.54 - 1.34s^{-1/2} - \exp(-19 + 3.775s^{-1/2}), \quad \text{for } s < 0.037 \qquad (2.131)$$

$$\gamma = 0.9 + \exp(18.86 - 3.67s^{-1/2}), \qquad \text{for } s < 0.037 \qquad (2.132)$$

which are valid for $0.03 < s < 0.15$ with standard error $= 0.06$.

The 1st and 2nd relations represent the relations of $f_m T_z$, and α/s^2 with γ respectively, while the last represents γ as function of slope s as shown in Fig. 2.7.

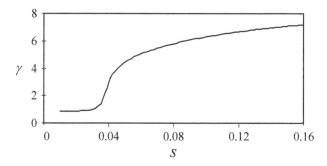

Fig. 2.7. Peakedness parameter versus wave slope.

All the above equations may be presented in graphical forms. Here we have the graph of γ versus s, indicating that if the real waves are steeper the spectrum will have higher peakedness. Since the slope s is determined given the wave spectrum, one can estimate the peakedness using Fig. 2.7. Since γ has been determined above one can determine f_m and α using Eqs. (2.129) and (2.130).

If the spectrum is used to simulate the wave in the time domain, the result will become large amplitude linear random wave. However, if it is used as the target spectrum for random wave in the wave tank, it will produce higher nonlinear random waves. Thus, this data may be useful for studying the effect of the peakedness on the responses of offshore structures in the wave tank. Given the significant wave height and average period for a sea state, good estimates of the three parameters of the corresponding JONSWAP spectrum can conveniently be made using the above formulas. For areas other than the North Sea, equation of γ as function of steepness s may require modification to fit empirical data for the sea area in question.

2.4 Family of Wave Spectra

There are families of B-spectrum and of Ochi-Hubble-6 parameter spectrum. We will review the former here. The latter can be referred to Ochi (1978).

2.4.1 Family of B-spectrum

Ochi (1978) derived families of two-parameter B-spectra and Ochi-Hubble six parameter spectra by analyzing the data of North Atlantic. He established a theory for finding the formula of conditional probability of the modal frequency $\omega_m(T_m)$ for any given specific significant wave height. It was based on the measured significant wave height and zero crossing period T_z at Weather Station I (Draper and Squire, 1967) and J (Draper and Whitaker, 1965) and observed data of waves at weather stations (A, B, C, D, E, I, J, K, and M) in the North Atlantic by Walden (1964). The observed data on the other hand was converted to the measured significant wave heights and modal periods. The conversion factors were established from an analysis of the measured and observed data taken at the stations I (Draper and Squire, 1967) and J (Draper and Whittaker, 1965). After the conversion, the five parameters $m_{lnx}, m_{lny}, \sigma_{lnx}, \sigma_{lny}, \rho_{xy}$ necessary for determining the conditional lognormal distribution for H_s and $T_z(T_m)$ at various locations A, B, C, D, E, J K, M, and I were determined using the method given in Sec. 1.3.2.3. The foregoing five parameters were averaged to represent the mean of the North Atlantic (Table 2.2). The most probable and upper and lower values of modal period T_m for a given confidence coefficient were derived from the conditional lognormal distribution in Sec. 1.3.2.3.

Table 2.2. Values of modal frequency ω_m as function of significant wave height (the mean of the North Atlantic) Ochi (1978).

		ω_m value
Lower ω_m	0.95	0.048 (8.75 – ln H_S)
	0.85	0.054 (8.44 – ln H_S)
	0.75	0.061 (8.07 – ln H_S)
	0.50	0.069 (7.77 – ln H_S)
Most probable		0.079 (7.63 – ln H_S)
Upper ω_m	0.50	0.099 (6.87 – ln H_S)
	0.75	0.111 (6.67 – ln H_S)
	0.85	0.119 (6.65 – ln H_S)
	0.95	0.134 (6.41 – ln H_S)

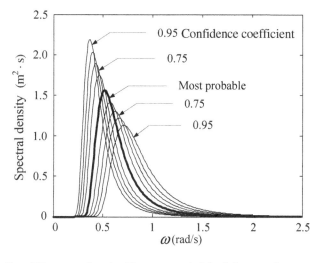

Fig. 2.8. Family of B-spectra for significant wave height 3.0 m as the mean of the North Atlantic.

The above formulas as shown in Table 2.2 were used to determine the relationship between H_s and modal frequencies ω_m.

Now we can generate the family of two parameter B-spectrum Eq. (2.107) consisting of nine modal frequencies for an arbitrarily chosen significant wave height. Figure 2.8 illustrates the family of B-spectra for H_s of 3 m as the mean of the North Atlantic.

In Fig. 2.8, there are four pairs of spectra with confidence coefficients in the lower and higher frequency region. These kinds of families of B-spectra can be generated for various significant wave height H_s from the low to phenomenal sea state using formulas in Table 2.2. Figure 2.8 illustrates that the B-spectra represent rising and falling seas.

References

Journal references:

Draper, L. and Squire, E.M. (1967). Waves at Ocean Weather Ship Station India, Trans. RINA, Vol. 109.

Hasselmann, K., Ross, D.B., Mueller, P. and Sell, W. (1976) A Parametric Wave Prediction Model, J. Phys. Oceanography Vol. 6, pp. 200–228.

Isherwood, R.M. (1987). Technical Note:A revised parameterization of the JONSWAP spectrum, Appl. Ocean Res., Vol. 9, No.1, pp. 47–50.

Michel, W. H. (1999). Sea Spectra Revisited, Marine Technology, Vol. 36, No. 4, pp. 211–227.

Ochi, M. K. (1978). Wave Statistics for the Design of Ships and Ocean Structures, SNAME Transactions, Vol. 86, pp. 47–76.

Pierson, W.J. and Moskowitz, L. (1964). A Proposed Spectral Form for Fully Developed Wind Seas Based on the Similarity Theory of S.A. Kitaigorodskii, J. Geophys. Res., Vol. 69, pp. 5181–5190.

Tucker, M.J., Challenor, P.G. and Carter, D.J.T. (1984). Numerical Simulation of a Random Sea: A Common Error and its Effect Upon Wave Group Statistics, Appl. Ocean Res.,Vol. 6, No. 2, pp.118–122.

Book references:

Barltrop, N.D.P and Adams, A.J. (1991) Dynamics of Fixed Marine Structures, Third Edition, Butterworth-Heinemann Ltd.

Bendat, J.S. and Piersol, A.G. (1986) Random Data Analysis and Measurement Procedures, Second Edition, John Wiley and Sons, New York.

Bendat, J.S. (1990) Nonlinear System Analysis and Identification from Random Data, John Wiley and Sons, New York.

Bretschneider, C.L. (1969) On Wind Generated Waves, Topics in Ocean Engineering, Gulf Publishing, Houston, pp. 37–41.

Blackman, R.B. and Tukey, J.W. (1956) The Measurement of Power Spectra from the Point of View of Communication Engineering, Dover Publications, New York.

Chakrabarti, K.S. (1990) Nonlinear Methods in Offshore Engineering, Elsevier.

Demirbilek, Z.(Ed) (1989) Tension Leg Platform — A State of the Art Review, ASCE, 345 East 47th Street, New York, NY 10017–2398.

Goda, Y. (2000) Random Seas and Design of Marine Structures, Advanced Series in Ocean Engineering — Vol. 15, World Scientific Publishing Co., Pte. Ltd.

Kinsman, B. (1965) Wind Waves, Dover Publications, Inc., New York.

Korvin-Kroukovsky, B.V. (1963) Theory of Seakeeping, The Society of Naval Architects and Marine Engineers, Jersey City, New Jersey.

Lewis, E.V. (1989) Motions in Waves and Controllability, Principles of Naval Architecture, Vol. III, The Society of Naval Architects and Marine Engineers, Jersey City, New Jersey.

Newland, D. E. (1986) An Introduction to Random Vibrations and Spectral Analysis, Longman Inc., New York.

Price, W.G. and Bishop, R.E.D. (1974) Probabilistic Theory of Ship Dynamics, Chapman and Hall Ltd.

Rice, S.O. (1954) Selected Papers on Noise and Stochastic Processes, Dover, New York.

Thomson, W.T. (1981) Theory of Vibration with Applications, Second Edition, Prentice-Hall, Inc., Englewood Cliffs, New Jersey 07632.

Tupper, E. (1996) Introduction to Naval Architecture, The Society of Naval Architects and Marine Engineers, Jersey City, New Jersey.

Proceedings references:

Bretschneider, C.L. (1959). Wave Variability and Wave Spectra for Wind-Generated Gravity Waves, Beach Erosion Board, US Army Corps of Engineers, Technical Memorandum, No. 118.

Draper, L. and Whittaker, M.A.B. (1965). Waves at Ocean Wheather Ship Station Juliett, Deutsche Hydrographische Zeitschrift, Band 18, Heft 1.

Houmb, O.G. and Overik, T. (1976). Parameterization of Wave Spectra and Long Term Joint Distribution of Wave Height and Period, Proc. BOSS'76, Vol. 1, pp. 144–169.

Lee W.T. and Bales, S.L. (1980). A Modified JONSWAP SPECTRUM Only on Wave Height and Period, David W. Taylor Naval Ship Research and Development Center, Bethesda, Maryland 20084, DTNSRDC/SPD-0918-01.

St. Denis, M. (1980). On the Statistical Description of Seaways of Moderate Severity, SNAME STAR Symposium.

Walden, H. (1964). Die Eigenschaften der Meerswellen in Nordatlandischen Ozean, Deutscher Wetterdienst, Einzelveroeffentlichungen, No. 41.

Chapter 3

Volterra Linear Model and Extreme Response

3.1 Introduction

The linear theory for response of offshore structures in the confused seas
employs the linear spectral method; using the square of response amplitude
operator (RAO) between the input and output energy density spectra (St.
Denis and Pierson, 1953). Two decades later, Cummins (1973) found that
theoretical RAO much overestimates the extreme deck-wettings, compared
to the experimental RAO, which may be attributed at least to the effect of
high nonlinear waves. Ever since the foregoing review, the questions about
the validity of theoretical RAO in various seas have remained unsolved.

The RAO square had been applied to predict extreme response of
structure to severe seas using linear spectral method, assuming that all the
waves are Gaussian regardless of the sea severity (Ochi, 1978).

3.1.1 Volterra linear model

The analysis of dynamic behavior of offshore structure in the seas applies a
linear system model with Gaussian excitation and response, both in the time
and frequency domain as shown in Fig. 3.1.

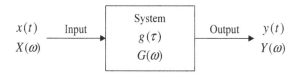

Fig. 3.1. Block diagram of a linear system, single input-output system.

93

The linear system is represented by the linear impulse response function (IRF), or linear frequency response function (FRF). We consider here the single input and single output linear system, and assume the input is Gaussian stationary zero-mean process.

3.1.2 Frequency response function

Linear system is characterized by FRF or IRF. As an example of FRF, one considers an equation of simple harmonic heaving motion $y(t)$ of a structure due to a simple harmonic wave-exciting force $x(t)$:

$$M(\omega)\ddot{y}(t) + N(\omega)\dot{y}(t) + By(t) = x(t) \tag{3.1}$$

where M and N are the sum of structure and added mass and wave-making or radiation damping coefficient in heaving motion that are functions of the wave frequency ω. And B is the heave restoring force coefficient (hydrostatic stiffness). The excitation $x(t)$ and response $y(t)$ are given in the complex forms:

$$x(t) = A_{in}(\omega)e^{i\omega t} \quad \text{and} \quad y(t) = A_{out}(\omega)e^{i\omega t} \tag{3.2}$$

assuming that one takes the real part of the RHS in Eq. (3.2).

Substituting Eq. (3.2) in Eq. (3.1), and omitting the time factor, one has:

$$(-M\omega^2 + iN\omega + B)A_{out}(\omega) = A_{in}(\omega) \tag{3.3}$$

from which the complex output amplitude:

$$A_{out}(\omega) = \frac{1}{B} \frac{1}{\left[1 - (\omega/\omega_n)^2 + i2\varsigma(\omega/\omega_n)\right]} A_{in}(\omega) \tag{3.4}$$

where $\omega_n = \sqrt{B/M}$ is the natural frequency of the system, and $\varsigma = N/(2M\omega_n)$ is the damping normalized by the critical damping $2M\omega_n$. FRF is defined as the ratio of the output amplitude to the input amplitude:

$$G(\omega) = \frac{A_{out}(\omega)}{A_{in}(\omega)} = \frac{1}{B} \frac{1}{\left[1 - (\omega/\omega_n)^2 + i2\varsigma(\omega/\omega_n)\right]} \tag{3.5}$$

One may obtain the same FRF by taking the ratio of FT of output and input in Eq. (3.2),

$$G(\omega) = \frac{Y(\omega)}{X(\omega)} \tag{3.6}$$

3.1.3 Impulse response function

The IRF represents the system response characteristics in the time domain, which may be derived considering a transient motion of the structure due to unit impulse:

$$M\ddot{y} + N\dot{y} + By = x(t) \tag{3.7}$$

where M, N, and B represent mass, damping and restoring force (stiffness). An impulsive force $x(t)$ is applied to the body for a very short time interval ε. Then the impulse I is in dimension of (force)×(time):

$$I = \int_{\varepsilon} x(t)dt \tag{3.8}$$

As ε approaches zero, such force tends to become infinite. However, the impulse I is in reality considered finite. When I is equal to unity, such force in the limiting $\varepsilon \to 0$ is called the unit impulse or the Dirac delta function ($\delta(t)$). A delta function at t is identified by $\delta(t - \tau)$ that represents the unit impulse at $t = \tau$:

$$\int_0^{\infty} \delta(t - \tau)dt = 1.0 \tag{3.9}$$

If $\delta(t-\tau)$ is multiplied by any time function $x(t)$ we have the excitation at $t = \tau$ in the form:

$$\int_0^{\infty} x(t)\delta(t - \tau)dt = x(\tau), \ \tau \le \infty \tag{3.10}$$

Since $I = x(t) \, dt = M \, dv$, where dv is the small velocity, the impulse acting on the structure, will result in a sudden change in its velocity $\dot{y}$ equal to I/M without an appreciable change in its displacement y. Under forced oscillation, the linear system with initial conditions $y(0) = 0$ and $\dot{y}(0) = I/M$ behaves according to:

$$y(t) = I \, g(t) \tag{3.11}$$

with

$$g(t) = \frac{1}{M\omega_n\sqrt{1-\zeta^2}} e^{-\zeta\omega_n t} \sin\sqrt{1-\zeta^2}\,\omega_n t \qquad (3.12)$$

where $g(t)$ is the response per unit impulse or unit impulse response function of the dynamic system as shown in Fig. 3.2.

The terminology impulse is carried over to the general case. For instance, if the input $x(t)$ is wave elevation, the impulse $x(t)dt$ is in the dimension of (length) $\times$ (time). If the input $x(t)$ is force, the impulse $x(t)dt$ is in the dimension of (force) $\times$ (time). The impulse response of a system is defined as the system response to an impulsive impact, $I\delta(t)$, where I has the proper dimension as above. The excitation $\delta(t)$ is described as a unit impulse when I is numerically unity in the above.

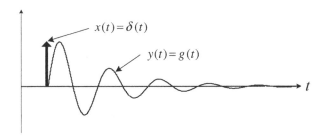

Fig. 3.2. Sketch of unit impulse $\delta(t)$ and unit impulse response function $g(t)$.

3.1.4 Relation between frequency response and impulse response

Either FRF or IRF fully defines the linear system characteristics. Therefore we can derive one from the other and vice versa. Because we are dealing with stable systems which are dormant before they are excited and for which motion dies away after an impulse, it is valid that:

$$\int_{-\infty}^{\infty} |g(t)|\, dt < \infty \qquad (3.13)$$

and we may therefore take Fourier transforms of both unit impulse of input $x(t) = \delta(t)$ and the transient output due to unit impulse response function $y(t) = g(t)$:

$$X(\omega) = \int_{-\infty}^{\infty} x(t)e^{-i\omega t}\,dt = \int_{-\infty}^{\infty} \delta(t)e^{-i\omega t}\,dt = 1 \tag{3.14}$$

$$Y(\omega) = \int_{-\infty}^{\infty} y(t)e^{-i\omega t}\,dt = \int_{-\infty}^{\infty} g(t)e^{-i\omega t}\,dt \tag{3.15}$$

Substituting Eqs. (3.14) and (3.15) into Eq. (3.6), one obtains the frequency response function FRF as the Fourier transform of the unit impulse response function IRF in the form:

$$G(\omega) = \int_{-\infty}^{\infty} g(t)e^{-i\omega t}\,dt \tag{3.16}$$

Consequently, the inverse Fourier transform of FRF in Eq. (3.16) gives the unit impulse response function:

$$g(t) = \frac{1}{2\pi} \int_{-\infty}^{\infty} G(\omega)e^{i\omega t}\,d\omega \tag{3.17}$$

Equations (3.16) and (3.17) constitute a Fourier transform pair in accordance to the convention A defined in Sec. 2.1.5. The FRF is more popular than IRF. However, the latter has better physical meaning because frequency response assumes a fully steady oscillatory state, whereas the real physical behavior of wave and structure is essentially unsteady. Thus the IRF can be applied to describe the arbitrary random excitation as shown below.

3.1.5 Impulse response to random excitation

In Fig. 3.3 we find that the impulse (shaded column) due to the input $x(\tau)$ between the time limits τ and $\tau + \Delta\tau$ has the magnitude $x(\tau)\Delta\tau$. The response at time t to this impulse alone is $x(\tau)\Delta\tau g(t-\tau)$. Hence we may obtain the total response $y(t)$ at t by linearly superposing all the separate responses to all the small impulses that make up the total time history of $y(t)$ back to $t = -\infty$:

$$y(t) = \int_{-\infty}^{t} g(t-\tau)x(\tau)\,d\tau \tag{3.18}$$

The superposition or convolution integral of the Eq. (3.18) is a very important input-output relationship for a linear system. However, in this relationship the system must be passive so that it only responds to past inputs and $g(t)$ decays eventually to static equilibrium so that magnitude $|g(t)|$ is in general bounded by a finite level, i.e.

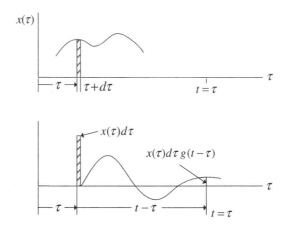

Fig. 3.3. Sketch of random excitation $x(\tau)$ and response to the random excitation.

$$\int_{-\infty}^{\infty} |g(t)dt| < \infty \qquad (3.19)$$

First recall that $g(t-\tau)$ is the response to a unit impulse at $t -\tau = 0$, that is at $t = \tau$. For $(t - \tau) < 0$, there has been no impulse applied. Hence for $t < \tau$, $g(t-\tau) = 0$. Thus we can extend the upper limit from $\tau = t$ to $\tau = \infty$:

$$y(t) = \int_{-\infty}^{\infty} g(t - \tau)x(\tau)d\tau \qquad (3.20)$$

Equation (3.20) represents the relation between Gaussian input x and Gaussian output y of a linear system. The IRF is frequently called linear filter.

Equation (3.20) is called Volterra linear system model in the time domain, and the input and output are zero-mean Gaussian process. The Volterra quadratic and cubic model will be discussed in Chaps. 7 through 11.

3.1.5.1 Linear filter for random wave simulation

The digital waves of arbitrarily long simulation are possible (Dalzell, 1971). The method simulates practically infinitely long unrepeatable Gaussian random waves given the target energy density spectrum. One may determine $G(\omega)$, from the energy density spectrum, and determine the IRF or linear filter:

$$g(t) = \frac{1}{2\pi} \int_{-\infty}^{\infty} G(\omega)e^{-i\omega t}d\omega \tag{3.21}$$

On the other hand we can determine the Gaussian white noise $x(t)$ with unit variance, which can be readily determined from MATLAB. Thus we can compute the random wave time series by the following convolution:

$$y(t) = \int_{-\infty}^{\infty} g(\tau)x(t-\tau)d\tau \tag{3.22}$$

The random wave simulated in the above manner may be utilized as the excitation to the 2nd-order impulse response function. The 2nd-order response needs much longer simulation than the corresponding linear response. This problem will be discussed in Chaps. 7, 8, and 9.

3.2 Relation Between Input and Output Energy Spectra

3.2.1 Relation between input and output auto-energy spectra

The input and output are stationary Gaussian processes with zero-mean. Thus the autocorrelation function of the output $y(t)$ is formed using Volterra linear model Eq. (3.20):

$$\begin{aligned} R_{yy}(\tau) &= E[y(t)y(t+\tau)] \\ &= E[\int_{-\infty}^{\infty} g(\theta_1)x(t-\theta_1)d\theta_1 \int_{-\infty}^{\infty} g(\theta_2)x(t+\tau-\theta_2)d\theta_2] \end{aligned} \tag{3.23}$$

Since we assume the input is stationary, the autocorrelation function is independent of absolute time t:

$$E[x(t-\theta_1)x(t+\tau-\theta_2)] = R_{xx}(\tau-\theta_2+\theta_1) \tag{3.24}$$

The autocorrelation function of output Eq. (3.23) is thus written:

$$R_{yy}(\tau) = \int_{-\infty}^{\infty} g(\theta_1)d\theta_1 \int_{-\infty}^{\infty} g(\theta_2)d\theta_2 R_{xx}(\tau-\theta_2+\theta_1) \tag{3.25}$$

Applying Fourier transform of convention A on both sides of Eq. (3.25), we have:

$$S_{yy}(\omega) = \int_{-\infty}^{\infty} R_{yy}(\tau)e^{-i\omega\tau}d\tau$$

$$= \int_{-\infty}^{\infty} g(\theta_1)e^{i\omega\theta_1}d\theta_1 \int_{-\infty}^{\infty} g(\theta_2)e^{-i\omega\theta_2}d\theta_2 \qquad (3.26)$$

$$\times \int_{-\infty}^{\infty} R_{xx}(\tau-\theta_2+\theta_1)e^{-i\omega(\tau-\theta_2+\theta_1)}d\tau = |G(\omega)|^2 S_{xx}(\omega)$$

The above relation is changed to the one-sided energy density spectra,

$$U_{yy}(\omega) = |G(\omega)|^2 U_{xx}(\omega) \qquad (3.27)$$

Given the wave energy spectrum and square of RAO $[G(\omega)^2]$ of the system, one can determine the response energy spectrum. The result above is a linear spectral relation between the input and output. The foregoing relation may be called linear spectral method for estimation of linear response in the Gaussian seas. This is used for estimating the statistical data such as the variance of the response, most probable extreme response and others.

3.2.2 RAO from cross- and auto-spectra

Similarly we assume the input and output are stationary Gaussian processes. Then the cross-correlation function between the input and output is formed using Volterra linear system model Eq. (3.20):

$$R_{xy}(\tau) = E[x(t)y(t+\tau)] = E[x(t)\int_{-\infty}^{\infty} g(\theta)x(t+\tau-\theta)d\theta]$$

$$= \int_{-\infty}^{\infty} g(\theta)d\theta E[x(t)x(t+\tau-\theta)d\theta] \qquad (3.28)$$

$$= \int_{-\infty}^{\infty} g(\theta)d\theta R_{xx}(\tau-\theta)$$

Similarly applying Fourier transform of convention A on the above cross- and auto-correlation, and referring to Eq. (3.16), we have:

$$G(\omega) = \frac{S_{xy}(\omega)}{S_{xx}(\omega)} \qquad (3.29)$$

Accordingly using the formula in Sec. 2.2.2 we can write Eq. (3.29) in the form,

$$G(\omega) = \frac{U_{xy}(\omega)}{U_{xx}(\omega)} \tag{3.30}$$

Equation (3.30) gives RAO (or LTF) of the system.

3.2.3 Linear coherency function

In order to examine the linearity of the system as a function of frequency, one may need to estimate the linear coherence function:

$$\gamma^2_{xy} = \frac{\left|U_{xy}(\omega)\right|^2}{U_{xx}(\omega)U_{yy}(\omega)} \tag{3.31}$$

This linear coherence function for all ω is a non-dimensional constant that is bounded between zero and one; $0 \le \gamma^2_{xy}(\omega) \le 1$. The coherency function is a measure of a possible linear relationship between $x(t)$ and $y(t)$. If a perfect linear relationship exists between $x(t)$ and $y(t)$ at some frequency ω, then the coherence function $\gamma^2_{xy}(\omega)$ will be equal to unity at the frequency. The linear coherence can provide a simple practical way to detect non-linearity without identifying its precise nature. Dalzell and Yamanouchi in Korvin-Kroukovski (1961) suggests that the linearity practically holds if $\gamma^2_{xy}(\omega) \ge 0.87$. Increasing the sample size, we may improve the linear coherency function.

Example 3.1 Coherency test of a barge test data

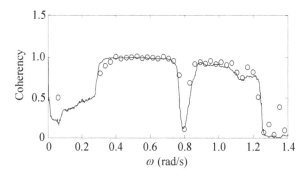

Fig. 3.4. Coherency function of force on a fixed barge in wave tank, circle (Blackman-Tukey spectrum); solid (direct FFT-spectrum), $H_s = 3$ m.

A model test was conducted to measure the surge-exciting force of a fixed barge in the random wave of H_s = 3 m. The auto- and cross-spectral densities were estimated to determine the RAOs (or LTFs). In order to investigate the linearity of the system the coherency function was computed as shown in Fig. 3.4. The indirect and direct methods are applied for estimation of the energy density spectra as given in Chap. 2. We see that these two methods give practically the same results.

The linearity holds in the range of frequency 0.3 rad/s–1.2 rad/s, where the averaged reflection coefficient, was 0.15 using a wave tank at Texas A & M University. Outside the foregoing frequency range the reflection coefficients drastically fluctuate.

3.2.4 Wave spectra for moving frame

Suppose a body-fixed coordinate system xyz moves at a constant speed U along the space fixed positive X-axis parallel to x-axis and the wave progresses in the direction making an angle μ with respect to positive X-axis as shown in Fig. 3.5, then we have the encounter frequency:

$$\omega_e = \omega - \frac{\omega^2}{g} U \cos \mu \qquad (3.32)$$

where ω is the wave frequency referred to the space-fixed frame XYZ. The wave spectrum with respect to the space-fixed frame is transformed to the moving frame, by considering that the energy remains same:

$$U_{xx}(\omega_e) d\omega_e = U_{xx}(\omega) d\omega \qquad (3.33)$$

$$U_{xx}(\omega_e) = U_{xx}(\omega) \frac{d\omega}{d\omega_e} \qquad (3.34)$$

Substituting Eq. (3.32) into Eq. (3.34), one has the wave spectrum with respect to the moving frame at constant velocity U:

$$U_{xx}(\omega_e) = U_{xx}(\omega) \left(1 - \frac{2\omega U}{g} \cos \mu \right)^{-1} \qquad (3.35)$$

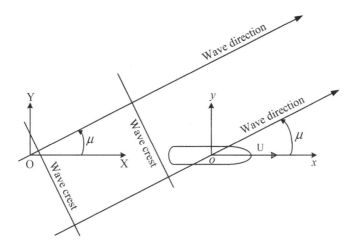

Fig. 3.5. Sketch of the wave heading, body-fixed and space-fixed coordinate system.

Thus the response spectrum in the encounter frequency domain is given by:

$$U_{yy}(\omega_e) = |G(\omega_e)|^2 U_{xx}(\omega_e)$$

(3.36)

which is the linear spectral relation similar to Eq. (3.27). In the above, the energy spectral densities are one-sided and the RAO is in the encounter frequency.

3.2.5 Response to short-crested seas

Wind-generated waves do not necessarily propagate toward one specific direction, such as the wind direction. Instead, the energy is distributed toward various directions. More specifically, wave energy associated with the frequencies close to the modal frequency ω_m is primarily propagated toward the wind direction, while wave energy associated with frequencies lower or higher than the modal frequency is distributed over a range of angular direction μ. Thus it is clear that we do need information on the directional dispersion of wave energy for a description of short-crested random seas. The general equation for the total wave system of components moving in different directions on the mean sea surface is:

$$\eta(x,y,t) = \sum_{i=1}^{\infty}\sum_{j=1}^{\infty} A_{ij}\cos\left[k_i(x\cos\mu_j + y\sin\mu_j) - \omega_i t + \varepsilon_{ij}\right] \qquad (3.37)$$

The wave energy density is defined by the variance of a component wave uniformly distributed over the elementary band area $\Delta\omega\Delta\mu$:

$$\frac{A_{ij}^2}{2} = U(\omega_i,\mu_j)\Delta\omega\Delta\mu \qquad (3.38)$$

The total energy E or variance m_0 of infinitely many wave components in multi-directions is given by:

$$E = m_0 = \frac{1}{2}\sum_i\sum_j A_{ij}^2 = \int_0^{\infty}\int_0^{\infty} U(\omega,\mu)d\omega d\mu \qquad (3.39)$$

The short-crested sea spectrum is assumed in the form:

$$U(\omega,\mu) = U(\omega)D(\mu) \qquad (3.40)$$

where we consider the widely accepted cosine square law for the spreading function:

$$\begin{cases} D(\mu) = \dfrac{2}{\pi}\cos^{2\nu}\mu; & \dfrac{-\pi}{2} \le \mu \le \dfrac{\pi}{2} \\ D(\mu) = 0 & ; \quad \text{otherwise} \end{cases} \qquad (3.41)$$

We assume $\nu = 1$ in the above, thus the total energy of the spectrum:

$$E = m_0 = \int_{-\pi/2}^{\pi/2}\int_0^{\infty} U(\omega)D(\mu)d\omega d\mu \qquad (3.42)$$

Since

$$\frac{2}{\pi}\int_{-\pi/2}^{\pi/2}\cos^2(\mu)d\mu = 1 \qquad (3.43)$$

the total energy of short-crested sea is reduced to that of the point spectrum:

$$E = m_0 = \int_0^{\infty} U(\omega)d\omega \qquad (3.44)$$

The response of structure is affected by the directional waves altogether in the short-crested sea (Lewis, 1989). The response spectral moment is given by:

$$m_n = \int_0^\infty \int_{-\pi/2}^{\pi/2} \omega^n U(\omega, \mu) \frac{2}{\pi} \cos^2 \mu \, d\mu \, d\omega \tag{3.45}$$

where $U(\omega, \mu)$ represents one-sided response energy density spectrum affected by the wave direction μ. Equation (3.45) may be integrated by separating the integral into two parts, in the variables of $\Delta\omega$ and $\Delta\mu$, respectively, and in a discrete form:

$$m_n = \sum_{j=1}^N m(n, j\Delta\mu) W(j\Delta\mu) \tag{3.46}$$

with

$$m(n, j\Delta\mu) = \int_0^\infty \omega^n U_R(\omega, j\Delta\mu) d\omega \tag{3.47}$$

$$\sum_{j=1}^N W(j\Delta\mu) = \int_{-\pi/2}^{\pi/2} \frac{2}{\pi} \cos^2 \mu \, d\mu = 1.0 \tag{3.48}$$

Equation (3.46) represents nth spectral moment of the response to each unidirectional wave in the direction μ. The spreading function weight Eq. (3.48) is determined by using the first Simpson's rule for odd number of ordinates, as shown in an example below.

Example 3.2 The effect of short crested sea on the ship motion
We determine the spreading function weight for 45° heading intervals by dividing the range between −90° and 90° by $\Delta\mu = 45°$. Then we obtain 4 intervals creating 5 ordinates. Accordingly the weight is determined by applying Simpson's first rule as shown in Table 3.1. Recall the sign convention of μ in Fig. 3.5.

Table 3.2 shows the estimation of variances of pitching motions at the dominant wave directions μ_a. Given the variances of the pitching response in the long-crested sea varying μ from 0° to 360° with 45° intervals, one can determine the wave direction μ_w for the spreading function weight varying from −90° to 90° according to $\mu_w = \mu - \mu_a$ and the variance at the dominant wave direction.

Table 3.1. Spreading function weight for 45° heading intervals.

μ_w (deg.)	$\frac{2}{\pi}\cos^2\mu_w$	SM	$W(\mu_w) = \left(\frac{2}{\pi}\cos^2\mu_w\right)(SM)$	μ_w (deg.)
−90	0.0000	1	0.0000	270
−45	0.3183	4	0.3331	315
0	0.6366	2	0.3331	0
45	0.3183	4	0.3331	45
90	0.0000	1	0.0000	90

Table 3.2. Computation of variance of the pitching response to the short-crested sea using wave spreading function for 45° heading intervals.

μ (deg)	$m(0,\mu)$	μ_w $=\mu-\mu_a$	$W(\mu_w)$	$m(0,\mu)$ $\times W(\mu_w)$	μ_w $=\mu-\mu_a$	$W(\mu_w)$	$m(0,\mu)$ $\times W(\mu_w)$	μ_w $=\mu-\mu_a$	$W(\mu_w)$	$m(0,\mu)$ $\times W(\mu_w)$
		$\mu_a = 0°$			$\mu_a = 45°$			$\mu_a = 90°$		
0.0	0.61	0.0	0.3331	0.2031	315.0	0.3331	0.2031	−90	0.0000	0.0000
45.0	0.65	45.0	0.3331	0.2165	0.0	0.3331	0.2165	−45	0.3331	0.2165
90.0	0.03	90.0	0.0000	0.0000	45.0	0.3331	0.0099	0	0.3331	0.0099
135.0	1.99	135.0	0.0000	0.0000	90.0	0.0000	0.0000	45	0.3331	0.6628
180.0	2.21	180.0	0.0000	0.0000	135.0	0.0000	0.0000	90	0.0000	0.0000
225.0	1.99	225.0	0.0000	0.0000	180.0	0.0000	0.0000	135	0.0000	0.0000
270.0	0.03	270.0	0.0000	0.0000	225.0	0.0000	0.0000	180	0.0000	0.0000
315.0	0.65	315.0	0.3331	0.2165	270.0	0.0000	0.0000	225	0.0000	0.0000
				$\Sigma =$ 0.6361			$\Sigma =$ 0.4295			$\Sigma =$ 0.8892

μ (deg)	$m(0,\mu)$	μ_w $=\mu-\mu_a$	$W(\mu_w)$	$m(0,\mu)\times$ $W(\mu_w)$	μ_w $=\mu-\mu_a$	$W(\mu_w)$	$m(0,\mu)$ $\times W(\mu_w)$
		$\mu_a = 135.0°$			$\mu_a = 180°$		
0.0	0.61	225.0	0.0000	0.0000	−180	0.0000	0.0000
45.0	0.65	270.0	0.0000	0.0000	−135	0.0000	0.0000
90.0	0.03	315.0	0.3331	0.0099	−90	0.0000	0.0000
135.0	1.99	0.0	0.3331	0.6628	−45	0.3331	0.6628
180.0	2.21	45.0	0.3331	0.7361	0	0.3331	0.7361
225.0	1.99	90.0	0.0000	0.0000	45	0.3331	0.6628
270.0	0.03	135.0	0.0000	0.0000	90	0.0000	0.0000
315.0	0.65	180.0	0.0000	0.0000	135	0.0000	0.0000
				$\Sigma =$ 1.4088			$\Sigma =$ 2.0617

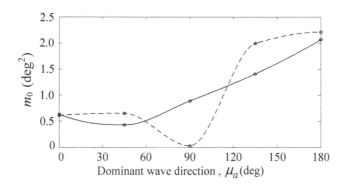

Fig. 3.6. The variance of the pitching response to short-crested sea (solid line) and long-crested sea (dotted line).

Figure 3.6 illustrates the variance of the pitching response of a ship as a function of dominant wave direction μ_a of a short-crested sea. It is to be noted that the variance of the pitch response at long-crested sea is quite different from the short-crested sea condition. At the quartering (45°) and bow (135°) seas, the long-crested sea effect on the pitching motion is higher than the short-crested seas, while at beam sea the pitching is much smaller than the short-crested sea. Thus, the extreme values that can be predicted from the variance of response in short crested-seas are smaller than those of the long-crested seas.

3.3 Extreme Responses to Severe Seas

The failure of structure may occur due to the fatigue of the structure in the seas as well as extreme response to the severest sea. Given the wave energy spectrum of a specific sea, one can determine the response energy spectrum, the variance, and zero-crossing period of the response, which were discussed in Chap. 1. Given the probable extreme and extreme amplitude among N observations, by employing the risk parameter, one may estimate the most probable responses given the response energy spectrum and expose time (Ochi, 1978).

Table 3.3 represents the frequency of occurrence of various sea states in the mean in the North Atlantic (Ochi, 1978).

Table 3.3. Frequency of occurrence of various sea states in the (mean) North Atlantic
(Ochi, 1978).

H_s (m)	Frequency of Occurrence	H_s (m)	Frequency of Occurrence
<1	0.0503	9–10	0.0079
1–2	0.2665	10–11	0.0054
2–3	0.2603	11–12	0.0029
3–4	0.1757	12–13	0.0016
4–5	0.1014	13–14	0.00074
5–6	0.0589	14–15	0.00045
6–7	0.0346	15–16	0.00020
7–8	0.0209	16–17	0.00012
8–9	0.0120	17<	0.00009

3.3.1 Operation time

The extreme amplitude depends on the number of encounters with waves, which is derived from the exposed time T. For instance, in the North Atlantic, we have a data of exposed time as function of sea severity as shown in Fig. 3.7. The significant wave heights between 6 m and 7.5 m are probably expected to occur for 40 hrs. The significant wave height around 14 m to 16 m may be sustained for 6 hrs.

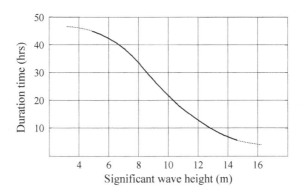

Fig. 3.7. Significant wave height and its persistence in the North Atlantic every 1.52 m interval of significant wave height (Ochi, 1978).

3.3.2 Design extreme response

The probable extreme amplitude was defined in Chap. 1:

$$\bar{y}_N = \sqrt{2\ln\left[\frac{3600T}{2\pi}\sqrt{\frac{m_2}{m_0}}\right]} \sqrt{m_0} \tag{3.49}$$

This is revised to include the factor k which is repeating number of the same sea state (frequency of occurrence in Tables 2.1 and 3.3) and risk factor α in the form:

$$\hat{y}_N(\alpha/k) = \sqrt{2\ln\left[\frac{3600T}{2\pi\alpha/k}\sqrt{\frac{m_2}{m_0}}\right]} \sqrt{m_0} \tag{3.50}$$

in which T is the longest duration of specified sea in hours, $\alpha = 0.01$, risk parameter, k is the repeating number of the same sea state during the life time. The revised extreme response may be called design extreme response (Ochi, 1978).

3.3.3 Short term prediction

Referring to the family of B-spectra in Table 2.2 in Sec. 2.4.1, we consider eight sea states H_s (m) of 2, 4, 6, 8, 10, 12, 14, 16, each of which has nine modal periods. Thus altogether there are 72 families of two parameter spectra. Nine spectra for $H_s = 3$ m are shown in Fig. 2.8. The foregoing environmental condition will produce responses of the offshore structure such as the wave-exciting force, motion and stress at the critical structural member. Here we consider the wave exciting transverse force on a semi-submersible fixed in the waves (Ochi, 1978).

The total service time is assumed to be 20 years. The total time to be exposed to the seas $T = $ (20 years) $\times$ (365 days) $\times$ (24 hrs) = 175200 hrs. These are used to obtain the 72 short-term predictions of the probable extreme force according to Eq. (3.49) as shown in Fig. 3.8. The effect of upper and lower bound of B-spectra with confidence coefficients 0.95 (Table 2.2) are computed as shown in Fig. 3.8.

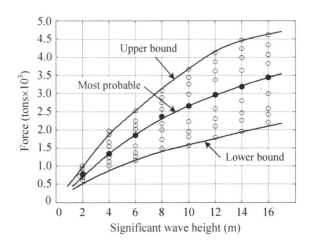

Fig. 3.8. Probable extreme amplitudes of transverse force in various sea states computed by using two parameter B-spectra (Ochi, 1978).

The short term prediction of the design extreme values against the sea severity is also obtained for the cases of upper and lower bounds using Eq. (3.50), where the repeating number of encounter k with a specified sea or the frequency of occurrence of the sea states as given in Table 3.3 is implemented. The highest design extreme force amounts to 6000 tons at the significant wave height of 16.5 m–17 m as shown in Fig. 3.9.

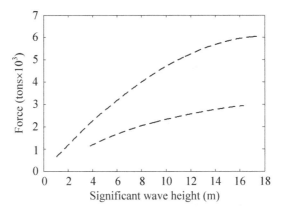

Fig. 3.9. Bounds of extreme values of transverse force for design consideration (Ochi, 1978).

3.3.4 Long-term prediction

The long-term prediction essentially depends on all of the responses predicted by the short-term predictions, as explained in the above section. Here we use a weighting factor approach to add up the weighting factors of the wave energy spectra in Table 3.4, frequency of occurrence of sea severity in Table 3.3, and the weighting factor for the wave direction.

3.3.4.1 Weighting factors for the energy spectra w_j

The conditional lognormal probability density function for significant wave height of 4.6 m at Station I in the North Atlantic and the modal periods were determined in Sec. 2.3.4, from which the weighting factors of the sea spectrum were determined as shown in Table 3.4. These weighting factors are applicable irrespective of sea severity in evaluating the long-term response of structure.

Table 3.4. Weighting factors for a sea spectrum (w_i) (Ochi, 1978).

Weighting factor	A wave energy spectrum
0.2500	most probable modal period
0.1875	modal period of confidence coefficient 0.50 (each)
0.0875	modal period of confidence coefficient 0.75 (each)
0.0500	modal period of confidence coefficient 0.85 (each)
0.0500	modal period of confidence coefficient 0.95 (each)

3.3.4.2 Weighting factor for frequency of encounter w_i

Information concerning the frequency of encounter for structure with each sea severity in its lifetime is necessary to evaluate the design extreme response. For this, information on the geographical location and information on the frequency in the service area of the marine structure should be considered. The frequency of occurrence of various sea severities for the mean of North Atlantic is given in Table 3.3.

3.3.4.3 Example of long term prediction

Here we will basically follow the same approach as used in 100 year return sea (Sec. 1.3.3). We need probability density distribution of H_s, from which we determine the cumulative probability $P(H_s)$. Then we apply it to the return period Eq. (1.128). In the long term prediction, we consider the response force instead of H_s. In this case one has to estimate the number of cycles to be encountered in the lifetime of the structure.

The long-term probability density function of the force applicable to the long-term response may be written as follows:

$$p(y) = \frac{\sum_i \sum_j \sum_k f_0^+ w_i w_j w_k \, p*(y)}{\sum_i \sum_j \sum_k f_0^+ w_i w_j w_k} \tag{3.51}$$

where

$p^*(y) = \dfrac{y}{m_0}\exp\left[-\dfrac{y^2}{2m_0}\right]$ = probability density function for short-term

response for a given spectrum (or Rayleigh probability density) and f_0^+ =

average zero-crossing frequency $= \dfrac{1}{2\pi}\sqrt{\dfrac{m_2}{m_0}}$

w_i = weighting factor for sea state (Table 3.3)
w_j = weighting factor for wave spectra (Table 3.4)
w_k = weighting factor for heading to waves in a given sea or spectrum
The total number of cycles of the responses expected in the lifetime of a marine system:

$$N = \left(\sum_i \sum_j \sum_k f_0^+ w_i w_j w_k\right)(60)^2 T \tag{3.52}$$

where T (hour) is the total exposed time to the sea.

Assume that the platform will be exposed to beam seas for one quarter of its life time, namely, $w_k = 1/4$ (constant), and the severest sea state has a significant wave height of 17 m. The number of response cycles in service of 20 years is estimated from Eq. (3.52), which amounts to $N = (1.864) \times (10^7)$. Analogous to Eq. (1.128), we have the design value, $\hat{y}_N$, with the risk parameter $\alpha = 0.01$

$$\left[\left(1-P(\hat{y}_N)\right)\right]^{-1}=\left(\frac{N}{\alpha}\right) \tag{3.53}$$

The cumulative distribution function involved in Eq. (3.53) is obtained by integrating the long term probability density function Eq. (3.51) with respect to y. Thus we determine the left hand side (LHS) of Eq. (3.53) in the logarithmic scale as shown in Fig. 3.10. Connecting these points we obtain a straight line. Since $N = (1.864) \times (10^7)$ and $\alpha = 0.01$, the total cycle is estimated as $(1.864) \times (10^9)$, the logarithm of which is 21.4, which intersects with the return period line in Fig. 3.10, and gives the design extreme load amplitude 6000 tons. The extreme design value agrees well with the extreme design value estimated by the short-term method for the significant wave height of 16.5 m–17 m as given in Fig. 3.9.

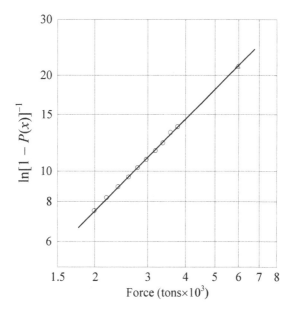

Fig. 3.10. Design extreme value of transverse force evaluated by using B-spectra family $(x = \hat{y}_N)$.

Hence it may safely be concluded (Ochi, 1978) that the extreme value for design consideration estimated from the long-term prediction method agrees well with the short-term prediction method. From the comparison of the two prediction methods, we conclude that as far as estimation of extreme values is concerned, it is appropriate to consider severest seas expected in the service area and to apply the short-term prediction method.

References

Journal references:

Ochi, M. K. (1978). Wave Statistics for the Design of Ships and Ocean Structures, SNAME Transactions, Vol. 86, pp. 47–76.

St. Denis, M. (1980). On the Statistical Description of Seaways of Moderate Severity, SNAME STAR Symposium.

St. Denis, M. and Pierson, W.J. (1953). On the Motions of Ships in Confused Seas, Transactions, SNAME, Vol. 61, pp. 280-357.

Book references:

Korvin-Kroukovsky, B.V. (1963) Theory of Seakeeping, The Society of Naval Architects and Marine Engineers, Jersey City, N.J.

Lewis, E.V. (1989) Motions in Waves and Controllability, Principles of Naval Architecture, Vol. 3, The Society of Naval Architects and Marine Engineers, Jersey City, New Jersey.

Newland, D.E. (1984) An Introduction to Random Vibrations and Spectral Analysis, Second Edition, Longman Inc., New York.

Thomson, W.T. (1981) Theory of Vibration with Applications, Second Edition, 1981, Prentice Hall, Inc., Englewood Cliffs, N.J.

Proceedings references:

Cummins, W.E. (1973). Pathologies of the Transfer Functions, The Seakeeping Symposium Commemorating the 20th Anniversary of the St. Denis-Pierson Paper.

Dalzell, J.F. and Yamanouchi, Y. (1958). Analysis of Model Test Results in Irregular Seas to Determine Motion Amplitudes and Phase Relationship to Waves, Ship Behavior of Ship at Sea, Second Summer Seminar, June 16-20, 1958, Experimental Towing Tank, Stevens Institute of Technology.

Dalzell, J.F. (1971). A Study of the Distribution of Maxima of Nonlinear Ship Rolling in a Seaway, Report SIT-DL-71-1562.

Division II

Linear Wave-Body Interaction

Chapter 4

Basics of Hydrodynamics and Linear Waves

4.1 Review of Vector Analysis

Basics of differential operations and integrals of vector and scalar functions are reviewed. The Gauss's, Stokes' and Green's theorems are frequently employed in the derivation of fundamental physical laws and other hydrodynamic theorems of waves and body motions in waves. (Sternburg and Smith, 1952; Milne-Thomson, 1955; Stoker, 1957; Sommerfeld in English, 1964; in German, 1949).

4.1.1 Differential operators

Let x, y, z be a set of variables and X, Y, Z and B be scalar functions of x, y, z. And let $\mathbf{A} = (X, Y, Z)$ be a vector function of the variables x, y, z with components X, Y, Z.

Define a vector differential operator ∇(del):

$$\nabla = \left(\frac{\partial}{\partial x}, \frac{\partial}{\partial y}, \frac{\partial}{\partial z} \right) = \mathbf{i}\frac{\partial}{\partial x} + \mathbf{j}\frac{\partial}{\partial y} + \mathbf{k}\frac{\partial}{\partial z} \tag{4.1}$$

Performing differential operations on B and $\mathbf{A}$ with ∇:

$$\nabla B = \left(\frac{\partial B}{\partial x}, \frac{\partial B}{\partial y}, \frac{\partial B}{\partial z} \right) = \operatorname{grad} B \tag{4.2}$$

$$\nabla \cdot \mathbf{A} = \frac{\partial X}{\partial x} + \frac{\partial Y}{\partial y} + \frac{\partial Z}{\partial z} = \operatorname{div} \mathbf{A} \tag{4.3}$$

$$\nabla \times \mathbf{A} = \left(\frac{\partial Z}{\partial y} - \frac{\partial Y}{\partial z}, \frac{\partial X}{\partial z} - \frac{\partial Z}{\partial x}, \frac{\partial Y}{\partial x} - \frac{\partial X}{\partial y} \right) = \text{curl} \, \mathbf{A} \qquad (4.4)$$

where, for instance, the x-components of the vectors ∇B and curl $\mathbf{A}$ are denoted by $\text{grad}_x B = \dfrac{\partial B}{\partial x}$ and $\text{curl}_x \mathbf{A} = \dfrac{\partial Z}{\partial y} - \dfrac{\partial Y}{\partial z}$.

4.1.2 Differential operators and surface integrals

Visualize the vector field $\mathbf{A}$ as a field of flow, i.e., let the magnitude and direction of the flow be given everywhere by the magnitude and the direction of $\mathbf{A}$. Let the field point P be surrounded by a closed surface σ and $\Delta \tau$ be the enclosing volume. If $\mathbf{n}$ is the unit outward normal from $\Delta \tau$ associated with the surface element $d\sigma$, $A_n = \mathbf{n} \cdot \mathbf{A}$ represents the outflow per unit surface.

On forming $\dfrac{1}{\Delta \tau} \int_\sigma \mathbf{n} \cdot \mathbf{A} d\sigma$, and taking limit with vanishing $\Delta \tau$ becomes:

$$\text{div} \, \mathbf{A} = \lim_{\Delta \tau \to 0} \frac{1}{\Delta \tau} \int_\sigma \mathbf{n} \cdot \mathbf{A} \, d\sigma \qquad (4.5)$$

The agreement between the definitions Eqs. (4.3) and (4.5) is evident in the special case that $\Delta \tau$ is chosen as a rectangular cell with center $P(x, y, z)$ and the sides $\Delta x, \Delta y, \Delta z$. For the two surfaces normal to the x-axis is shown in Fig. 4.1.

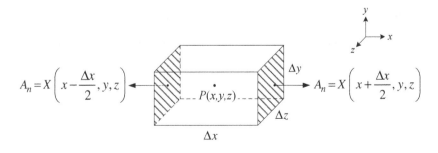

Fig. 4.1. Rectangular element.

Then:

$$\sum A_n \Delta \sigma = \{X(x + \frac{\Delta x}{2}, y, z) - X(x - \frac{\Delta x}{2}, y, z)\}\Delta y \Delta z$$

$$= \frac{\partial X}{\partial x}\Delta x \Delta y \Delta z = \frac{\partial X}{\partial x}\Delta \tau.$$

(4.6)

Similarly for the two pairs of surfaces normal to y- and z-axis, we have:
$\frac{\partial Y}{\partial y}\Delta \tau$ and $\frac{\partial Z}{\partial z}\Delta \tau$.

Substituting the sum $(\frac{\partial X}{\partial x} + \frac{\partial Y}{\partial y} + \frac{\partial Z}{\partial z})\Delta \tau$ into Eq. (4.5), one has the identity between Eqs. (4.5) and (4.3).

Similarly grad B and curl $\mathbf{A}$ are defined by surface integral operations:

$$\nabla B = \lim_{\Delta \tau \to 0} \frac{1}{\Delta \tau} \int_\sigma \mathbf{n} B d\sigma$$

(4.7)

$$\nabla \times \mathbf{A} = \lim_{\Delta \tau \to 0} \frac{1}{\Delta \tau} \int_\sigma \mathbf{n} \times \mathbf{A} d\sigma$$

(4.8)

It is noted that an operational similarity holds between $(\nabla \cdot, \nabla, \nabla \times)$ and $(\mathbf{n} \cdot, \mathbf{n}, \mathbf{n} \times)$ on both sides of Eqs. (4.5), (4.7), and (4.8).

Let an arbitrarily chosen oriented line a pass through the point P in the plane that contains P and is normal to a. Draw a closed curve surrounding P, and denote the enclosed area by $\Delta \sigma$. Let A_s be the component of $\mathbf{A}$ in the direction of the arc element ds, taken in that direction that forms a right-hand screw with the axis a. Consider now the line integral for which Lord Kelvin introduced the appropriate name circulation. The limit of the circulation to the area $\Delta \sigma$ is the component of the curl $\mathbf{A}$ in the direction of a:

$$\text{curl}_a \mathbf{A} = \lim_{\Delta \sigma \to 0} \frac{1}{\Delta \sigma} \oint_s A_s ds$$

(4.9)

Suppose one wants to verify for the x-component of the curl. The line a then coincides with the $+x$ direction, and s may be chosen as a rectangle with center $P(x, y, z)$ and sides $\Delta y, \Delta z$ in Fig. 4.2.

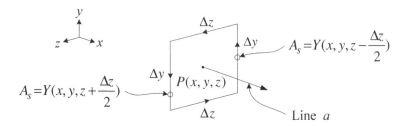

Fig. 4.2. Surface element.

The contribution of the pair, which is parallel to the y-axis, yields:

$$\sum A_s \Delta s = \left\{ -Y\left(x, y, z + \frac{\Delta z}{2}\right) + Y\left(x, y, z - \frac{\Delta z}{2}\right) \right\} \Delta y$$

$$= -\frac{\partial Y(x, y, z)}{\partial z} \Delta y \Delta z \qquad (4.10)$$

And the second pair, which is parallel to the z-axis

$$\sum A_s \Delta s = \left\{ -Z\left(x, y - \frac{\Delta y}{2}, z\right) + Z\left(x, y + \frac{\Delta y}{2}, z\right) \right\} \Delta z$$

$$= \frac{\partial Z(x, y, z)}{\partial y} \Delta y \Delta z \qquad (4.11)$$

Altogether, one obtains according to Eqs. (4.10) and (4.11):

$$\text{curl}_x \, \mathbf{A} = \frac{\partial Z}{\partial y} - \frac{\partial Y}{\partial z} \qquad (4.12)$$

4.1.3 Gauss's theorem

Case 1:
Let $\mathbf{A} = (X, Y, Z)$ be a vector function with $X(x, y, z)$, $Y(x, y, z)$, $Z(x, y, z)$ defined in the volume τ and on the surface σ and continuously differentiable. Slicing the τ, we obtain an infinite number of $\Delta \tau$. Then, repeatedly applying Eq. (4.5) to each $\Delta \tau$ one obtains:

$$\int_\tau \nabla \cdot \mathbf{A} d\tau = \int_\sigma \mathbf{n} \cdot \mathbf{A} d\sigma \qquad (4.13)$$

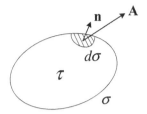

Fig. 4.3. Volume τ enclosed by surface σ.

Case 2:
Let $B = B(x, y, z)$ be a scalar function, defined in the volume τ and on the surface σ and continuously differentiable. Then, slicing the τ into $\Delta\tau$ and using repeatedly Eq. (4.7), one has:

$$\int_\tau \nabla B d\tau = \int_\sigma \mathbf{n} B d\sigma \qquad (4.14)$$

4.1.4 Stokes' theorem

Let $\mathbf{A} = (X, Y, Z)$ be a vector function with $X(x, y, z)$, $Y(x, y, z)$ and $Z(x, y, z)$ defined on the surface σ to which the boundary contour s belongs and $\mathbf{A}$ is continuously differentiable. Dividing σ into an infinite number of $\Delta\sigma$ and applying Eq. (4.9) to each $\Delta\sigma$, we obtain circulation Γ:

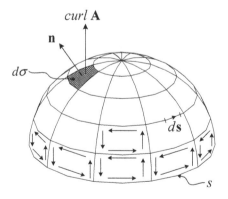

Fig. 4.4. Domain of surface σ and contour s.

$$\Gamma = \int_{\sigma} \text{curl}\,\mathbf{A} \cdot \mathbf{n} d\sigma = \oint_{s} \mathbf{A} \cdot d\mathbf{s} \tag{4.15}$$

The sum of the surface integral on $\Delta\sigma$ inside the contour s cancels each other except the contour integrals along s.

4.1.5 Green's theorem

Green's theorem and Green's function are used in modeling the fluid flows in waves as well as the flows due to body motion in waves. Let $\varphi_1(x, y, z)$ and $\varphi_2(x, y, z)$ be two scalar functions which are defined in the domain τ and on the surface σ, differentiable up to the second order and the derivatives are continuous. Replacing a vector $\mathbf{A}$ by $\varphi_1 \nabla \varphi_2$ in Gauss's theorem Eq. (4.13), we have the first form of Green's theorem:

$$\int_{\tau} \left(\nabla \varphi_1 \cdot \nabla \varphi_2 + \varphi_1 \nabla^2 \varphi_2 \right) d\tau = \int_{\sigma} \varphi_1 \frac{\partial \varphi_2}{\partial n} d\sigma \tag{4.16}$$

In the above equation, $\partial \varphi_2 / \partial n = \mathbf{n} \cdot \nabla \varphi_2$ and $\mathbf{n}$ is unit outward normal on σ from domain τ.

Interchanging φ_1 and φ_2, we have the first form of Green's theorem again:

$$\int_{\tau} \left(\nabla \varphi_2 \cdot \nabla \varphi_1 + \varphi_2 \nabla^2 \varphi_1 \right) d\tau = \int_{\sigma} \varphi_2 \frac{\partial \varphi_1}{\partial n} d\sigma \tag{4.17}$$

Subtracting Eq. (4.16) from Eq. (4.17), we have the second form of Green's theorem:

$$\int_{\tau} \left(\varphi_2 \nabla^2 \varphi_1 - \varphi_1 \nabla^2 \varphi_2 \right) d\tau = \int_{\sigma} \left(\varphi_2 \frac{\partial \varphi_1}{\partial n} - \varphi_1 \frac{\partial \varphi_2}{\partial n} \right) d\sigma \tag{4.18}$$

If φ_1 and φ_2 are harmonic or regular ($\nabla^2 \varphi = 0$) everywhere in τ, we have:

$$\int_{\sigma} \left(\varphi_2 \frac{\partial \varphi_1}{\partial n} - \varphi_1 \frac{\partial \varphi_2}{\partial n} \right) d\sigma = 0 \tag{4.19}$$

The above equation is frequently called Green's 2nd identity or Green's theorem. We will extend the Green's 2nd identity to Green's third identity when one of the two scalar functions is singular.

Suppose $\varphi_1(x, y, z)$ is harmonic in τ and φ_2 is singular at a point in τ and σ in the form:

$$\varphi_2(x,y,z;x_0,y_0,z_0)=1/r(P;P_0)$$

$$r=\sqrt{(x-x_0)^2+(y-y_0)^2+(z-z_0)^2}$$

(4.20)

$P_0(x_0, y_0, z_0)$ and $P(x, y, z)$ may be defined as fixed and moving points, and vice versa. If these two points are endlessly close to each other φ_2 becomes singular. We will take here P_0 as a moving point (differentiation and integration variable) in τ and σ and consider three cases depending on where P lies in Fig. 4.5. The positive unit normal is taken outward from the domain τ, as defined in the beginning of this section.

Case 1:
$P(x, y, z)$ lies outside $\tau + \sigma$
then,

$$\left.\begin{array}{l}\nabla^2(1/r)=0\\ \nabla^2\varphi_1=0\end{array}\right\} \text{ in } \tau$$

(4.21)

Thus Green's 2nd identity Eq. (4.19) holds:

$$\int_\sigma\left(\frac{1}{r(P;P_0)}\frac{\partial\varphi_1(P_0)}{\partial n(P_0)}-\varphi_1(P_0)\frac{\partial(1/r(P;P_0))}{\partial n(P_0)}\right)d\sigma(P_0)=0$$

(4.22)

Case 2:
$P(x, y, z)$ lies in τ
Then, we consider a small sphere of radius ε whose origin is at P in $\tau = \tau^* + \tau'$ as shown in Fig. 4.5. The domain of a small sphere is denoted by τ' and σ'. And consider another domain $\tau^* = \tau - \tau'$ bounded by σ and σ'. When P_0 is in the domain τ^*:

$$\nabla^2\varphi_1=0, \quad \nabla^2\frac{1}{r}=0$$

(4.23)

Hence Green's 2nd identity Eq. (4.19) holds:

$$\int_{\sigma'} \left(\frac{1}{r(P;P_0)} \frac{\partial \varphi_1(P_0)}{\partial n(P_0)} - \varphi_1(P_0) \frac{\partial(1/r)}{\partial n(P_0)} \right) d\sigma'(P_0)$$

$$+ \int_{\sigma} \left(\frac{1}{r(P;P_0)} \frac{\partial \varphi_1(P_0)}{\partial n(P_0)} - \varphi_1(P_0) \frac{\partial(1/r)}{\partial n(P_0)} \right) d\sigma(P_0) = 0 \qquad (4.24)$$

In view of Gauss's theorem Eq. (4.13), the first term of the LHS of the first integral in Eq. (4.24) is written in the following form:

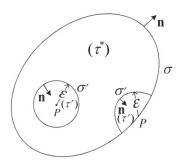

Fig. 4.5. Domain of volume τ' and τ^* and surfaces σ and σ' with unit outward normals.

$$\int_{\sigma'} \frac{1}{r(P;P_0)} \frac{\partial \varphi_1(P_0)}{\partial n(P_0)} d\sigma'(P_0) = \frac{1}{\varepsilon} \int_{\sigma'} \mathbf{n} \cdot \nabla \varphi_1 d\sigma' = \frac{1}{\varepsilon} \int_{\tau'} \nabla^2 \varphi_1 d\tau' = 0 \quad (4.25)$$

Hence, Eq. (4.24) is reduced to:

$$\int_{\sigma'} \varphi_1(P_0) \frac{\partial(1/r)}{\partial n(P_0)} d\sigma'(P_0) = \int_{\sigma} \left(\frac{1}{r} \frac{\partial \varphi_1(P_0)}{\partial n(P_0)} - \varphi_1(P_0) \frac{\partial(1/r)}{\partial n(P_0)} \right) d\sigma(P_0) \quad (4.26)$$

Since the unit normal $\mathbf{n}$ is opposite to r, $\dfrac{\partial}{\partial n}$ on $\sigma' = -\dfrac{\partial}{\partial r}$, we have

$$\int_{\sigma'} \varphi_1(P_0) \frac{\partial(1/r)}{\partial n(P_0)} d\sigma'(P_0) = \int_{\sigma'} \frac{1}{r^2} \varphi_1(P_0) d\sigma'(P_0) \qquad (4.27)$$

In Eq. (4.27) we take r as small as possible so that the integral converges to a constant $\varphi_1(P)$ at the center of the shrunken sphere bounded by σ', then we have:

$$\int_{\sigma'} \frac{1}{r^2(P; P_0)} \varphi_1(P_0) d\sigma'(P_0) = 4\pi\varphi_1(P) \tag{4.28}$$

Thus Eq. (4.26) takes the form:

$$4\pi\varphi_1(P) = \int_\sigma \left(\frac{1}{r} \frac{\partial \varphi_1(P_0)}{\partial n(P_0)} - \varphi_1(P_0) \frac{\partial(1/r)}{\partial n(P_0)} \right) d\sigma(P_0) \tag{4.29}$$

Equation (4.29) is the Green's third identity, stating that, given the potential φ_1 and its normal derivative on the boundary surface σ', outward from $\tau*$, the surface integral will determine the value of potential φ_1 at a point in τ'.

Case 3:
$P(x, y, z)$ lies on σ
The solid angle 4π of the sphere in Eq. (4.28) now becomes 2π because the sphere of radius ε is reduced to a hemisphere:

$$\int_{\sigma'} \frac{1}{r^2} \varphi_1(P_0) d\sigma' = 2\pi\varphi_1(P) \tag{4.30}$$

Thus,

$$2\pi\varphi_1(P) = \int_\sigma \left(\frac{1}{r} \frac{\partial \varphi_1(P_0)}{\partial n(P_0)} - \varphi_1(P_0) \frac{\partial(1/r)}{\partial n(P_0)} \right) d\sigma(P_0), \ P \in \sigma \tag{4.31}$$

where P lies on the smooth surface σ.

In summary, we have the formulas depending on where P lies:

$$\begin{pmatrix} 4\pi \\ 2\pi \\ 0 \end{pmatrix} \varphi_1(P) = \int_\sigma \left(\frac{1}{r} \frac{\partial \varphi_1(P_0)}{\partial n(P_0)} - \varphi_1(P_0) \frac{\partial(1/r)}{\partial n(P_0)} \right) d\sigma(P_0) \tag{4.32}$$

4π, 2π, 0 correspond to P lying in the domain τ, on the boundary σ and outside of $\tau + \sigma$, respectively.

The Green's third identity in 2-D case is similarly derived. Accordingly the 2-D Green's theorems, all along the smooth contour s, are given in the form:

$$\left.\begin{matrix} 2\pi \\ \pi \\ 0 \end{matrix}\right\} \varphi_1(P) = \int_s \left(\log\frac{1}{r}\frac{\partial \varphi_1(P_0)}{\partial n(P_0)} - \varphi_1(P_0)\frac{\partial}{\partial n(P_0)}\log\frac{1}{r} \right) ds(P_0) \qquad (4.33)$$

where $r(P;P_0) = \sqrt{(x-x_0)^2 + (y-y_0)^2}$.

In Eq. (4.33), 2π is applied when P lies at an interior point, π is used when P is on the smooth surface contour s and 0 is employed when P lies outside the region τ.

4.1.6 Green's function

Let $g(x, y, z; x_0, y_0, z_0)$ be harmonic in each set of variables $P(x, y, z)$ and $P_0(x_0, y_0, z_0)$. Then Green's function is defined in the form:

$$G(P;P_0) = \frac{1}{r(P;P_0)} + g(P;P_0) \qquad (4.34)$$

Substituting Eq. (4.34) in the Green's third identity Eq. (4.29) and making use of Green's 2nd-identity Eq. (4.19), we derive:

$$\varphi(P) = \frac{1}{4\pi}\int_\sigma \left(\frac{\partial \varphi(P_0)}{\partial n(P_0)}G(P;P_0) - \varphi(P_0)\frac{\partial G(P;P_0)}{\partial n(P_0)} \right) d\sigma(P_0) \qquad (4.35)$$

If the point P is taken on σ, 4π is replaced by 2π according to Eq. (4.31):

$$\varphi(P) = \frac{1}{2\pi}\int_\sigma \left(\frac{\partial \varphi(P_0)}{\partial n(P_0)}G(P;P_0) - \varphi(P_0)\frac{\partial G(P;P_0)}{\partial n(P_0)} \right) d\sigma(P_0), \quad P \in \sigma \quad (4.36)$$

It is to be noted that the unit normal on the boundary surface is directed outward from the domain.

The corresponding 2-D Green's function is defined:

$$G(P;P_0) = \log\frac{1}{r(P;P_0)} + g(P;P_0) \qquad (4.37)$$

The Green's third identity for 2-D becomes:

$$\varphi(P) = \frac{1}{2\pi}\int_s \left(\frac{\partial \varphi(P_0)}{\partial n(P_0)}G(P;P_0) - \varphi(P_0)\frac{\partial G(P;P_0)}{\partial n(P_0)} \right) ds(P_0) \qquad (4.38)$$

If P is on contour s, 2π is replaced by π according to Eq. (4.33):

$$\varphi(P) = \frac{1}{\pi} \int_s \left(\frac{\partial \varphi(P_0)}{\partial n(P_0)} G(P; P_0) - \varphi(P_0) \frac{\partial G(P; P_0)}{\partial n(P_0)} \right) ds(P_0), \quad P \in s \qquad (4.39)$$

4.2 Basics of Hydrodynamics

The basics of hydrodynamics is reviewed herein (Plandtl and Tietjens, 1934; Milne-Thomson, 1955; Stoker, 1957; Wehausen and Laitone, 1960; Kinsman, 1965; Le Mehaute, 1976; Newman, 1977; Schlichting, 1979; Munson et al., 1998).

4.2.1 Kinematics of fluid flow

Description of the particles kinematics in the flow field is made by the Lagrangian and Eulerian method.

4.2.1.1 Lagrangian method

The Lagrangian method may be used to answer the question: what occurs to a given fluid particle as it moves about its own path? The path of a specific particle of the fluid is denoted by its position as a function of time. If we denote the initial position of a given particle at time t_0 by $\mathbf{r}_0 = \{x_0, y_0, z_0\}$, then the position vector $\mathbf{r}$ at time t is given:

$$\mathbf{r}(t) = \mathbf{r}(\mathbf{r}_0, t - t_0)$$

It is noted that the coordinates $\mathbf{r}_0 = \{x_0, y_0, z_0\}$ indicate the name of the particle given at time t_0. How do we determine the paths? Suppose the velocity field of the particles with the names designated by the coordinates of the initial positions is defined:

$$\mathbf{V} = \left(\frac{d\mathbf{r}}{dt} \right)_{\mathbf{r}_0} \qquad (4.40)$$

or

$$u = \left(\frac{dx}{dt}\right)_{r_0}, v = \left(\frac{dy}{dt}\right)_{r_0}, w = \left(\frac{dz}{dt}\right)_{r_0}$$

then we can obtain the new positions at t_1 after $\Delta t = t_1 - t_0$, by integrating the velocities:

$$x(t_1) - x(t_0) = \int_{t_0}^{t_1} u(x_0, y_0, z_0, t)dt$$

$$y(t_1) - y(t_0) = \int_{t_0}^{t_1} v(x_0, y_0, z_0, t)dt \qquad (4.41)$$

$$z(t_1) - z(t_0) = \int_{t_0}^{t_1} w(x_0, y_0, z_0, t)dt$$

Continuing step by step integration from t_0 to t_1, t_1 to t_2 and so forth, we may obtain the path of the particle named by $r_0 = \{x_0, y_0, z_0\}$. This procedure can be carried out for the entire fluid particles.

4.2.1.2 Eulerian method

What happens at a given time t at the various points r in a space filled by a fluid? The velocity at any point and at any time can be written:

$$V = V(r, t) \qquad (4.42)$$

$$u = u(x, y, z, t), v = v(x, y, z, t), w = w(x, y, z, t) \qquad (4.43)$$

In order to find out subsequently what is happening to the individual particle according to the Lagrangian method, we consider a set of equations for each particle:

$$\frac{dx}{dt} = u, \frac{dy}{dt} = v, \frac{dz}{dt} = w \qquad (4.44)$$

Combining with Eulerian description, i.e., Eq. (4.43), a set of three simultaneous equations is given as follows:

$$\frac{dx}{dt} = u(x, y, z, t), \qquad \frac{dy}{dt} = v(x, y, z, t), \qquad \frac{dz}{dt} = w(x, y, z, t) \qquad (4.45)$$

The solution of the set of equations Eq. (4.45) involves the requirement at a given time $t = t_0$ to determine the coordinates of the particle become $x = x_0$, $y = y_0$, $z = z_0$. These three integration constants can be considered as initial coordinates in Lagrangian description. Therefore in principle the Lagrangian method of description can always be derived from Eq. (4.45) by the Eulerian method. A streamline is defined as a line which is tangential at every point to the velocity vector at a given time t_0:

$$\mathbf{V} = k d\mathbf{r}, \quad k = \text{real const} \tag{4.46}$$

or

$$\frac{dx}{u(x, y, z, t_0)} = \frac{dy}{v(x, y, z, t_0)} = \frac{dz}{w(x, y, z, t_0)} \tag{4.47}$$

Integration of Eq. (4.47) gives the path as obtained by Eq. (4.41). The streamlines and paths are identical in the steady flow.

4.2.1.3 Material derivative

Let $f(x, y, z, t)$ be a scalar or vector function representing property of fluid (velocity, pressure, density etc). Then the total differential is given in the form:

$$df = \frac{\partial f}{\partial t} dt + \frac{\partial f}{\partial x} dx + \frac{\partial f}{\partial y} dy + \frac{\partial f}{\partial z} dz \tag{4.48}$$

The rate of change of the property with respect to time is given by:

$$\frac{df}{dt} = \frac{\partial f}{\partial t} + (\mathbf{V} \cdot \nabla) f \tag{4.49}$$

Eq. (4.49) may be expressed in the form:

$$\frac{Df}{Dt} = \left(\frac{\partial}{\partial t} + (\mathbf{V} \cdot \nabla) \right) f \tag{4.50}$$

The differential operator D/Dt is called material or essential derivative; $\partial f / \partial t$ the rate of change of the property at a local point designated as the local term. $\mathbf{V} \cdot \nabla f$ is the rate of change of the property associated with the same fluid particle as it moves about, and called the convective term. If the particle is at rest, then the property experiences its local change only.

4.2.1.4 The Reynolds transport theorem

A system of flowing fluid is defined as a collection of the same fluid particles. The system of flowing fluid carries fluid properties such as mass, linear momentum, kinetic energy etc. as it moves about. Such property is called extensive property. The amount of the extensive property contained in the system is given in the form:

$$B_{sys}(t) = \int_{sys} \rho b d\tau \qquad (4.51)$$

where ρ (x, y, z, t) $d\tau$ is the elementary mass of the fluid and b is the intensive property of the system, or the amount of property per unit mass of the fluid. For instance $b = 1$, $\mathbf{V}$, and $V^2/2$ are the intensive properties of mass, linear momentum and kinetic energy, respectively. We prescribe an arbitrary control (geometrical) and non-deforming volume in the fluid domain. If we let the system of flowing fluid coincide with a prescribed non-deforming control volume CV at time t, the time rate of change of the amount of the extensive property of the system at t is given by:

$$\frac{DB_{sys}}{Dt} = \frac{\partial}{\partial t} \int_{CV} \rho b d\tau + \int_{CS} \rho b \mathbf{V} \cdot \mathbf{n} d\sigma \qquad (4.52)$$

where CS denotes the control surface. $\mathbf{V}$ is fluid velocity and $\mathbf{n}$ is unit outward normal from the fluid domain τ on $d\sigma$ of the control surface CS. The first term indicates the local rate of change of the property contained in the CV. The second term indicates the total flux of the system property across the CS at time t. Equation (4.52) is called Reynolds transport theorem for the non-deforming fixed control volume. This system volume may be finite as well as infinitesimal.

The second term is converted to the volume integral using Gauss's theorem:

$$\int_{CS} \rho b \mathbf{V} \cdot \mathbf{n} d\sigma = \int_{CS} \mathbf{n} \cdot \rho b \mathbf{V} d\sigma = \int_{CV} \nabla \cdot (\rho b \mathbf{V}) d\tau \qquad (4.53)$$

Since $\nabla \cdot (\rho b \mathbf{V}) = \nabla(\rho b) \cdot \mathbf{V} + (\rho b) \nabla \cdot \mathbf{V}$, we have:

$$\frac{DB_{sys}}{Dt} = \int_{CV} \left[(\rho b)_t + \mathbf{V} \cdot \nabla(\rho b) + (\rho b) \nabla \cdot \mathbf{V} \right] d\tau \qquad (4.54)$$

Making use of the definition of material derivative on ρb, we may write the Reynolds transport theorem in a modified form:

$$\frac{DB_{sys}}{Dt} = \int_{CV} \left(\frac{D(\rho b)}{Dt} + (\rho b)\nabla \cdot \mathbf{V} \right) d\tau \qquad (4.55)$$

4.2.2 Conservation laws

4.2.2.1 Mass conservation law

The mass in a system of the same flowing fluid particles is represented by Eq. (4.51) with $b = 1$:

$$M_{sys}(t) = \int_{sys} \rho(x, y, z, t) d\tau \qquad (4.56)$$

Since the amount of mass system is supposed to be unchanged while it is moving with the system flowing fluid, we have:

$$\frac{DM_{sys}}{Dt} = 0 \qquad (4.57)$$

Applying the Reynolds transport theorem Eq. (4.55) to the system mass of the same flowing fluid with $b = 1$:

$$\int_{CV} \left(\frac{D\rho}{Dt} + \rho \nabla \cdot \mathbf{V} \right) d\tau = 0 \qquad (4.58)$$

Since Eq. (4.58) holds for an arbitrary volume, the mass conservation is defined at a point in the form:

$$\frac{D\rho}{Dt} + \rho \nabla \cdot \mathbf{V} = 0 \qquad (4.59)$$

If the density is constant, and the flow is steady or unsteady, the above is reduced to:

$$\nabla \cdot \mathbf{V} = 0 \qquad (4.60)$$

Equation (4.60) is the continuity equation for a steady or unsteady incompressible fluid flow. If the above holds in the velocity field, the flow is continuous and Eq. (4.60) is called continuity equation. It is noted that continuity condition may be used to check if an assumed flow is possible.

4.2.2.2 Momentum conservation law

The general momentum equation will give the equation of motion of a particle in the viscous fluid (Schlichting, 1957). However, we assume our immediate interest is to derive equation of motion of fluid particle from the momentum equation for inviscid flow. The linear momentum of the moving system of fluid is given in the form:

$$\mathbf{M}_{sys}(t) = \int_{sys} \rho \mathbf{V} d\tau \qquad (4.61)$$

According to Newton's second law, the rate of change of the linear momentum of the system of fluid equals the external force on the contents in the system. Since we assume inviscid flow, the external force consists of the pressure force on the control surface and the body force on the particles due to gravity. The rate of change of the linear momentum of the system is equal to the modified form of the Reynolds transport theorem Eq. (4.55), where $b = \mathbf{V}$:

$$\frac{D}{Dt} \int_{sys} \rho \mathbf{V} d\tau = \int_{CV} \left(\frac{D(\rho \mathbf{V})}{Dt} + (\rho \mathbf{V}) \nabla \cdot \mathbf{V} \right) d\tau = \int_{CV} \rho \mathbf{g} d\tau - \int_{CS} p \mathbf{n} d\sigma \qquad (4.62)$$

In Eq. (4.62), p is the pressure acting on the surface in the direction opposite to the unit normal $\mathbf{n}$ outward from the CV domain, $\mathbf{g}\{0, -g, 0\}$ is the body force per unit mass, and g is acceleration due to gravity in y-direction.

In Eq. (4.62), div·$\mathbf{V}$ = 0, thus we convert the surface integral into the volume integral by Gauss's theorem Eq. (4.14):

$$\int_{CV} \left(\frac{D(\rho \mathbf{V})}{Dt} + \nabla p - \rho \mathbf{g} \right) d\tau = 0 \qquad (4.63)$$

The foregoing relation is valid for an arbitrary volume. If we assume incompressible flow:

$$\frac{D\mathbf{V}}{Dt} = -\frac{\nabla p}{\rho} + \mathbf{g} \qquad (4.64)$$

In components:

$$\frac{\partial u}{\partial t} + u\frac{\partial u}{\partial x} + v\frac{\partial u}{\partial y} + w\frac{\partial u}{\partial z} = -\frac{1}{\rho}\frac{\partial p}{\partial x}$$

$$\frac{\partial v}{\partial t} + u\frac{\partial v}{\partial x} + v\frac{\partial v}{\partial y} + w\frac{\partial v}{\partial z} = -\frac{1}{\rho}\frac{\partial p}{\partial y} - gy \qquad (4.65)$$

$$\frac{\partial w}{\partial t} + u\frac{\partial w}{\partial x} + v\frac{\partial w}{\partial y} + w\frac{\partial w}{\partial z} = -\frac{1}{\rho}\frac{\partial p}{\partial z}$$

Equation (4.65) is called Euler equation of motion of a fluid particle that holds at any point in the flow field when the fluid is inviscid and incompressible.

Using both Euler equation and continuity Eq. (4.60) will determine the four unknowns p, u, v, and w namely pressure and velocity field. Integration of Euler equation is difficult because the acceleration is made of the products of two linear terms. However, there is a way to integrate the Euler equation along a streamline, the result of which is called Bernoulli's equation, which will be shown shortly. We may obtain Navier-Stokes equation of motion in the laminar viscous flow if we consider the shear stress due to viscosity:

$$\frac{D\mathbf{V}}{Dt} = -\frac{\nabla p}{\rho} + \mathbf{g} + \nu\,\nabla^2\mathbf{V} \qquad (4.66)$$

where ν is kinematic viscosity of the fluid. Equations (4.66) and (4.60) constitute 4 equations to solve four unknowns three velocity components and pressure. A few of the exact analytical solutions of simplified Stokes' problems are known (Schlichting, 1957) but the solution of Navier-Stokes (N-S) equation in general may only be obtained by applying the numerical method such as finite difference scheme applied to the entire fluid domain. There are two N-S equations, namely one for the laminar and the other for turbulent boundary layer flow. Prandtl's hypothesis is used to determine the pressure force on the body surface due to inviscid flow by potential theory outside the layer, while the friction force due to the viscous boundary layer flow around the body surface by solving N-S equation. Computational fluid dynamics has recently been developed for application to solve some realistic problems.

4.2.2.3 Conservation of circulation

When the external forces are conservative and are derived from single-valued potential, the circulation Γ in any circuit that moves with the fluid is conserved. This is proved as follows: let s be a closed circuit which moves with the fluid, i.e. a circuit moves with the same fluid particles always. Then we have:

$$\frac{D\Gamma}{Dt} = \frac{D}{Dt} \oint_s \mathbf{V} \cdot d\mathbf{r} = \oint_s \left(\frac{D\mathbf{V}}{Dt} \cdot d\mathbf{r} + \mathbf{V} \cdot d\mathbf{V} \right) = 0 \tag{4.67}$$

To prove this we use Euler's Eq. (4.64) in the gravitational field. Since $\nabla p \cdot d\mathbf{r} = dp$; $\mathbf{g} \cdot d\mathbf{r} = d(-gy)$; and $\mathbf{V} \cdot d\mathbf{V} = d(V^2/2)$, we have:

$$\frac{D\mathbf{V}}{Dt} \cdot d\mathbf{r} + \mathbf{V} \cdot d\mathbf{V} = d\left(-\left(\frac{p}{\rho} + gy \right) + \frac{V^2}{2} \right) \tag{4.68}$$

Each quantity in the RHS of Eq. (4.68) is single-valued:

$$\frac{D\Gamma}{Dt} = \int_A^A d\left(-\left(\frac{p}{\rho} + gy \right) + \frac{V^2}{2} \right) = 0 \tag{4.69}$$

Therefore the circulation is independent of time. The constancy of circulation of the inviscid compressible or incompressible flow is due to Lord Kelvin.

If $\Gamma = 0$ at a time, then according to Lord Kelvin's theorem, $\Gamma = 0$ always holds. Thus according to Stokes' theorem Eq. (4.15), we have always irrotational motion when the motion started from rest ($\mathbf{V} = 0$), since the motion is initially irrotational:

$$\nabla \times \mathbf{V} = 0 \tag{4.70}$$

If the motion is irrotational, there exists a single-valued scalar function φ, from which the velocity field is determined by:

$$\mathbf{V} = \nabla \varphi \tag{4.71}$$

To derive the above Eq. (4.71), we consider a line integral in a simply connected region. According to Stoke's theorem Eq. (4.15) we may write:

$$\int_{OAP} \mathbf{V} \cdot d\mathbf{s} + \int_{PBO} \mathbf{V} \cdot d\mathbf{s} = \oint (\nabla \times \mathbf{V}) \, d\sigma \tag{4.72}$$

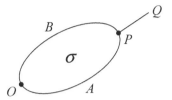

Fig. 4.6. Simply connected region.

Since the motion is irrotational, the term on RHS of Eq. (4.72) vanishes and the equation gives a point-dependent scalar function:

$$\int_{OAP} \mathbf{V} \cdot d\mathbf{s} = \int_{OBP} \mathbf{V} \cdot d\mathbf{s} = \varphi_P \tag{4.73}$$

Take a point Q so near to P that the velocity may be assumed nearly constant along PQ ($= \boldsymbol{\eta}$), then approximately:

$$\int_{PQ} \mathbf{V} \cdot d\mathbf{s} = \mathbf{V} \cdot \boldsymbol{\eta} \tag{4.74}$$

Since φ_P in Eq. (4.73) is a point-dependent scalar function:

$$\int_{PQ} \mathbf{V} \cdot d\mathbf{s} = \varphi_Q - \varphi_P = d\varphi = \boldsymbol{\eta} \cdot \nabla \varphi \tag{4.75}$$

which leads to:

$$\mathbf{V} = \nabla \varphi \tag{4.76}$$

The above φ is called velocity potential. If the flow is steady or unsteady and the fluid is inviscid, incompressible and irrotational, the velocity potential satisfies the Laplace's equation:

$$\nabla \cdot \mathbf{V} = \nabla \cdot \nabla \varphi = \nabla^2 \varphi = 0 \tag{4.77}$$

This is the governing equation to be satisfied for the ideal (inviscid and incompressible) fluid, irrotational and steady or unsteady flow in the entire fluid domain, as will be used in solving the wave and body interaction problems throughout the text.

4.2.3 Bernoulli's equations

Since the acceleration in unsteady flow is given by:

$$\frac{D\mathbf{V}}{Dt} = \frac{\partial \mathbf{V}}{\partial t} + \nabla \frac{1}{2}V^2 + \nabla \times \mathbf{V} \times \mathbf{V} \tag{4.78}$$

Euler equation (4.64) of the particle takes the form:

$$\frac{\partial \mathbf{V}}{\partial t} + \nabla \times \mathbf{V} \times \mathbf{V} + \nabla \left(\frac{p}{\rho} + gy + \frac{1}{2}V^2 \right) = 0 \tag{4.79}$$

Before integrating the above Euler equation Eq. (4.79), we need to consider two cases. Case 1: steady and possibly rotational flow, and Case 2: unsteady and irrotational flow.

Case 1:
If the flow is steady and possibly rotational, then

$$\frac{\partial \mathbf{V}}{\partial t} = 0, \ \nabla \times \mathbf{V} \times \mathbf{V} \neq 0 \tag{4.80}$$

Scalar product of $\mathbf{V}$ on the convective acceleration gives:

$$\mathbf{V} \cdot (\mathbf{V} \cdot \nabla)\mathbf{V} = \mathbf{V} \cdot \nabla \left(\frac{V^2}{2} \right) + \mathbf{V} \cdot \nabla \times \mathbf{V} \times \mathbf{V} = \mathbf{V} \cdot \nabla \left(\frac{V^2}{2} \right) \tag{4.81}$$

And the scalar product of $\mathbf{V}$ on the terms in Euler equation in steady flow:

$$\mathbf{V} \cdot \nabla \left(\frac{p}{\rho} + gy + \frac{1}{2}V^2 \right) = 0 \tag{4.82}$$

Substituting the streamline equation Eq. (4.46) into the velocity in Eq. (4.82) we obtain the differential of the expression in the parenthesis along the streamline:

$$d \left(\frac{p}{\rho} + gy + \frac{1}{2}V^2 \right) = 0 \tag{4.83}$$

The integration of the above differential equation along the streamline gives:

$$\frac{p}{\rho} + gy + \frac{1}{2}V^2 = C \tag{4.84}$$

C is constant along the same streamline, which holds only along the same streamline in the steady, ideal fluid flow, that is the familiar conventional Bernoulli's equation from the fluid mechanics.

Case 2:
If the flow is unsteady and irrotational, we have $\partial V / \partial t \neq 0$ and $\nabla \times V = 0$, accordingly the velocity field is given by $V = \nabla \varphi$ and Eq. (4.79) becomes:

$$\nabla \left(\frac{\partial \varphi}{\partial t} + \frac{p}{\rho} + gy + \frac{1}{2}V^2 \right) = 0 \tag{4.85}$$

Scalar multiplication of $d\mathbf{r}$ to ∇ in Eq. (4.85) gives:

$$d \left(\frac{\partial \varphi}{\partial t} + \frac{p}{\rho} + gy + \frac{1}{2}V^2 \right) = 0 \tag{4.86}$$

Since $d\mathbf{r}$ is an arbitrary vector, not specific along the streamline as it was for the rotational steady flow in the Case 1, the differential equation is to be considered along any direction in the irrotational unsteady flow. Integrating the above:

$$\frac{\partial \varphi}{\partial t} + \frac{p}{\rho} + gy + \frac{1}{2}V^2 = C(t) \tag{4.87}$$

The constant $C(t)$ called Bernoulli constant is same throughout the fluid at an instant time t but independent of space variable. Addition of this to pressure p has no effect on the motion of the fluid since no pressure gradients result from such an addition to the pressure. Thus, we may put the $C(t) = 0$. Stoker (1957) proved that the constant $C(t)$ can be put to zero without loss of generality. The general Bernoulli's equation of ideal fluid for unsteady irrotational flow is:

$$\frac{\partial \varphi}{\partial t} + \frac{p}{\rho} + gy + \frac{1}{2}V^2 = 0 \tag{4.88}$$

When the motion is steady and irrotational we obtain the same Bernoulli's equation as Eq.(4.84), where the constant C is same at any point in the fluid.

4.2.4 Boundary value problems

The problem in determining a function Φ in a domain τ given the value of normal derivative $\partial\Phi/\partial n$ on the boundary surface of the domain is called Neumann boundary value problem (BVP). The problem to determine a function Φ in the same τ given the value of Φ on the boundary surface of the domain is called Dirichlet BVP. The wave-body interaction problem frequently employs Green's third identity which contains Φ and $\partial\Phi/\partial n$ defined in a mixed manner on the boundary.

The BVP generally considers the kinematical and dynamical boundary conditions.

4.2.4.1 Kinematical boundary condition

We assume the fluid under consideration to have a boundary σ, fixed or moving, which separates it from some other medium, and which has the property that any particle that is once on the surface remains on it[a].

Examples of such boundary surfaces of importance for us are those in which σ is the surface of a fixed rigid body in contact with the fluid — the bottom of the sea, for example — or the free surface of the water in contact with the air.

Assume such a surface is given, for example by $F(x, y, z, t) = 0$. Then, since F moves with the same fluid particles:

$$\frac{DF}{Dt} = \frac{\partial F}{\partial t} + \mathbf{V} \cdot \nabla F = 0, \text{on } F = 0 \tag{4.89}$$

Since $\nabla F = \mathbf{n}|F|$, and $\mathbf{n}$ is unit outward normal from the fluid, covered by the surface $F = 0$ the foregoing kinematical boundary condition becomes:

$$\frac{\partial \Phi}{\partial n} = \frac{-F_t}{|\nabla F|} = V_n, \text{ on } F = 0 \tag{4.90}$$

---

[a] Actually (Stoker, 1957), the property is a consequence of the basic assumption in continuum mechanics that the motion of the fluid can be described mathematically as a topological deformation which depends on the time t.

V_n means the common velocity of fluid and boundary surface in the direction normal to the surface.

In the important special case in which the boundary surface is fixed, i.e., it is independent of the time t, we have the condition:

$$\frac{\partial \Phi}{\partial n} = 0 \text{ on } \sigma \tag{4.91}$$

If the boundary surface moves at V_n the boundary condition is obtained from Eq. (4.90) as:

$$\frac{\partial \Phi}{\partial n} = V_n \text{ on } \sigma \tag{4.92}$$

4.2.4.2 Dynamical boundary condition

On the boundary of two inviscid fluids presenting a surface of separation, the pressure shall be continuous. On the free surface boundary between air and water, referring to Bernoulli's Eq. (4.88), we have:

$$p_{water} = -\rho \frac{\partial \Phi}{\partial t} - \rho g y - \frac{\rho}{2}(\nabla \Phi)^2 = p_{air} = 0 \tag{4.93}$$

On the inviscid fluid in contact with rigid boundary fixed or moving, the condition is such that the fluid force shall be normal to the boundary surface.

4.2.5 Energy flux

The total mechanical energy E contained in the system of flowing fluid is identical to the sum of the potential and kinetic energies of the flowing particles. Taking the intensive property b or mechanical energy per unit mass of the flowing fluid (where heat is not involved):

$$b = gy + \frac{1}{2}V^2 \tag{4.94}$$

In Eq. (4.52), we write the energy contained in the system of fluid at time t:

$$E_{sys}(t) = \int_{sys} \rho b d\tau \tag{4.95}$$

Since the energy of the system moves with the same fluids at velocity V, the rate of increase of energy of the system with respect to time t is according to Reynolds transport theorem in the form:

$$\frac{DE_{sys}}{Dt} = \frac{\partial}{\partial t}\left(\int_{CV(t)} \rho b d\tau\right) + \int_{CS(t)} \rho b V_n \, d\sigma \qquad (4.96)$$

Here we wish to calculate the energy flux, having in mind that the control volume and control surface are not necessarily fixed, but may vary depending on time. And we assume that the control surface moves with the same fluid particles. In Eq. (4.96) V_n denotes the normal velocity of the boundary of $CS(t)$ that may or may not be identical to the fluid normal velocity Φ_n.

We will consider these two cases below.

Use of Bernoulli's Eq. (4.88) in Eq. (4.96) gives:

$$\frac{DE_{sys}}{Dt} = \rho \int_{CV(t)} \nabla\Phi \cdot \nabla\Phi_t d\tau - \int_{CS(t)} \rho\left(\frac{p}{\rho} + \Phi_t\right) V_n d\sigma \qquad (4.97)$$

Referring to Gauss's theorem Eq. (4.13), we can express the volume integral in Eq. (4.97) in the form:

$$\int_{CV(t)} \nabla\Phi \cdot \nabla\Phi_t \, d\tau = \int_{CS(t)} \Phi_t \Phi_n d\sigma \qquad (4.98)$$

Substitution of Eq. (4.98) into Eq. (4.97) gives:

$$\frac{DE_{sys}}{Dt} = \int_{CS(t)} \left(\rho \, \Phi_t(\Phi_n - V_n) - p V_n\right) d\sigma \qquad (4.99)$$

If the control surface $CS(t)$ is moving or being fixed consisting of the same fluid particles $\Phi_n = V_n$, the energy flux through $CS(t)$ is given by:

$$\left(\frac{DE_{sys}}{Dt}\right)_{CS(physical)} = \int_{CS(physical)} (-p\Phi_n) d\sigma \qquad (4.100)$$

The above relation indicates that no energy flux is obtained if either the boundary is fixed, i.e., $V_n = 0$, or the pressure p on the boundary vanishes. Such boundary surface as being fixed or in motion consisting of the same fluid particles is called the physical or material boundary. The free surface is an example of physical boundary.

If *CS* is the geometric surface fixed in space but not consisting of the same fluid particles, i.e., $V_n = 0$, then Eq. (4.99) becomes:

$$\left(\frac{DE_{sys}}{Dt}\right)_{CS(geometric)} = \int_{CS(geometric)} \rho \Phi_t \Phi_n d\sigma \qquad (4.101)$$

From the above one may determine the average energy flux of sinusoidal propagating wave crossing the fixed geometric surface.

4.3 Linear Wave Theory

The potential flow of a linear wave in a given domain is determined by satisfying the linearized boundary conditions. Such potential flow is a solution of a typical BVP introduced in Sec. 4.2.4. We will set up a linear BVP in a general form and apply it to the standing waves. The solution of linear standing waves can be linearly superimposed to make progressing waves. A similar approach will be applied to obtain the circular waves.

4.3.1 General formulation

We consider the fluid contained in a basin which is bounded by the free surface and bottom. The fluids may be disturbed in various modes due to wind and currents. The fluids may also be disturbed by the motion of ships on the calm and wavy free surface. Other type of disturbances may be due to falling bodies on the free surface, movement of crust of the ocean bottom due to volcano, steadily pulsating source near the free surface etc. All such disturbances of fluids are called surface waves contained in arbitrary shape in the water basin.

We seek for the basic formulation for determining the velocity potential of the wave in the basin referring to the fundamental theory of boundary value problem in Sec. 4.2.4. The fluid must satisfy the continuity in the fluid domain, and satisfy the prescribed boundary conditions. On the free surface one must satisfy the kinematical and dynamical boundary conditions, while on the bottom surface non-penetration is allowed. In addition, the most fundamental condition is that the fluids motions are all bounded in the far-field. These conditions are expressed in the forms:

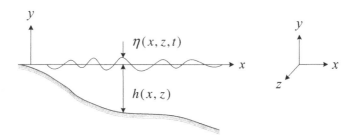

Fig. 4.7. Sketch of the coordinate system.

$$\nabla^2 \Phi = 0, \quad 0 < x < \infty, -\infty < z < \infty, 0 \le y \le h \tag{4.102}$$

$$\Phi_n = 0, \quad \text{on } y = -h(x,z) \tag{4.103}$$

$$-\eta_t - \Phi_x \eta_x + \Phi_y - \Phi_z \eta_z = 0, \text{ on } y = \eta(x,z,t) \tag{4.104}$$

$$\Phi_t + \frac{1}{2}(\nabla \Phi)^2 + gy = 0, \quad \text{on } y = \eta(x,z,t) \tag{4.105}$$

$$\Phi, \eta \text{ are bounded; } |x| \mapsto \infty, |z| \mapsto \infty$$
$$\eta(x,z,t) = 0, \Phi_x = \Phi_y = \Phi_z = 0, \text{ for } t = 0 \tag{4.106}$$

Equations (4.104) and (4.105) are the nonlinear (or exact) boundary conditions on the free surface because these contain the products of derivatives of two unknown functions Φ and η. It is difficult to analytically determine the velocity potential that satisfies the exact boundary conditions. Thus we have to resort to linearization of the nonlinear boundary conditions. Perturbation procedure is applied in linearizing the boundary conditions, as shown in Chap. 9 for the 2nd-order waves, that gives the 1st- and 2nd-order boundary conditions separately.

4.3.2 Formulation of linear wave

Consider 2-D waves horizontally unbounded and vertically bounded by the free surface of uniform atmospheric pressure and the horizontal bottom of constant depth h. Assume that the perturbed fluid motion from the equilibrium is small, irrotational and oscillates harmonically in time. Then the linearized BVP is written:

$$\nabla^2 \Phi(x, y, z, t) = 0, \; -\infty \le (x, z) \le \infty, \; 0 \le y \le \infty \tag{4.107}$$

$$\Phi_{tt} + g\Phi_y = 0 \text{ on } y = 0 \tag{4.108}$$

$$\Phi_n = 0 \text{ on } y = -h(x, z) \tag{4.109}$$

$$\eta(x, z, t) = -\frac{1}{g}\Phi_t(x, y, z, t) \text{ on } y = 0 \tag{4.110}$$

where the velocity potential and wave elevation are unknown.

4.3.2.1 The linear wave in uniform current

The same perturbation procedure is applied to the nonlinear wave on the uniform current $-U$ along x-axis. Omitting the repeated presentation of the continuity and bottom condition, we write the linear free surface boundary condition and linear wave elevation in the forms:

$$\left(\frac{\partial}{\partial t} - U\frac{\partial}{\partial x}\right)^2 \Phi + g\Phi_y = 0, \; y = 0 \tag{4.111}$$

$$\Phi_t - U\Phi_x + g\eta = 0, \; y = 0 \tag{4.112}$$

Equations (4.111) and (4.112) represent the linear free surface boundary condition and linear wave elevation with respect to the ship advancing at U in the waves.

For the steady fluid motion at U, we have the linear free surface boundary condition and linear wave elevation, respectively in the forms:

$$U^2\Phi_{xx} + g\Phi_y = 0, \quad y = 0 \tag{4.113}$$

$$\eta = \frac{U}{g}\Phi_x, \quad y = 0 \tag{4.114}$$

The wave elevation due to the uniform motion of a ship in calm water can be estimated by Eq. (4.114).

4.3.3 Formulation of standing waves

Consider the velocity potential for 2-D standing wave in the water bounded by the bottom and free surface and horizontally infinite in the form:

$$\Phi(x,y,t) = \varphi(x,y)\cos\omega t \qquad (4.115)$$

The above fulfills the continuity equation in the fluid domain and linear boundary conditions:

$$\nabla^2\varphi = \frac{\partial^2\varphi}{\partial x^2} + \frac{\partial^2\varphi}{\partial y^2} = 0, \; -h \leq y \leq 0, -\infty \leq x \leq \infty \qquad (4.116)$$

$$\frac{\partial\varphi}{\partial y} = 0, \quad y = -h \qquad (4.117)$$

$$\frac{\partial^2\varphi}{\partial t^2} + g\frac{\partial\varphi}{\partial y} = 0, \quad y = 0 \qquad (4.118)$$

$$\eta = -\frac{1}{g}\Phi_t = \frac{\omega}{g}\varphi\sin\omega t, \quad y = 0 \qquad (4.119)$$

where the velocity potential and wave elevation are unknown. We look for the solution of φ and η, assuming that the velocity potential $\varphi(x, y)$ in Eq. (4.115) is given in the form of separation of variables:

$$\varphi(x,y) = X(x)Y(y) \qquad (4.120)$$

Substituting the above in the continuity equation Eq. (4.116), we have:

$$\frac{X''}{X} = -\frac{Y''}{Y} \qquad (4.121)$$

For an arbitrary pair of x and y, there exists an arbitrary k such that:

$$\frac{X''}{X} = -\frac{Y''}{Y} = -k^2 \text{ or } +k^2, \, (k > 0 \text{ and real}) \qquad (4.122)$$

Thus we will have two solutions as shown next section.

4.3.3.1 Solution of standing wave for $-k^2$

$$X'' + k^2 X = 0 \tag{4.123}$$

$$Y'' - k^2 Y = 0 \tag{4.124}$$

The general solutions of Eqs. (4.123) and (4.124) are written in the forms:

$$X = c_1 \cos kx + c_2 \sin kx \tag{4.125}$$

and

$$Y = a_1 e^{ky} + a_2 e^{-ky} \tag{4.126}$$

where c_i, and a_i are arbitrary constants. Accordingly the general solution for φ in Eq. (4.120) is given in the form:

$$\varphi(x, y) = (a_1 e^{ky} + a_2 e^{-ky}) \begin{Bmatrix} \cos \\ \sin \end{Bmatrix} kx \tag{4.127}$$

Equation (4.127) has three undetermined constants a_1, a_2 and k. These are determined by satisfying the bottom condition Eq. (4.117), free surface boundary condition Eq. (4.118), and linear wave elevation Eq. (4.119). Expressing the exponential functions in terms of the hyperbolic functions, $e^{\pm ky} = \cosh ky \pm \sinh ky$ and fulfilling the bottom condition Eq. (4.117), we have:

$$\varphi(x, y) = C \frac{\cosh k(y + h)}{\sinh kh} \cos kx \tag{4.128}$$

where C is an arbitrary constant. It is noted that we have taken the $\cos kx$ term here. One may take $\sin kx$ term as well. Let the above φ in Eq. (4.128) satisfy the linear free surface condition Eq. (4.118), then we have:

$$\omega^2 = gk \tanh kh \tag{4.129}$$

The arbitrary number k in Eq. (4.122) has now been determined by the above dispersion relation, that is, k is defined in terms of wave frequency and water depth. In fact Eq. (4.129) gives the natural frequency of the free oscillation. Because we satisfy the homogeneous free surface boundary condition Eq. (4.118), the standing wave is called free wave. There is 2nd-order free surface boundary condition that is not homogeneous. Then the solution is called phase-locked or bounded wave as discussed in Chap. 9.

To determine the arbitrary constant C in φ in Eq. (4.128), we substitute the φ into linear wave elevation Eq. (4.119) and let the amplitude of η be A, then the arbitrary constant C:

$$C = \frac{gA}{\omega} \tanh kh \qquad (4.130)$$

Thus the final solutions of the velocity potential and wave elevation of the free standing wave are given, respectively by:

$$\Phi(x, y, t) = \frac{gA}{\omega} \frac{\cosh k(y+h)}{\cosh kh} \cos kx \, \cos \omega t \qquad (4.131)$$

$$\eta(x, t) = A \cos kx \, \sin \omega t, \quad y = 0 \qquad (4.132)$$

We may similarly derive the above equations with the terms of sin kx and sin ωt instead of cos kx and cos ωt, respectively.

The above Φ and η have sinusoidal profiles along the x-axis and these profiles harmonically oscillate in time. The k is called the wavenumber, from which the wavelength is determined by $\lambda = 2\pi/k$. The period of oscillation is given by $T = 2\pi/\omega$. The points of zero elevation are obtained from Eq. (4.132) by putting cos $kx = 0$ at $x = \pm(2m - 1)(\pi/2) \, k$ for $m = 1, 2, \cdots$. These points are called nodes. Points of the maximum and minimum elevations are called crest and trough. It should be noted that there are infinite numbers of such standing waves along the x-axis. However, the formula may be used as a finite sized rectangular wave tank bounded by walls.

4.3.3.2 Solution for standing wave for $+k_2$

$$X'' - k^2 X = 0 \qquad (4.133)$$

$$Y'' + k^2 Y = 0 \qquad (4.134)$$

The general solutions of the above are:

$$X = a_1 e^{kx} + a_2 e^{-kx} \qquad (4.135)$$

$$Y = c_1 \cos ky + c_2 \sin ky \qquad (4.136)$$

$$\varphi(x, y) = (a_1 e^{kx} + a_2 e^{-kx}) \begin{Bmatrix} \cos \\ \sin \end{Bmatrix} ky \qquad (4.137)$$

There are three arbitrary constants including k. These will be determined by satisfying the bottom and free surface boundary conditions and the boundedness of the solution.

To satisfy the constant depth condition in Eq. (4.117) we need:

$$Y'(-h) = 0 \qquad (4.138)$$

which is satisfied by:

$$Y = B \cos k(y+h) \qquad (4.139)$$

where B is an arbitrary constant. Let $\varphi(x,y) = (a_1 e^{kx} + a_2 e^{-kx}) B \cos k(y+h)$ satisfy the free surface boundary condition Eq. (4.118), then we obtain:

$$\frac{\omega^2}{g} = -k \tan kh \qquad (4.140)$$

Equation (4.140), in contrast to the single solution Eq. (4.129), has many roots k_i ($i = 1, 2, 3 \cdots$) given the frequency ω and depth h.

The roots may be graphically determined by writing Eq. (4.140) in the following form as shown in Fig. 4.8:

$$-\frac{\dfrac{\omega^2}{gh}}{kh} = \tan kh \qquad (4.141)$$

Thus, the solution is given in the form:

$$\Phi(x, y, t) = \sum_{i=1}^{\infty} B_i (a_1 e^{k_i x} + a_2 e^{-k_i x}) \cos k_i(y+h) \cos \omega t \qquad (4.142)$$

The bounded solution of the above is obtained in the form:

$$\Phi(x, y, t) = \sum_{i=1}^{\infty} C_i (e^{-k_i |x|}) \cos k_i(y+h) \cos \omega t \qquad (4.143)$$

Equation (4.143) contains arbitrary constants C_i which implies that many components of standing waves fluctuate horizontally with exponentially decreasing amplitudes. Such waves are called evanescent standing waves.

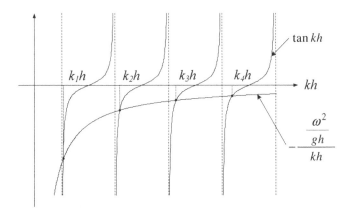

Fig. 4.8. Graphycal determination of roots k_i for the given frequency ω and depth h.

4.3.3.3 Seiches in lakes and harbors

Consider the 2-D standing waves in a rectangular tank with a uniform section of breadth b and depth h in xy plane. And let the potential in Equation (4.131) satisfy the kinematical boundary condition $\partial\Phi/\partial x = 0$ on the walls at $x = 0$ and b, namely:

$$\sin kb = 0 \tag{4.144}$$

which holds when $k_m = m\pi/b$ and mode m is a positive integer. This indicates there exists an infinite number of solutions. By substituting the above in Eq. (4.132), one obtains:

$$\eta = \sum_m A_m \cos\left(\frac{m\pi}{b}x\right)\sin\omega_m t \tag{4.145}$$

with

$$\omega_m = \sqrt{k_m g \tanh k_m h} \quad \text{and} \quad k_m = \frac{m\pi}{b} \tag{4.146}$$

The component standing waves are illustrated in Fig. 4.9.

When an excitation frequency is in coincidence with one of the natural frequencies ω_m, the motion of that frequency will become large. When the depth h is very small relative to the breadth, b, the period T_1 of the first mode oscillation is given by

$$T_1 = \frac{2b}{\sqrt{gh}} \qquad (4.147)$$

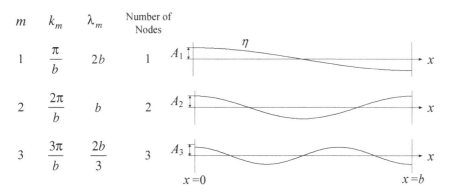

m	k_m	λ_m	Number of Nodes
1	$\dfrac{\pi}{b}$	$2b$	1
2	$\dfrac{2\pi}{b}$	b	2
3	$\dfrac{3\pi}{b}$	$\dfrac{2b}{3}$	3

Fig. 4.9. Modal motions of standing waves.

In the case of Lake Baikal; $b = 665$ km, $h = 680$ m gives $T_1 = 4.5$ hrs. Here the observed period is 4.64 hrs. In the case of Lake Geneva, which is not rectangular, the formula does not give such good agreement. Here, $b = 70$ km and $h = 160$ m and the calculated period is 59 min against an observed period of 73.5 min (Davis, 1971). This kind of phenomenon which concerns motions of rather long periods called seiches occurs in lakes in various parts of the world. Wave motions in harbors are often of the type of seiches. In these cases, oscillations of water in the harbor are also commonly excited by the motion at the mouth of the harbor, which of course is due to the wave motions generated in the open sea.

4.3.3.4 Standing wave on model test in wave tank

The wavemaker produces not only the progressing waves due to the flap but also resonance standing wave motion of the water due to the rectangular tank. This may affect the acquisition of the data of incident wave and response.

4.3.3.5 Streamlines of standing waves (Eulerian method)

The standing wave potential Eq. (4.131) provides the wave elevation and the differential equation of streamline equation (4.47), respectively:

$$\eta(x,t) = A \cos kx \sin \omega t \tag{4.148}$$

$$\frac{dx}{-\cosh k(y+h)\sin kx} = \frac{dy}{\sinh k(y+h)\cos kx} \tag{4.149}$$

Separation of variables in Eq. (4.149) makes an integrable form:

$$\frac{k \cosh k(y+h)}{\sinh k(y+h)} dy = -\frac{k \cos kx}{\sin kx} dx \tag{4.150}$$

Integration of the above gives the stream function:

$$\psi = \sinh k(y+h) \, |\sin kx| = \text{const} \tag{4.151}$$

As an example we consider the standing wave of length $2\pi/k$ = 10 m, amplitude 1 m and water depth of 10 m. Incorporating these values in Eq. (4.151) while varying the constant value of ψ gives the streamlines as shown in Fig. 4.10. The three vertical lines at x = 0 m, 5 m, and 10 m represent the streamlines of ψ = 0, regardless of the y-value since sin kx = 0 on the lines, whereas the wave elevations at the same abscissa are +1.0 m or −1.0 m. The bottom line presents the streamlines of ψ = 0 since sinh $k(y + h)$ = 0 for y = − h. The points satisfying $\partial \psi/\partial x$ = 0, i.e., cos kx = 0, regardless of any y-value, indicate the lowest points of the streamline. Along the vertical axis, for instance, at x = 2, ψ = 0 is the lowest, and increases to ψ = 0.5, ψ = 4 and so forth.

4.3.3.6 Paths of particles of standing waves

The velocity potential for the standing wave in Eq. (4.131) may be written in the form:

$$\Phi(x,y,t) = \alpha \cosh k(y+h) \cos kx \cos \omega t \tag{4.152}$$

from which one obtains the following differential equations of the path of a particle,

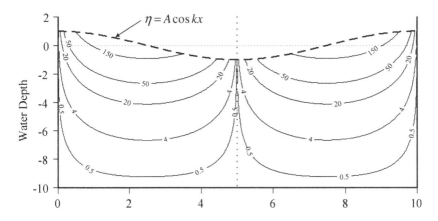

Fig. 4.10. Streamlines of the flow of a standing wave in a rectangular tank.

$$\frac{dx}{dt} = -\alpha\,k\cosh k(y+h)\sin kx\cos\omega t$$
$$\frac{dy}{dt} = \alpha\,k\sinh k(y+h)\cos kx\cos\omega t \qquad (4.153)$$

Equation (4.153) is nonlinear. Therefore it is difficult to integrate. Thus we assume that the motion of the particle is small and the particle's velocity at (x, y) is identical to that of the particle at equilibrium position, or the mid-position of oscillation (x_0, y_0), i.e.:

$$\frac{dx}{dt} = -\alpha\,k\cosh k(y_0+h)\sin kx_0\cos\omega t$$
$$\frac{dy}{dt} = \alpha\,k\sinh k(y_0+h)\cos kx_0\cos\omega t \qquad (4.154)$$

Integrating the above equations we have:

$$x = x_0 - \left(\frac{\alpha k}{\omega}\right)\cosh k(y_0+h)\sin kx_0\sin\omega t$$
$$y = y_0 + \left(\frac{\alpha k}{\omega}\right)\sinh k(y_0+h)\cos kx_0\sin\omega t \qquad (4.155)$$

where (x_0, y_0) are the integral constants or the name of the particle given at time $t = 0$ in Lagrangian method. Eliminating the time factor:

$$y - y_0 = -\frac{\sinh k(y_0 + h)\cos kx_0}{\cosh k(y_0 + h)\sin kx_0}(x - x_0) \qquad (4.156)$$

The path of the particle is rectilinear and depends on the Lagrangian coordinate (x_0, y_0). The slope of the path at (x_0, y_0) is identical to the slope of the streamline at (x_0, y_0) in Eq. (4.149). This demonstrates the connection of Lagrangian to Eulerian method.

4.3.4 Progressing wave

The 2-D standing wave in water of constant depth has been determined by satisfying the continuity and linear boundary conditions. We consider the sum of two such velocity potentials in the form in Sec. 4.3.3.6:

$$\Phi(x,y,t) = \begin{cases} \alpha\cosh k(y+h)\sin kx\cos \omega t \\ \mp\alpha\cosh k(y+h)\cos kx\sin \omega t \end{cases} \qquad (4.157)$$

which gives:

$$\Phi(x,y,t) = \alpha\cosh k(y+h)\,\sin(kx \mp \omega t) \qquad (4.158)$$

Since each standing wave satisfies continuity and all the boundary conditions given in Eqs. (4.116) through (4.119), the sum of any two standing wave potentials gives another potential satisfying the same boundary conditions. The velocity potential Eq. (4.158) represents the 2-D progressing wave in the direction of the positive or negative x-direction, which gives the wave elevation in the form:

$$\eta = -\frac{1}{g}\frac{\partial \Phi}{\partial t}\Big|_{y=0} = \pm\frac{\omega\alpha}{g}\cosh kh\,\cos(kx \mp \omega t) \qquad (4.159)$$

Denoting the wave amplitude by A, we determine the arbitrary constant α:

$$\alpha = \pm\frac{gA}{\omega}\frac{1}{\cos kh} \qquad (4.160)$$

Substituting the constant in Eqs. (4.158) and (4.159), we have the velocity potential and wave elevation:

$$\Phi(x, y, t) = \pm \frac{gA}{\omega} \frac{\cosh k(y+h)}{\cosh kh} \sin(kx \mp \omega t) \tag{4.161}$$

$$\eta(x,t) = A \cos(kx \mp \omega t) \tag{4.162}$$

The wavenumber k was given in the dispersion relation Eq. (4.129). The argument of the trigonometric functions in the above is called the phase or phase function. If it is kept constant, say zero, that is, if one looks for the locus of equal phase in positive x-direction as time proceeds, one obtains the condition:

$$kdx - \omega dt = 0 \tag{4.163}$$

Thus:

$$\frac{dx}{dt} = \frac{\omega}{k} = V_p = c \tag{4.164}$$

V_p is called phase velocity or wave celerity as denoted by c. For waves of invariable frequency or for monochromatic wave, this is the only velocity that can come into consideration.

The phase velocity for finite water is derived from the dispersion relation:

$$c = \sqrt{g/k \tanh kh} = \sqrt{g\lambda/2\pi \tanh 2\pi h/\lambda} \tag{4.165}$$

Taking limits, one has the celerity:

$$c = \sqrt{g\lambda/2\pi} \qquad \text{for deep water} \tag{4.166}$$

$$c = \sqrt{gh} \qquad \text{for shallow water} \tag{4.167}$$

For the depth-to-wavelength ratio $h/\lambda > 0.5$ and $h/\lambda < 0.1$, Eqs. (4.166) and (4.167) can be used with negligibly small errors. For the ratio $0.1 < h/\lambda < 0.5$; Eq. (4.165) can be approximately used. The shallow water progressing speed $c = (gh)^{1/2}$ is the maximum physically possible celerity. Considering the ocean as a thin film covering the surface of the earth, we can apply the shallow water celerity for the Tsunami's propagation speed. If the volcano had started on the bottom of 5000 m deep sea, the Tsunami will propagate at 430.5 knots.

The deep water phase velocity depends on the wavelength alone, whereas the shallow water phase velocity depends on the water depth. In deep water, long waves travel faster than short waves. When the propagation velocity of a wave depends on the wavelength, we speak of dispersion, using expression borrowed from optics. The dispersion in a medium is normal when longer (red) waves have larger velocities than shorter (violet) waves. The behavior of gravity waves in deep water thus corresponds to the case of normal dispersion in optics. In shallow water there is no dispersion. For the finite water depth lying between deep and shallow water, it is seen that the propagation speed c increases as λ increases. The progressing wave satisfies the linear free surface boundary condition or homogeneous differential equation. Thus the progressing wave is called free wave.

4.3.4.1 The streamlines of progressing wave

The streamlines of a progressing wave in the positive x-direction are obtained in the same manner as the streamlines of a standing wave expressed by differential equation $dx / \Phi_x = dy / \Phi_y$:

$$\frac{dx}{\cosh k(y+h)\cos(kx-\omega t)} = \frac{dy}{\sinh k(y+h)\sin(kx-\omega t)} \qquad (4.168)$$

Integrating Eq. (4.168) we have:

$$\psi = \sinh\ k(y+h)\cos(kx-\omega t) = \text{const} \qquad (4.169)$$

and

$$\eta(x) = A\cos(kx-\omega t) \qquad (4.170)$$

When the fluid motion is observed relative to the frame moving at the phase speed with the progressing wave, the streamlines will appear exactly the same as those of the standing wave. In other words, with respect to the moving frame, the above ψ and η in Eqs. (4.169) and (4.170) will take the following forms:

$$\psi = \sinh k(y+h)\cos kx \qquad (4.171)$$

$$\eta = A\cos kx \qquad (4.172)$$

4.3.4.2 Paths of particles in progressing wave

Making use of the velocity potential of progressing wave in the positive x-direction, in Eq. (4.161) we have:

$$\frac{dx}{dt} = \Phi_x(x,y,t) = \frac{gAk}{\omega}\frac{\cosh k(y+h)}{\cosh kh}\cos(kx-\omega t)$$

$$\frac{dy}{dt} = \Phi_y(x,y,t) = \frac{gAk}{\omega}\frac{\sinh k(y+h)}{\cosh kh}\sin(kx-\omega t)$$

(4.173)

Equation (4.173) is a nonlinear differential equation system. Linearization gives the velocity of a fluid particle identical to that of the particle at the mean position of the oscillation, namely:

$$\frac{dx}{dt} = \Phi_x(x_0,y_0,t), \quad \frac{dy}{dt} = \Phi_y(x_0,y_0,t)$$

(4.174)

Integrating the above with respect to time t we obtain a new position x and y displaced from x_0 and y_0. A simple manipulation leads to:

$$\frac{(x-x_0)^2}{A^2} + \frac{(y-y_0)^2}{B^2} = 1$$

(4.175)

$$A = \frac{gAk}{\omega^2}\frac{\cosh k(y_0+h)}{\cosh kh}, \quad B = \frac{gAk}{\omega^2}\frac{\sinh k(y_0+h)}{\cosh kh}$$

(4.176)

Equation (4.175) represents a closed elliptic path or orbit. This is only valid under the assumption of linearity. The orbit becomes a circle as the depth becomes infinite while it becomes flatter as the depth decreases.

4.3.5 Wave energy and energy flux

The total energy E_λ contained in the oscillating fluid column under a wavelength λ, is the energy surplus in one wavelength over the fluid at rest:

$$E_\lambda = \rho \int_{CV}\left(\frac{1}{2}(\Phi_x^2+\Phi_y^2)+gy\right)dxdy = \rho\int_0^h\int_0^\lambda\left(\frac{1}{2}(\Phi_x^2+\Phi_y^2)+gy\right)dxdy \quad (4.177)$$

The calculation shows that the total energy consists of the potential and kinetic energy and each component is identical to each other:

$$E_{p_\lambda} = E_{k_\lambda} = \frac{1}{4}\rho g A^2 \lambda$$

(4.178)

Hence the total energy contained in the water column of one wavelength is:

$$E_\lambda = \frac{1}{2}\rho g A^2 \lambda$$

(4.179)

Thus the energy per unit wavelength or energy of sinusoidal wave per unit area of ocean surface is given:

$$\frac{E_\lambda}{\lambda} = \frac{1}{2}\rho g A^2 \tag{4.180}$$

Using Eq. (4.161) we evaluate the time average of the energy flux crossing the geometrical control surface x:

$$\overline{\frac{DE}{Dt}}_{CS(\text{geometric})} = \frac{1}{T}\int_0^T \left(-\int_{-h}^0 \Phi_t \Phi_n dy\right)dt = \frac{\rho g A^2}{2}c_g \tag{4.181}$$

where

$$c_g = \frac{c}{2}\left(1 + \frac{2kh}{\sinh 2kh}\right) \tag{4.182}$$

which is called group velocity. Reynolds and Lord Rayleigh interpreted the group velocity c_g as the ratio of the average energy flux to the energy contained per unit wavelength. Taking the limit for deep water:

$$c_g = \frac{1}{2}c, \qquad h \to \infty \tag{4.183}$$

This indicates clearly that c_g in deep water is one half the phase speed. The group velocity is also interpreted kinematically by observing the progressing wave group formed by the interference of two progressing waves of similar height and wavelength as shown in Sec. 5.1.1.

4.3.6 Circular waves

4.3.6.1 Circular standing waves

The water is horizontally unbounded and vertically bounded by the free surface of uniform atmospheric pressure and the deep water bottom. Assuming that the perturbed flow is small, irrotational and oscillates harmonically in time, we can determine the linear potential of the circular waves in the following form:

$$\Phi(r,\theta,y,t) = \varphi(r,\theta,y)\sin \omega t \tag{4.184}$$

where

$$r = \sqrt{x^2 + z^2}, \quad \theta = \tan^{-1}\frac{z}{x} \qquad (4.185)$$

Assume the solution $\varphi(r,\theta,y)$ in the following form:
$$\varphi(r,\theta,y) = F(r,\theta)\,Y(y) \qquad (4.186)$$

Applying continuity law to the above equation, we have:
$$Y'' - k^2 Y = 0$$
$$\nabla^2 F + k^2 F = 0, \quad (k \text{ posivite, real}) \qquad (4.187)$$

and assume F in the form of separation of variable:
$$F(r,\theta) = R(r)\,\Theta(\theta) \qquad (4.188)$$

Then, since
$$\nabla^2 F = \frac{1}{r}\frac{\partial}{\partial r}\left(r\frac{\partial F}{\partial r}\right) + \frac{1}{r^2}\frac{\partial^2 F}{\partial \theta^2} \qquad (4.189)$$

the second equation of Eq. (4.187) is reduced to:
$$r^2\frac{R''}{R} + r\frac{R'}{R} + r^2 k^2 + \frac{\Theta''}{\Theta} = 0 \qquad (4.190)$$

Assume:
$$\frac{\Theta''}{\Theta} = -n^2, \quad n\text{: real and positive} \qquad (4.191)$$

Then we have:
$$r^2 R'' + r R' + (r^2 k^2 - n^2)R = 0 \qquad (4.192)$$

Transforming the above differential equation by using $\rho = rk$, one has:
$$\rho^2 R''(\rho) + \rho R'(\rho) + (\rho^2 - n^2)R(\rho) = 0 \qquad (4.193)$$

The above is the Bessel equation of order n. If n is zero or positive integer, the general solution of the above is given by:
$$R(\rho) = c_1 J_n(\rho) + c_2 Y_n(\rho) \qquad (4.194)$$

The solutions of Eqs. (4.187) and (4.191) are $Y(y) = Ae^{ky}$, and $\Theta = B\cos(n\theta + \delta)$, thus the general solution is:
$$\Phi(r,\theta,y,t) = e^{ky}\left[A J_n(kr) + B Y_n(kr)\right]\begin{Bmatrix}\cos\\\sin\end{Bmatrix}(n\theta)\,\sin\omega t \qquad (4.195)$$

where A and B are arbitrary constants. Satisfying the linearized free surface condition we can determine the dispersion relation $k = \omega^2/g$.

Consider the Bessel functions of 0th order J_0 and Y_0 as illustrated in Fig. 4.11. J_0 is regular everywhere while Y_0 is singular at the origin behaving logarithmically:

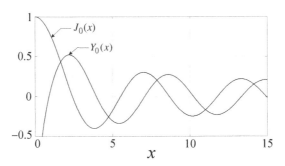

Fig. 4.11. The Bessel function of zeroth order.

$$J_0\big|_{kr\to0} \approx 1, \quad Y_0\big|_{kr\to0} \approx \frac{2}{\pi}\log\frac{kr}{2} \tag{4.196}$$

The asymptotic expression of 0th-order Bessel functions for a large kr behave as sinusoidal waves with diminishing amplitudes:

$$J_0(kr) \approx \sqrt{\frac{2}{\pi kr}} \, \cos\left(kr - \frac{\pi}{4}\right)$$
$$Y_0(kr) \approx \sqrt{\frac{2}{\pi kr}} \, \sin\left(kr - \frac{\pi}{4}\right) \tag{4.197}$$

We introduce Hankel functions of the 1st and 2nd kind of 0th-order:

$$H_0^{(1)}(kr) = J_0(kr) + iY_0(kr) \tag{4.198}$$

$$H_0^{(2)}(kr) = J_0(kr) - iY_0(kr) \tag{4.199}$$

The non-oscillatory Bessel functions of 0th-order are:

$$I_0\big|_{kr\to0} \approx 1, \qquad\qquad K_0\big|_{kr\to0} \approx -\log kr$$
$$I_0\big|_{kr\to\infty} \approx \frac{e^{kr}}{\sqrt{2\pi kr}}, \qquad K_0\big|_{kr\to\infty} \approx \sqrt{\frac{\pi}{2kr}}e^{-kr} \tag{4.200}$$

4.3.6.2 Circular progressing waves

A progressing circular wave potential is obtained by linearly superposing two standing circular waves:

$$\Phi(r, y, t) = \alpha e^{ky} \left[J_0(kr) \cos \omega t + Y_0(kr) \sin \omega t \right]$$
$$= \text{Re} \left[\alpha e^{ky} H_0^{(1)}(kr) e^{-i\omega t} \right] \tag{4.201}$$

where α is an arbitrary positive constant.

The asymptotic expression of the above Φ is:

$$\Phi(r, y, t)\big|_{kr \to \infty} \approx \alpha e^{ky} \sqrt{\frac{2}{\pi kr}} \cos\left(kr - \omega t - \frac{\pi}{4} \right) \tag{4.202}$$

Thus the circular progressing wave elevation outward from the origin is given by:

$$\eta(r, t)\big|_{kr \to \infty} \approx -\frac{\omega \alpha}{g} \sqrt{\frac{2}{\pi kr}} \sin\left(kr - \omega t - \frac{\pi}{4} \right) \tag{4.203}$$

The amplitudes of the Eqs. (4.202) and (4.203) decrease according to $1/r^{1/2}$.

Similarly we can show that the potential for the circular progressing wave in the finite water in the form:

$$\Phi(x, y, z, t) = \text{Re} \left[\sum_{n=0}^{\infty} \alpha_n \cos n\theta \left[H_n^{(1)}(kr) \right] \frac{\cosh k(h + y)}{\cosh kh} e^{-i\omega t} \right] \tag{4.204}$$

$$H_n^{(1)}(kr) = J_n^{(1)}(kr) + i Y_n^{(1)}(kr)$$

$$\left. \begin{array}{c} J_n^{(1)}(kr)\big|_{kr \to \infty} \\ Y_n^{(1)}(kr)\big|_{kr \to \infty} \end{array} \right\} \approx \sqrt{\frac{2}{\pi kr}} \begin{array}{c} \cos \\ \sin \end{array} \left(kr - \frac{\pi}{4} - \frac{n\pi}{2} \right) \tag{4.205}$$

$$\Phi(x, y, z, t)\big|_{kr \to \infty} \approx \text{Re} \left[\begin{array}{c} \sum_{n=0}^{\infty} \alpha_n \cos n\theta \sqrt{\dfrac{2}{\pi kr}} e^{i\left(kr - \frac{\pi}{4} - \frac{n\pi}{2} \right)} \times \\ \dfrac{\cosh k(h + y)}{\cosh kh} e^{-i\omega t} \end{array} \right] \tag{4.206}$$

References

Book references:

Davis, T. (1971) Applied Oceanography, SIT-OE-71-Report, Lecture Note in Ocean Engineering Department, Stevens Institute of Technology, Hoboken, New Jersey.

Kinsman, B. (1965) Wind Generated Waves, Dove Publications, Inc., New York.

Le Mehaute, B. (1976) Introduction to Hydrodynamics and Water Waves, Spinger-Verlag New York, Heidelberg, Berlin.

Milne-Thomson. L.M. (1955) Theoretical Hydrodynamics, Macmillan & Co., Ltd. NewYork.

Munson, B.R., Young, D.F. and Okiishi, T.H. (2002) Fundamentals of Fluid Mechanic, John Wiley & Sons, Inc.

Newman, J.N. (1977) Marine Hydrodynamics, The MIT Press, Cambridge, Massachusetts.

Prandtl, L. and Tietjens, O.G. (1934) Fundamentals of Hydro- and Aeromechanics, Dover Publications Inc., New York.

Schlichting, H. (1979) Boundary Layer Theory, McGraw-Hill Book Co.

Sommerfeld., A. (1964) Mechanics of Deformable Bodies, Lectures on Theoretical Physics, Vol. II, Academic Press, New York and London (in English).

Sommerfeld, A. (1949) Vorlesungen ueber theortetische Physik, II Mechanik der Deformierbaren Medien, Akademie Verlag, Leipzig.

Sternburg, W.J. and Smith, T.L. (1952) The Theory of Potential and Spherical Harmonics, Mathematical Expositions, No. 3, University of Toronto Press.

Stoker, J.J. (1957) Water Waves, Interscience Publishers, Inc., New York.

Wehausen, J.V. and Laitone, E.V. (1960) Surface Waves, Encyclopedia of Physics Vol. 9, Springer-Verlag, Berlin, Gottingen, Heidelberg.

Chapter 5

Propagation of Wave Group and Pulsating Source Wave

5.1 Propagation of Wave Groups

The waves near the beach periodically coming into the beach line appear to be in regular forms. The crest of these waves is regarded as a result of superimposed sinusoidal waves. The grouped waves are usually larger than ordinary linear sinusoidal waves. We observe also that the grouped waves come to the shore line slowly while the component wave in the group moves forward faster. Such behavior can be modeled by superposition of many sinusoidal waves. When the wave components are distributed in a packed form over the narrow frequency band, the narrow-banded amplitude spectrum can be used for simulation of large progressing wave groups (Kinsman, 1965).

The knowledge of the large wave groups is essential in the analysis of wave effects on the large motion of vessels in seas including slowly-varying drift motion of moored vessels. The transient wave groups due to the impulsive pressure applied on the free surface initially at rest has similar property as above.

5.1.1 Amplitude-modulated wave

By superposing two similar sinusoidal waves we obtain another linear wave:

$$\eta(x,t) = A\cos(k_1 x - \omega_1 t) + A\cos(k_2 x - \omega_2 t) \tag{5.1}$$

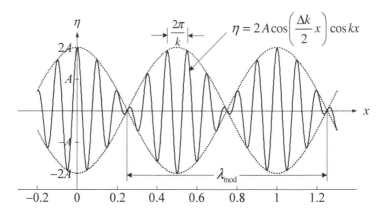

Fig. 5.1. Amplitude modulated wave formed by the addition of two sinusoids of different wave numbers ($k = 20\pi$, $\Delta k = 4\pi$).

If the two wavenumbers or wave frequencies are nearly same

$$k_1 = k_2 + \Delta k, \quad \omega_1 = \omega_2 + \Delta \omega \tag{5.2}$$

Then, Eq. (5.1) may be written in the form:

$$\eta(x,t) = 2A\cos\left(k_1 x - \omega_1 t + \frac{\Delta k}{2}x - \frac{\Delta \omega}{2}t\right)\cos\left(\frac{\Delta k}{2}x - \frac{\Delta \omega}{2}t\right) \tag{5.3}$$

Assuming that $\Delta k << k$, $\Delta \omega << \omega$, we may write $k_1 = k_2 = k$, $\omega_1 = \omega_2 = \omega$:

$$\eta(x,t) \approx B(x,t)\cos(kx - \omega t) \tag{5.4}$$

where

$$B(x,t) = 2A\cos\left(\frac{\Delta k}{2}x - \frac{\Delta \omega}{2}t\right) \tag{5.5}$$

$B(x, t)$ is the envelope that modulates the amplitudes of carrier or basic wave. Consider Eq. (5.4) for two cases. When $t = 0$,

$$\eta(x) \approx 2A\cos\left(\frac{\Delta k}{2}x\right)\cos(kx) \tag{5.6}$$

and when $x = 0$,

$$\eta(t) \approx 2A\cos\left(\frac{\Delta \omega}{2}t\right)\cos(\omega t) \tag{5.7}$$

These two equations represent amplitude-modulated waves in space and time, respectively. The carrier wave $\cos(kx)$ or $\cos(\omega t)$ is modulated by the function $2A\cos(\Delta k / 2)x$ or $2A\cos(\Delta\omega / 2)t$. Because of the assumption, i.e., $\Delta k \ll k$, $\Delta\omega \ll \omega$, the wavelength and period of modulation, L_{mod} and T_{mod} are very much greater than the length and period of the carrier wave.

$$\lambda_{mod} \approx \frac{4\pi}{\Delta k}, \ T_{mod} \approx \frac{4\pi}{\Delta\omega} \tag{5.8}$$

This implies that the carrier wave oscillates frequently while the maximum amplitude slowly decreases and then increases.

Figure 5.1 shows an amplitude-modulated wave in space formed by the addition of two sinusoids of different wavenumber. The modulated wavelength is 1.0 m, while the carrier wavelength is 0.1 m. Similarly we may have an amplitude-modulated wave in time domain.

The velocity of propagation of the group amplitude $B(x, t)$ is found by setting

$$c_{mod} \approx \frac{\lambda_{mod}}{T_{mod}} = \frac{\Delta\omega}{\Delta k} \tag{5.9}$$

In the limit,

$$\lim_{\Delta\omega \to 0, \Delta k \to 0} c_{mod} = \frac{d\omega}{dk} \tag{5.10}$$

which can be obtained by taking the derivative from the dispersion relation:

$$\frac{d\omega}{dk} = \frac{1}{2}c\left(1 + \frac{2kh}{\sinh 2kh}\right) \tag{5.11}$$

where c is the phase velocity. The group velocity was defined as the ratio of the average wave energy flux to the wave energy of a sinusoidal progressing wave per unit area in Sec. 4.3.5. Here we define the group velocity as the propagation velocity of the modulated amplitude, which is a kinematical definition. The average speed of the group amplitude approaches the group velocity as the two component waves become more and more nearly alike in both length and period. The group velocity in the finite water is greater than in the deep water. The group and carrier waves are graphically symmetric about the MWL and vertical axis through the crest, indicating they are linear waves.

5.1.2 Propagation of Gaussian transient

The ocean waves come in groups of high and low waves, and the groups move with a speed approaching group velocity. However, the model with two components is still far too simple to simulate the irregularity of the sea surface. We can go beyond the very simple model (sum of two sinusoids) where we consider an arbitrary group of waves whose frequencies are continuously spread over a small frequency band. The small frequency band contains infinitely many sinusoids. Such an arbitrary group is called wave packet or transient. For convenience, we employ the complex representation, $\exp[i(kx - \omega t)]$ for $\cos(kx - \omega t)$ or $\sin(kx - \omega t)$ for propagation of wave transient,

$$\eta(x,t) = \text{Re} \int_{k_0-\varepsilon}^{k_0+\varepsilon} a(k)e^{i(kx-\omega t)}dk \qquad (5.12)$$

where $a(k)$ is the amplitude density spectrum and the amplitude of the partial wave is $a(k)dk$. The small band width of the whole group is 2ε, if measured in wavenumber unit, so as to be concentrated about the central wavenumber k_0. We rewrite the exponent in Eq. (5.12):

$$kx - \omega t = k_0 x - \omega_0 t + (k - k_0)x - (\omega - \omega_0)t \qquad (5.13)$$

Then Eq. (5.12) assumes the form:

$$\eta(x,t) = \text{Re}\left[B(x,t)\, e^{i(k_0 x - \omega_0 t)} \right] \qquad (5.14)$$

with

$$B(x,t) = \text{Re} \int_{k_0-\varepsilon}^{k_0+\varepsilon} a(k)\, e^{i[(k-k_0)x-(\omega-\omega_0)t]}dk \qquad (5.15)$$

To find the velocity of propagation of the group amplitude $B(x,t)$ or the sums of many sinusoids in this more general case, we have to set the exponential in B, which alone contains x and t and constants. This yields, for the entire wave group, an approximately constant value:

$$\frac{dx}{dt} = \frac{\omega - \omega_0}{k - k_0} = \frac{\Delta\omega}{\Delta k} \tag{5.16}$$

which is in agreement with Eq. (5.9). Taking the limit $\Delta k \to 0$ the above constant value becomes the group velocity. The simple group is illustrated as an approximately simple wave train.

By superposing infinitely many sinusoids on a finite band, instead of infinitesimally small 2ε, we have the progressing wave transient in general form:

$$\eta(x,t) = \mathrm{Re} \int_{-\infty}^{\infty} a(k)e^{i[kx - \omega t]}dk \tag{5.17}$$

with

$$a(k) = \frac{a_0}{\sqrt{2\pi s}} \, e^{[-(k-k_0)^2/2s^2]} \tag{5.18}$$

The above Gaussian amplitude density spectrum is an even function of k. a_0 and k_0 are the maximum amplitude and wavenumber, and s is the std of Eq. (5.17). Substituting Eq. (5.18) into Eq. (5.17), one obtains the progressing surface of Gaussian transient:

$$\eta(x,t) = \mathrm{Re} \int_{-\infty}^{\infty} \frac{a_0}{\sqrt{2\pi s}} \, e^{\left[\frac{-(k-k_0)^2}{2s^2}\right]} e^{i(kx - \omega t)}dk \tag{5.19}$$

Integration of Eq. (5.19) in the deep water is given by expanding $\omega(k)$ in Taylor series about k_0:

$$\omega(k) = \omega(k_0) + A(k - k_0) + B/2(k - k_0)^2 + \cdots \tag{5.20}$$

with

$$A = \frac{d\omega}{dk}\bigg|_{k=k_0}, \quad B = \frac{d^2\omega}{dk^2}\bigg|_{k=k_0}, \quad \cdots \tag{5.21}$$

Retaining the first two terms of the Taylor series, one will have a linear approximation for the frequency as a function of wavenumber, yielding:

$$\eta_1(x,t) = a_0 \exp\{-(s^2/2)(x - At)^2\} \exp\{i[k_0 x - \omega(k_0)t]\} \tag{5.22}$$

Equation (5.22) shows that the individual wave runs at the phase speed of the dominant central component, but the group as a whole, moves steadily with the group velocity $A = (d\omega/dk)|_{k=k0}$ without changing the amplitude.

If one takes all the three terms in Eq. (5.20), the solution becomes quadratic in approximation for the frequency as a function of wavenumber. The analysis results in:

$$\eta_2(x,t) = a_0 \sqrt{\frac{1}{(1+is^2Bt)}} \exp\left\{-\frac{1}{2}\frac{s^2}{(1+s^4B^2t^2)}(x-At)^2\right\}$$

$$\cdot \exp\left\{i\left[k_0x - \omega(k_0)t + \frac{1}{2}\frac{s^2Bt}{(1+s^4B^2t^2)}(x-At)^2\right]\right\}$$

(5.23)

5.1.2.1 Propagation of even Gaussian transient

Clauss and Bergmann (1986) extended Eq. (5.23) to the finite water condition in the form;

$$\left\{{Re \atop Im}\right\}\eta_2(x,t) = a_0\left[\frac{1}{(1+s^4B^2t^2)}\right]^{1/4} \cdot \exp\left\{-\frac{1}{2}\frac{s^2}{(1+s^4B^2t^2)}(x-At)^2\right\}\cdot\left\{{\cos \atop \sin}\right\}$$

$$\cdot \exp\left\{k_0x - \omega_0t + \frac{1}{2}\tan^{-1}(-s^2Bt) + \frac{1}{2}\frac{s^4Bt}{(1+s^4B^2t^2)}(x-At)^2\right\}$$

(5.24)

where

$$A = \frac{d\omega}{dk}\bigg|_{k=k_0} = \frac{c_0}{2}\left[1 + \frac{2k_0h}{\sinh 2k_0h}\right]$$

$$B = \frac{d^2\omega}{dk^2}\bigg|_{k=k_0} = c_0h\left[\frac{1 - 2k_0h\coth 2k_0h}{\sinh 2k_0h}\right.$$

$$\left. - \frac{1}{2\sinh 2k_0h}\left(\frac{\sinh 2k_0h}{2k_0h} - \frac{2k_0h}{\sinh 2k_0h}\right)\right]$$

(5.25)

with

$$c_0 = \sqrt{\frac{g}{k_0} \tanh k_0 h} \qquad (5.26)$$

The three terms of the products in Eq. (5.24) are called the damping term, modulation function, and oscillation term, respectively. The modulation exponentially decreases with the distance x. Figure 5.2(a) shows an even Gaussian transient wave.

The evolution of the transient is illustrated in the time domain at a number of fixed locations in Fig. 5.2(b), and in space at a number of time instants as shown in Fig. 5.2(c). The ratio of the position x to time t of the peaks of each wave group gives the group velocity of 1.56 m/s, which is approximately equal to $\Delta\omega/\Delta k$. It is to be noted that the transient at $x = 0$ in Fig. 5.2(b) is precisely identical to the original Gaussian transient in Fig. 5.2(a).

Figure 5.2(b) shows the evolution of the transients whose amplitudes are diminishing with the elongated group period. Figure 5.2(c) shows that both individual wavelength and amplitude decrease with increase of distance, while the modulated wavelength increases with decrease of amplitude as the distance increases.

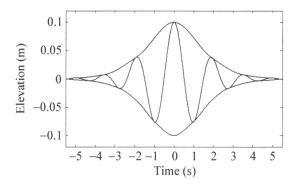

Fig. 5.2(a) Gaussian transient at $x = 0$, even function of t ; $a_0 = 0.1$ m , $T_0 = 2$ s, std s = 0.4, $h = 9$ m.

Clauss and Bergmann (1986) generated Gaussian wave packet in the wave tank to measure the response of offshore structures.

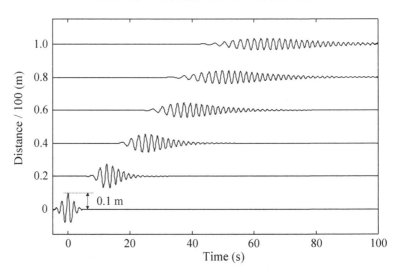

Fig. 5.2(b) Propagation of even Gaussian transient as function of time at various locations.

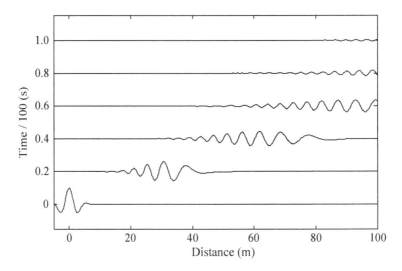

Fig. 5.2(c) Propagation of even Gaussian transient in space at various instants.

5.1.3 Propagation of odd Gaussian transient

Pierson et al. (1992) generated Gaussian and various types of transients in a wave tank and observed the evolutions to investigate the nonlinearity of the waves. With the low wave condition the linearity was confirmed, while with the higher wave, nonlinear waves were observed. The study found the experimental reversibility of the linear wave group and suggested it to be used for checking the degree of linearity of the laboratory wave.

Here we consider the Gaussian envelope defined by:

$$E(t) = A\exp(-B^2t^2) \qquad (5.27)$$

which is used to make an odd transient wave in the form:

$$\eta(t) = E(t)\sin\omega_0 t \qquad (5.28)$$

as illustrated in Fig. 5.3(a). It assumes $A = 0.05$ m, $B = 1/\sqrt{3}\,\text{s}^{-1}$, $\omega_0 = 4\pi/3$ rad/s. Equation (5.28) apparently satisfies the condition for the existence of Fourier transform:

$$\int_{-\infty}^{\infty} |\eta(t)|\,dt < \infty = 0 \qquad (5.29)$$

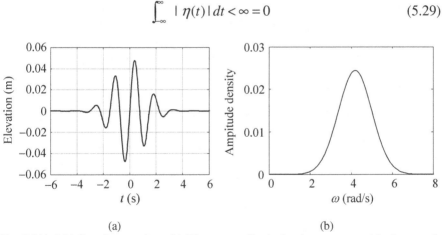

(a)   (b)

Fig. 5.3(a) Odd Gaussian transient. (b) Discrete amplitude density spectrum with phase angle of constant 90°.

The DFT of the transient wave Eq. (5.27) is carried out with $T = 85.33$ s, $\Delta t = 1/24$ s, $N = 2048$, $\Delta\omega = 2\pi/T = 0.0736$ rad/s, and $h = 10$ m, which gives the amplitude density spectrum $b(\omega)$ with constant phase of $\pi/2$ as shown in Fig. 5.3(b). The propagation of the transient is given by:

$$\eta(x,t) = \text{Re} \int_{-\infty}^{\infty} b(\omega) e^{i(kx-\omega t+\pi/2)} d\omega = -\int_{-\infty}^{\infty} b(\omega) \sin(kx - \omega t) d\omega \qquad (5.30)$$

where the wave satisfies the linear dispersion:

$$\frac{\omega^2}{g} = k \tanh kh \qquad (5.31)$$

Figures 5.4(a) and (b) illustrate the evolution of the transient at few locations and time instants. Figure 5.4(a) may be used to determine approximate values of group velocities whose average is closely identical to the theoretical group velocity. It should be noted that the simulated transient at $x = 0$ in Fig. 5.4(a) is precisely identical to the original Gaussian transient in Fig. 5.3(a).

Given the wave elevation time series measured in the wave tank, one may simulate the progressing random sea wave. Since the propagation assumes linear dispersion, if the wave is weakly nonlinear, the nonlinearity of the simulated wave will be slightly weakened compared to the original measured progressing wave. However, it might still be useful for investigation of the effects of large wave groups propagating over the ship hull on the responses such as deck-wetting.

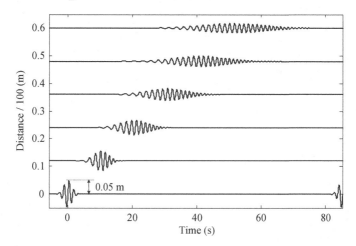

Fig. 5.4(a) Propagation of odd Gaussian transient as function of time at various locations.

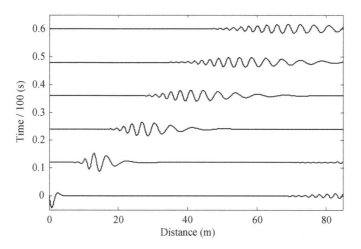

Fig. 5.4(b) Propagation of odd Gaussian transient in space at various instants.

5.1.4 Transient wave due to impulse on the free surface

The foregoing wave is obtained based on the linear superposition of infinitely many free waves satisfying the steady state linear BVP. Here we consider the linear BVP for waves generated by concentrated disturbance acting like a source emitting waves of all wavelengths and frequencies. This sort of wave has to satisfy the linear initial boundary conditions. The initial state of the fluid may be determined when we know the initial form of the boundary and the boundary values of the normal velocity $\partial\Phi/\partial n$ or the velocity potential Φ. Hence two forms of the problem naturally present themselves. We may start with the initial elevation of the free surface without initial velocity, or we may start with the surface undisturbed and an initial distribution of surface impulse ($\rho\,\Phi_0$). Solutions of these problems are based on Cauchy (1827) and Poisson (1816). Thus, it is known as the Cauchy-Poisson problem. Fourier transform technique was applied to solve afore-mentioned problem (Stoker, 1956). The linear initial BVP is formulated as follows:

$$\nabla^2\Phi = 0, \quad -\infty \le x \le \infty, -\infty \le y \le 0, t \ge 0 \qquad (5.32)$$

$$\Phi_{tt} + g\Phi_y = 0, \quad y = 0, \quad t > 0 \qquad (5.33)$$

$$\Phi_y = 0, \quad y = -\infty, \quad t \geq 0 \tag{5.34}$$

The potential and the derivatives are bounded for

$$\sqrt{x^2 + y^2} \rightarrow \infty \tag{5.35}$$

The initial elevation is applied on the free surface,

$$\Phi = 0, \quad y = 0, \quad t = 0 \tag{5.36}$$

$$\Phi_t = -g\alpha(x), \quad y = 0, \quad t = 0 \tag{5.37}$$

where $\alpha(x)$ is the elevated volume per unit area such that,

$$\alpha(x) = \begin{cases} \alpha_0 = \text{const}, |x| < a \\ 0 \qquad\qquad |x| > a \end{cases} \tag{5.38}$$

The above systems of equations are Fourier transformed and the solution for the velocity potential in the transformed domain is reversed to the original physical domain. Defining a unit total elevated volume:

$$2\alpha_0 a \rightarrow 1 \text{ when } \alpha_0 \rightarrow \infty \text{ and } a \rightarrow 0 \tag{5.39}$$

one obtains the solution,

$$\Phi = -\frac{\sqrt{g}}{\pi} \int_0^\infty e^{ky} \cos kx \sin\left(\sqrt{gk}\ t\right) \frac{dk}{\sqrt{k}} \tag{5.40}$$

$$\eta = -\frac{1}{g}\Phi_t = \frac{1}{\pi} \lim_{y \rightarrow 0} \int_0^\infty e^{ky} \cos kx \cos\left(\sqrt{gk}\ t\right) dk \tag{5.41}$$

Note that $\eta(x,t)$ is taken for $y \rightarrow 0$ since the integral obviously diverges for $y = 0$. When $gt^2/4x$ is large, an approximate integration of Eq. (5.41) can be carried out using Kelvin's method of stationary phase (Stoker, 1956), which can calculate the major contribution of the slowly varying phase to the main or significant motion in certain regions. The motion is created by the superposition of waves whose wavelength and amplitude differ insignificantly from a certain given value (Kelvin's method). The final result is given in the form:

$$\eta(x,t) = \frac{1}{x\sqrt{\pi}} \sqrt{\frac{gt^2}{4x}} \cos\left(\frac{gt^2}{4x} - \frac{\pi}{4}\right) \tag{5.42}$$

Equation (5.42) is numerically computed for $x = 4.9$ m varying t as shown in Fig. 5.5(a). With fixed time $t = 0.45$ s, x is varied from a small value toward zero as shown in Fig. 5.5(b). The non-dimensional definitions are according to Lamb (1943): $\tilde{t} = t / \sqrt{2x / g}$, $\tilde{\eta} = \eta / \pi x$.

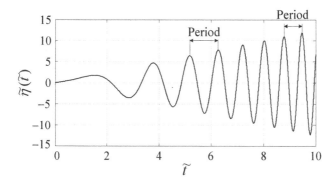

Fig. 5.5(a) The transient wave elevation @ $x = 4.9$ m versus $\tilde{t}$.

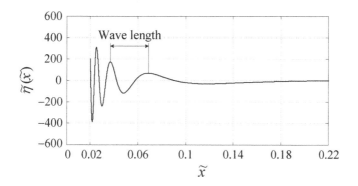

Fig. 5.5(b) The transient wave on the free surface $\overline{\eta}$ versus $\overline{x}$ @ $t = 0.45$ sec.

If the phase is fixed by setting $gt^2/4x = c$, then the phase moves in accordance with the relation $x = gt^2/4c$. Equation (5.42) holds only where the quantity $gt^2/4x$ is large, and hence the individual phases move in such region at nearly constant velocity. Also, for small changes in x or t the waves behave very nearly like simple harmonic waves of a certain fixed period and wavelength. This can be seen as follows. Suppose that we vary t alone from t_0 to $t_0 + \Delta t$. Then the phase φ at $t = t + t_0$ may be written:

$$\varphi = \frac{gt_0^2}{4x_0}\left[1 + \frac{2\Delta t}{t_0} + \left(\frac{\Delta t}{t_0}\right)^2\right] \tag{5.43}$$

Thus if $\Delta t / t_0$ is small, i.e., if the change Δt in the time is small compared with the total lapse of time since the motion is initiated, we have the following change in phase:

$$\Delta\varphi = \varphi - \varphi_0 \simeq \frac{2\Delta t}{t_0}\left(\frac{gt_0^2}{4x_0}\right) \tag{5.44}$$

Consequently the period $T = \Delta t$ of the motion corresponding to the change $\Delta\varphi = 2\pi$ is given approximately by the formula:

$$T \simeq \frac{4\pi x_0}{gt_0} \tag{5.45}$$

Accuracy of this formula is good if $T/t_0 \approx 4\pi x_0 / gt_0^2$ is small and this is the case since $gt_0^2/4x_0$ is always assumed to be large. Thus the period varies slowly for a fixed position and it decreases with the increase of time as shown in Fig. 5.5(a).

In the same way one finds for the local wavelength the approximate formula:

$$\lambda \simeq \frac{8\pi x_0^2}{gt_0^2} \tag{5.46}$$

by varying with respect to x alone, and this is shown to be accurate if $gt_0^2/4\pi x_0$ is large. Thus $\lambda / x_0 \approx 8\pi x_0/gt_0^2$ is small since $gt_0^2/4x_0$ is always assumed to be large. Also, the wavelength at any fixed time varies slowly and increases with the distance from the source as shown in Fig. 5.5(b).

The velocity of the individual wave in the neighborhood of t and x, i.e., for instance, the velocity of a zero of η (x, t), for example from $gt^2/4x = c$ when x and t both vary independently, is determined by using Eq. (5.46):

$$\frac{dx}{dt} = \frac{2x}{t} \tag{5.47}$$

If one fixes attention on the group rather than on an individual wave of the group, the velocity of the group will differ from that of its component waves. The phase velocity for the present case can be obtained in terms of the local wavelength readily from Eq. (5.47) by expressing the right hand side in terms of the local wavelength through use of Eq. (5.46). The result is:

$$\frac{dx}{dt} = \frac{2x}{t} = \sqrt{\frac{g\lambda}{2\pi}} \qquad (5.48)$$

On the other hand, the position x of a group of waves of fixed wavelength λ at time t in Eq. (5.48) is given by the formula:

$$x = \frac{1}{2}\sqrt{\frac{g\lambda}{2\pi}}\, t \qquad (5.49)$$

so that the velocity of the group is just half the phase speed of the component waves. In other words, the component waves in a particular group move forward through the group with a speed twice that of the group. The group does not, however, maintain constant amplitude as it proceeds. It is easily seen from Eq. (5.42) in combination with Eq. (5.46) that the group amplitude is proportional inversely with $x^{0.5}$ for the fixed wavelength λ. The transient wave elevation Fig. 5.5(a) indicates that the amplitude increases while the period decreases as the time increases. Figure 5.5(b) shows that the amplitude increases, while the wavelength decreases as x approaches to origin, in both respects without limit.

The kinematic definition of group velocity in Sec. 5.1.1 was obtained by the superposition of trains of simple harmonic waves of slightly different wave- length and amplitude, while in the present case the waves are the result of superposition of waves of all wavelengths and periods.

It appears that the region in the immediate neighborhood of the origin may be an endless succession of waves of continually increasing amplitude and frequency, whose activity of the source is not paradoxical, for our assumed initial accumulation of a finite volume of elevated fluid on an infinitely narrow base implies an unlimited store of energy. The storm wave [Fig. 5.5(b)] has a shorter wavelength while in the far-field it becomes swell.

The drop of a pebble on calm water surface creates a similar wave pattern as the 2-D transient wave but in the form of circular waves. Likewise the amplitude of the generated wave by the pebble decreases as it moves out as the 2-D wave.

5.2 Waves Due to 2-D Pulsating Source

We will derive the velocity potential due to the 2-D pulsating source.

In an infinitely extended 2-D fluid domain, we consider the velocity potential of a harmonically pulsating source (Holstein, 1937; Stoker, 1957; Wehausen and Laitone, 1960; Frank, 1967) in the form:

$$\Phi_1(x, y, t) = \frac{Q}{2\pi} \log r \cos \omega t \qquad (5.50)$$

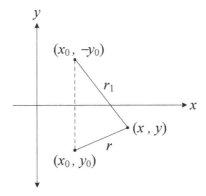

Fig. 5.6. (x_0, y_0) source point and (x, y) field point in the flow field.

where r is the radius vector from the source point (x_0, y_0) to a field point (x, y), t indicates time, ω the circular frequency, and Q strength of the 2-D source that is in dimension of radiating flow rate per unit length of a long cylinder. The source point is evidently a singular point of the potential Φ_1. The pulsating source flow may be regarded as a periodically varying in- and ex-haling circle whose center is at the point (x_0, y_0). Designate the amplitude of radial displacement of a point on the periphery of radius r by A_0. Then the maximum radial velocity of the peripheral point is $A_0 \omega$. Thus we determine the radial velocity at the peripheral point of a circle of radius r:

$$\frac{\partial \Phi_1}{\partial r} = \frac{Q}{2\pi r} \cos \omega t, \qquad Q = 2\pi r A_0 \omega \tag{5.51}$$

Our task is to find a pulsating source induced flow when the fluid has a free surface and the gravitational field acts on the fluid. In particular, we are interested in the manner of the wave formation by the action of the periodically oscillating source below the free surface. Using potential Φ_1, only the fluid motion near the source can be studied. The motion of the remaining fluid may be studied by adding potentials. We set a plane rectangular coordinate system as shown in Fig. 5.6. As usual, the x-axis lies on the calm free surface and the y-axis points vertically upward. The radius vector r is then given by:

$$r = \sqrt{(x - x_0)^2 + (y - y_0)^2} \tag{5.52}$$

Let the final solution Φ be given by:

$$\Phi(x, y, t) = \Phi_1(x, y, t) + \Phi_2(x, y, t) + \cdots + \Phi_n(x, y, t)$$

$$= \sum_{i=1}^{n} \mathrm{Re}[\phi_i(x, y) e^{-i\omega t}] \tag{5.53}$$

$$\phi_i(x, y) = \phi_{ic}(x, y) + i\phi_{is}(x, y)$$

The foregoing Φ must satisfy the following conditions:

$$\nabla^2 \Phi = 0, \ -\infty \leq x \leq \infty, \ y \leq 0 \tag{5.54}$$

$$\Phi_y - \nu \, \Phi = 0, \ \text{on } y = 0 \quad \nu = \frac{\omega^2}{g} : \text{deep water wavenumber} \tag{5.55}$$

$$\frac{\partial \Phi}{\partial y} = 0, \ \ y \to -\infty \tag{5.56}$$

$$\eta(x, t) = -\frac{1}{g} \frac{\partial \Phi}{\partial t}, \ \ y = 0 \tag{5.57}$$

$$\Phi_n(x, y, t) = \phi_n(x, y) \begin{Bmatrix} \cos \omega t \\ \sin \omega t \end{Bmatrix} \tag{5.58}$$

$$\lim_{x \to \pm\infty} \left(\frac{\partial \phi_c}{\partial x} \pm \nu \phi_s \right) = 0, \text{ or } \lim_{x \to \pm\infty} \left(\frac{\partial \phi_s}{\partial x} \mp \nu \phi_c \right) = 0 \qquad (5.59)$$

Equations (5.54) through (5.57) are the familiar formulas for the linear BVP as given in Chap. 4, while Eq. (5.58) represents standing waves that may be used to make combined flows to satisfy the required boundary conditions. The last Eq. (5.59) is called the radiation condition, which enables the source induced waves propagate outward from the disturbance. We will build up the 2-D pulsating source potential Eq. (5.53) by following five steps:

Step 1: Φ_1 is already given in Eq. (5.50).

Step 2: Add Φ_2 to the above basic potential.

We assume that the fluid domain extends over the entire xy-plane. Mathematically, the potential is analytically continued to the upper half-plane. The continuation is obtained by locating a periodic sink of the same strength at the mirror point which acts in the opposite direction to the source. In other words, one is in an inhale, while the other is in an exhale mode of breath. Thus the periodic sink potential Φ_2 is given in the form:

$$\Phi_2 = -\frac{Q}{2\pi} \log r_1 \cos \omega t \qquad (5.60)$$

$$r_1 = \sqrt{(x - x_0)^2 + (y + y_0)^2} \qquad (5.61)$$

Dropping the time factors from the potential Eq. (5.58), we obtain the following result:

$$\phi_1 + \phi_2 = 0, \ y = 0 \qquad (5.62)$$

Thus

$$\frac{\partial \phi_1}{\partial y} + \frac{\partial \phi_2}{\partial y} = -\frac{Q}{\pi} \frac{y_0}{(x - x_0)^2 + y_0^2}, \ y = 0 \qquad (5.63)$$

Step 3: Add $\Phi_3 = \phi_3 \cos \omega t$.

Because the foregoing Eqs. (5.62) and (5.63) indicate that ϕ_1 with ϕ_2 alone cannot satisfy the linear free surface condition Eq. (5.55), we add another potential. Substituting $\phi_1 + \phi_2 + \phi_3$ in the free surface boundary condition Eq. (5.55), we have:

$$\frac{\partial \phi_3}{\partial y} - v\phi_3 = \frac{Q}{\pi} \frac{y_0}{(x-x_0)^2 + y_0^2}, \quad y=0 \tag{5.64}$$

In order to determine $\phi_3(x, y)$ in the above boundary condition, we employ the formula representing the right-hand side of Eq. (5.64) in an integral form, known as Laplace's transform:

$$\frac{-y_0}{(x-x_0)^2 + y_0^2} = \int_0^\infty e^{ky_0} \cos k(x-x_0)dk, \ y_0 < 0 \tag{5.65}$$

Equations (5.64) and (5.65) together suggest that $\phi_3(x, y)$ may be assumed in the form:

$$\phi_3(x, y) = \int_0^\infty F(k)e^{k(y+y_0)} \cos k(x-x_0)dk \tag{5.66}$$

Substituting Eqs. (5.65) and (5.66) into Eq. (5.64), we have:

$$F(k) = -\frac{Q}{\pi} \frac{1}{k-v} \tag{5.67}$$

Thus, substitution of Eq. (5.67) in Eq. (5.66) yields the solution ϕ_3 in the form:

$$\phi_3(x, y) = -\frac{Q}{\pi} \text{PV} \int_0^\infty \frac{e^{k(y+y_0)}}{k-v} \cos k(x-x_0)dk \tag{5.68}$$

where PV $\int_0^\infty$ indicates the Cauchy principal value integral. Hence using Eq. (5.58):

$$\Phi_1 + \Phi_2 + \Phi_3$$

$$= \left[\frac{Q}{2\pi} \log \frac{r}{r_1} - \frac{Q}{\pi} \text{PV} \int_0^\infty \frac{e^{k(y+y_0)}}{k-v} \cos k(x-x_0)dk \right] \cos \omega t \tag{5.69}$$

Step 4 : Radiation condition.

Now we examine if Eq. (5.69) satisfies the radiation condition in the far-field. Since the log terms vanish in the far-field, we need to determine the asymptotic expression of the Cauchy PV integral in Eq. (5.69), for $x \to \pm\infty$, the definition of which is given by:

$$J = \text{PV} \int_0^\infty \frac{G(k)}{k-v} dk = \lim_{\rho \to 0} \left\{ \int_0^{v-\rho} \frac{G(k)}{k-v} dk + \int_{v+\rho}^\infty \frac{G(k)}{k-v} dk \right\} \tag{5.70}$$

where $G(k)$ is a regular function of real variable k and ρ is a small real number. In our case,

$$G(k) = e^{k(y+y_0)}\cos k(x-x_0) \qquad (5.71)$$

Eq. (5.70) is evaluated by contour integral in the complex k-plane as shown in Fig. 5.7, where we consider a closed contour integral by indenting along c_2 (clockwise) over the singular point $k = v$ and according to Cauchy integral theorem that the closed contour integral vanishes:

$$\oint \frac{e^{k(y+y_0)}}{k-v}e^{ik(x-x_0)}dk = \int_{c_1} + \int_{c_2} + \int_{c_3} + \int_{c_4} = 0 \qquad (5.72)$$

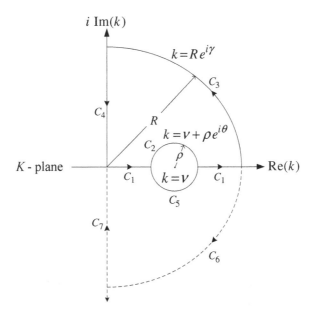

Fig. 5.7. Contour integral in k-complex domain.

The real k in Eqs. (5.70) and (5.71) now represents a complex variable in the complex k-plane. The contour integral of k along the circular arcs in the first quadrant indicates that it is real along the real axis, a pure imaginary along the imaginary axis, and along the semi-circular and quadrant arcs c_2 and c_3, $k = v + \rho\,e^{i\theta}$ and $k = Re^{i\gamma}$, respectively. The real part of the first integral along c_1–c_1 is identical to the Cauchy PV integral J in Eq. (5.70), which is obtained from Eq. (5.72):

$$\int_{c_1} = -\int_{c_2} - \int_{c_3} - \int_{c_4} \tag{5.73}$$

Since the integrals along c_3 and c_4 vanish for $R \to \infty$ and $x \to +\infty$, respectively, the integral along c_2 (clockwise) only remains. According to the residue theorem (positive for counter-clockwise integral),

$$-\int_{c_2} = i\pi e^{v(y+y_0)} e^{iv(x-x_0)}, \quad x \to +\infty \tag{5.74}$$

In the foregoing calculation, x is considered to be positive. For $x \to -\infty$, we take the path c_1–c_5–c_1–c_6–c_7 in Fig. 5.7, where indent is taken along c_5 (counter-clockwise) below the real k-axis. The integrals along the quadrant arc and imaginary axis vanish for $R \to \infty$ and $x \to -\infty$. Thus the above contour integral is reduced to:

$$-\int_{c_5} = -i\pi e^{v(y+y_0)} e^{iv(x-x_0)}, \quad x \to -\infty \tag{5.75}$$

Since the Cauchy PV integral J in Eq. (5.73) is real, only the real parts of Eqs. (5.74) and (5.75) are substituted in Eq. (5.73). Consequently:

$$\text{PV} \int_0^\infty \frac{e^{k(y+y_0)}}{k-v} \cos k(x-x_0)dk$$
$$= \mp\pi e^{v(y+y_0)} \sin v(x-x_0), \quad x \to \pm\infty \tag{5.76}$$

Equation (5.76) represents the asymptotic expression of the Cauchy PV integral for $x \to \pm\infty$. Use of Eq. (5.76) gives the far-field expression of $\Phi_1 + \Phi_2 + \Phi_3$ Eq. (5.69):

$$\Phi_1 + \Phi_2 + \Phi_3 |_{x \to \pm\infty} = \pm Qe^{v(y+y_0)} \sin v(x-x_0)\cos \omega t \tag{5.77}$$

Equation (5.77) indicates that the summed potential $\Phi_1 + \Phi_2 + \Phi_3$ behaves as a standing wave in the far field $x \to \pm\infty$.

Step 5: Add the fourth potential Φ_4.
As Eq. (5.77) shows, the summed potential $\Phi_1 + \Phi_2 + \Phi_3$ obtained in step 4 evidently does not satisfy the radiation condition (outgoing progressing wave in the far-field). Thus, we add the following fourth potential Φ_4, in order to make it outgoing progressive waves:

$$\Phi_4(x,y,t) = -Qe^{v(y+y_0)}\cos v(x-x_0)\sin \omega t \tag{5.78}$$

The addition of Eq. (5.78) to Eq. (5.77) yields outgoing progressing waves in the far-field:

$$\Phi_{\pm} = \pm Qe^{v(y+y_0)}\sin\{v(x-x_0) \mp \omega t\} \tag{5.79}$$

Equation (5.79) indicates that the resulting wave in the far-field satisfies the radiation condition.

Hence the resultant velocity potential satisfying the conditions of the flow due to 2-D pulsating source in the deep water is given in the form:

$$\Phi(x,y,t) = \left\{ \frac{Q}{2\pi}\log\frac{r}{r_1} - \frac{Q}{\pi}PV\int_0^\infty \frac{e^{k(y+y_0)}}{k-v}\cos k(x-x_0)dk \right\}\cos \omega t$$
$$- Qe^{v(y+y_0)}\cos v(x-x_0)\sin \omega t \tag{5.80}$$

5.2.1 Remarks on the radiation condition

The 2-D pulsating source potential in Eq. (5.80) may be written in the form:

$$\Phi(x,y,t) = \text{Re}[\phi(x,y)e^{-i\omega t}] = \text{Re}[(\phi_c + i\phi_s)e^{-i\omega t}] \tag{5.81}$$

with

$$\phi_c = \frac{Q}{2\pi}\log\frac{r}{r_1} - \frac{Q}{\pi}PV\int_0^\infty \frac{e^{k(y+y_0)}}{k-v}\cos k(x-x_0)dk$$

$$\phi_s = -Q\, e^{v(y+y_0)}\cos v(x-x_0) \tag{5.82}$$

Thus, referring to Eq. (5.76), we may write the asymptotic expression $\Phi_{\pm}$ in the form:

$$\Phi_{\pm}(x,y,t) = \text{Re}\left[\lim_{x\to\pm\infty}(\phi_c + i\phi_s)e^{-i\omega t} \right]$$
$$= \text{Re}\left[\left(\pm Qe^{v(y+y_0)}\sin v(x-x_0) - iQe^{v(y+y_0)}\cos v(x-x_0) \right)e^{-i\omega t} \right] \tag{5.83}$$

Equation (5.83) evidently satisfies the prescribed radiation condition in Eq. (5.59).

We may also write the radiation condition in complex form:

$$\lim_{x\to\pm\infty}\frac{\partial\phi}{\partial x} \mp iv\phi = 0 \quad \text{with } \phi = \phi_c + i\phi_s \tag{5.84}$$

In Sec. 5.2.1, the radiation condition was imposed in order to derive the velocity potential due to 2-D pulsating source. However, it is rather artificial (Stoker, 1957). If we solve the initial boundary value problem due to a pulsating source of a given frequency, we may obtain a steady outgoing progressing wave in the far-field as the time increases infinitely.

5.2.2 2-D Green's function

The 2-D pulsating source Eq. (5.80) is also called 2-D Green's function $G(P; P_0)$ which is written in the form:

$$\Phi = \text{Re}\left[\frac{Q}{2\pi}G(P;P_0)e^{-i\omega t}\right]$$

$$= \text{Re}\left[\frac{Q}{2\pi}\{\log r(P;P_0) + g(P;P_0)\}e^{-i\omega t}\right]$$

with   (5.85)

$$g(P;P_0) = \left\{\begin{bmatrix}-\log r_1 - 2\text{PV}\int_0^\infty \frac{e^{k(y+y_0)}}{k-\nu}\cos k(x-x_0)dk\end{bmatrix} + \\ i\begin{bmatrix}-2\pi e^{\nu(y+y_0)}\cos\nu(x-x_0)\end{bmatrix}\right\}$$

where $g(P;P_0)$ is a complex function in the lower x y-plane. The Green's function $G(P; P_0)$ in Eq. (5.85) is used, for instance, in building the Green's third identity as defined in Sec. 4.1.6 and is applied to the close-fit or source distribution method (Frank,1967) in Chap. 6.

5.2.3 Far-field waves due to 2-D pulsating source

Referring to Eq. (5.85) we find the wave potential due to 2-D pulsating source in the far-field $x \to \pm\infty$ in the form:

$$\Phi_\pm(x,y,t) = \text{Re}\left[\frac{Q}{2\pi}G_\pm e^{-i\omega t}\right] = \pm Q e^{\nu(y+y_0)}\sin\{\nu(x-x_0)\mp\omega t\}   (5.86)$$

Thus, the progressing wave η in the far-field is given by:

$$\eta_\pm(x,t) = \frac{\omega Q}{g}e^{\nu y_0}\cos\{\nu(x-x_0)\mp\omega t\}, \quad y=0 \tag{5.87}$$

It is to be noted that the two waves from the same source propagate in opposite directions.

In Eq. (5.87), νx_0 is the phase difference between the maximum radial velocity of the pulsating source at (x_0, y_0) and the maximum wave elevation at the origin of the coordinate system. Substituting the source strength $Q = 2\pi A_0\omega$ in Eq. (5.51) into Eq. (5.87), we consider the following amplitude ratio:

$$\bar{A} = \frac{A}{A_0} = 2\pi r \nu e^{\nu y_0} \tag{5.88}$$

where A is the pulsating source-induced wave amplitude and A_0 is the radial displacement amplitude of a circle of radius r, whose origin is at the source point (x_0, y_0). The foregoing amplitude ratio may be regarded as LTF as defined in Chap. 3. The input is the maximum radial displacement of the periphery of a circle of radius r, while the output is the propagated wave amplitude. The LTF is dependent on the non-dimensional wave frequency $\nu |y_0|$ associated with the position of the source. Equation (5.88) is also written in the following form;

$$\frac{\bar{A}|y_0|}{2\pi r} = \nu|y_0|e^{-\nu|y_0|} \tag{5.89}$$

For a simple illustration we assume unit radius $r = 1.0$ in Eq. (5.89) as shown in Fig. 5.8. If the non-dimensional frequency $\nu |y_0|$ equals 1, the radiated wave amplitude ratio takes its maximum value, as given in Eq. (5.90):

$$\left.\frac{\bar{A}|y_0|}{2\pi \times 1}\right|_{max} = \frac{1}{e} \tag{5.90}$$

Accordingly,

$$\omega = \sqrt{\frac{g}{|y_0|}}, \quad \lambda = 2\pi|y_0| \tag{5.91}$$

When the source is located deep below the free surface, the wave becomes long.

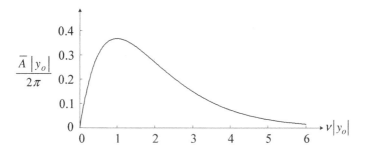

Fig. 5.8. Ratio of radial displacement amplitude of pulsating source to wave amplitude in the far-field.

5.2.4 Cauchy PV integral in the near-field

In Sec. 5.2.1 we have derived the asymptotic expression of the Cauchy PV integral Eq. (5.70) for $x \to \pm \infty$, which was necessary to derive the relation to satisfy the radiation condition. Evaluation of the same PV integral in the near-field is necessary to compute the velocity potential containing the PV integral near the source point employing the Green's function. The PV integral in Eq. (5.80) is written in complex form:

$$PV \int_0^\infty \frac{e^{-ik(z-\bar{z}_0)}}{k-v}dk$$

$$= I \pm i\pi e^{-iv(z-\bar{z}_0)}, x-x_0 > 0 \quad \text{and} \quad x-x_0 < 0; y+y_0 \le 0 \tag{5.92}$$

with

$$I = \int_0^\infty \frac{e^{-ik(z-\bar{z}_0)}}{k-v}dk \Big|_{\text{contours } c_1 \to c_2 \to c_1 \text{ and } c_1 \to c_5 \to c_1}$$

$$x-x_0 > 0 \text{ and } x-x_0 < 0; y+y_0 \le 0 \tag{5.93}$$

where $z = x + iy$ and $\bar{z}_0 = x_0 - iy_0$. The contour integral is shown in Fig. 5.7. The integral is taken along the real k-axis ($c_1 \to c_1$) and the semi-circular arcs c_2 indented upward (clockwise) and c_5 indented downward (counter-clockwise) respectively in the k-plane, where θ in $\rho e^{i\theta}$ represents positive for counter-clockwise.

We transform the contour integral Eq. (5.93); first translate the contour integral I in the k-plane Eq. (5.93) to w-plane by $w = k - v$:

$$I = e^{-iv(z-\overline{z}_0)} \int_{-v}^{\infty} \frac{e^{-iw(z-\overline{z}_0)}}{w} dw \tag{5.94}$$

Next we rotate the integral Eq. (5.94) by the transform:

$$u = iw(z - \overline{z}_o)$$

then, we obtain I in the form in the u-plane:

$$I = e^{-iv(z-\overline{z}_0)} \int_{-iv(z-\overline{z}_0)}^{\infty(z-\overline{z}_0)} \frac{e^{-u}}{u} du \tag{5.95}$$

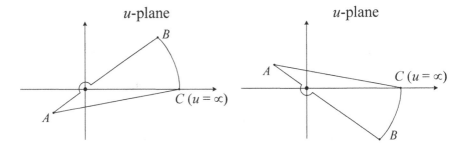

Fig. 5.9. Closed contours for integration of I in Eq. (5.95) in u-plane for $x-x_0>0$ and $x-x_0<0$.

The integral in Eq. (5.95) can be evaluated by employing the formula (Abramowitz and Stegun, 1972):

$$E_1[Z] = \int_z^{\infty} \frac{e^{-t}}{t} dt, \ \arg(Z) < \pi \tag{5.96}$$

where Z is complex.

Figure 5.9 represents the closed contours. The left figure is for $x - x_0 > 0$ while the right figure is for $x - x_0 < 0$, respectively. The integral I is now modified by taking closed contour integrals connecting $Z = -iv(z - \overline{z}_0)$ and $Z = \infty$ for both cases $x - x_0 > 0$ and $x - x_0 < 0$. Both closed contour integrals are formed along $A \rightarrow B \rightarrow C \rightarrow A$ and then reversed along $A \rightarrow C \rightarrow B$ to obtain the original contour $A \rightarrow B$. Then we have the residues $- i2\pi$ and $+ i2\pi$ due to the closed contour integrals in the clock-wise and anti-clock-wise mode, respectively. Hence, the integral I is determined in the form:

$$I = e^{-iv(z-\overline{z}_0)}\left\{E_1\left[-iv(z-\overline{z}_0)\right]\mp i2\pi\right\}$$
$$x - x_0 > 0 \quad \text{and} \quad x - x_0 < 0; y + y_0 \leq 0$$

(5.97)

Consequently the real part of the Cauchy PV integral Eq. (5.80) is given by:

$$\text{PV}\int_0^\infty \frac{e^{k(y+y_0)}}{k-v}\cos k(x-x_0)dk = \text{Re}\left[e^{-iv(z-\overline{z}_0)}\{E_1[-iv(z-\overline{z}_0)]\mp \pi i\}\right]$$
$$x - x_0 > 0 \text{ and } x - x_0 < 0; y + y_0 \leq 0$$

(5.98)

In the above, $E_1[-iv(z-\overline{z}_0)]$ is represented in terms of the exponential integral of complex argument:

$$E_1(Z) = -\gamma - \log Z - \sum_{n=1}^{\infty}\frac{(-1)^n Z^n}{n \cdot n!}$$

(5.99)

where $\gamma = 0.5773$ an Euler constant.

If Z is a pure imaginary and x is real positive:

$$E_1[ix] = -Ci(x) + i\ si(x)$$

(5.100)

$$si(x) = Si(x) - \frac{\pi}{2}$$

(5.101)

where Ci and Si are the cosine and sine integrals that are defined by:

$$Ci(x) = -\int_x^\infty \frac{\cos t}{t}dt$$

(5.102)

$$Si(x) = \int_0^x \frac{\sin t}{t}dt$$

(5.103)

or

$$Ci(x) = \gamma + \log x + \sum_{n=1}^{\infty}\frac{(-1)^n x^{2n}}{2n(2n)!}$$

(5.104)

$$Si(x) = \sum_{n=1}^{\infty}\frac{(-1)^n x^{2n+1}}{(2n+1)(2n+1)!}$$

(5.105)

The graphs of cosine and sine integrals are shown in Fig. 5.10(a). The behavior of the exponential integral $E_1[x]$ of a pure real argument is shown in Fig. 5.10(b). For the exponential integral of a "non-pure real argument", we have:

$$E_1[-x \pm i0] = \int_{-x \pm i0}^{\infty} \frac{e^{-t}}{t} dt = PV \int_{-x}^{\infty} \frac{e^{-t}}{t} dt \mp i\pi \qquad (5.106)$$

where $\mp$ indicates indentations are taken above and below the origin of the real t axis, respectively. We designate the above Cauchy type PV integral by:

$$-Ei[x] = PV \int_{-x}^{\infty} \frac{e^{-t}}{t} dt \qquad (5.107)$$

The $Ei(x)$ is the real part of the exponential integral of the "non-pure real argument". The behavior of the real component Ei is illustrated in Fig. 5.10(b).

The foregoing formulas were applied for computation of the wave due to oscillating and moving pressure patch on the free surface (Kim and Tsakonas, 1981). A similar analysis of Cauchy PV integral is given in Frank (1967).

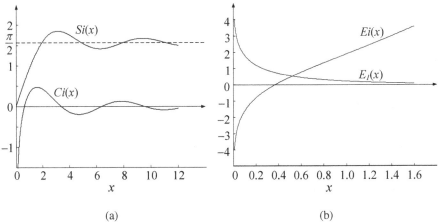

(a)   (b)

Fig. 5.10(a) Sine and cosine integrals of real argument. (b) Exponential integral of real argument and the real part of exponential integral.

5.3. Waves Due to 3-D Pulsating Source

The steadily pulsating 2-D source generates 2-D progressing wave as shown in Sec. 5.2.1. Similarly the potential flow due to 3-D pulsating source at a fixed source point (x_0, y_0, z_0) in the deep water bounded by free surface S_F, bottom surface S_B, and far-field cylindrical surface S_∞ will produce 3-D progressing wave.

Let the 3-D source potential be written in the form:

$$\Phi(x, y, z; x_0, y_0, z_0; t) = \mathrm{Re}[(\phi_1 + i\phi_2)e^{-i\omega t}] \tag{5.108}$$

that satisfies the continuity and boundary conditions as described below:

$$\nabla^2 \phi_i = 0, \ -\infty \le x \le \infty, \ y \le 0 \tag{5.109}$$

$$\phi_{iy} - v\phi_i = 0, \ \ y = 0 \ \text{ or } \text{ on } S_F, \ \ v = \frac{\omega^2}{g} \tag{5.110}$$

$$\phi_{iy} = 0, \ \ y = -\infty, \ \text{ or on } S_B \tag{5.111}$$

$$\lim_{R \to \infty} \sqrt{R}\left(\frac{\partial \phi_1}{\partial R} + v\phi_2\right) = 0, \ \text{ on } S_\infty \tag{5.112}$$

where

$$R = \sqrt{(x - x_0)^2 + (z - z_0)^2} \tag{5.113}$$

Similar expression to the 2-D radiation condition given in Eq. (5.59) is applied for 3-D radiation condition in the above considering:

$$\lim_{x \to \infty} \frac{\partial \phi_c}{\partial x} + v\phi_s = 0, \ x = R, \ \phi_c = \phi_1, \ \phi_s = \phi_2 \tag{5.114}$$

Since the sum of 3-D simple source $1/r$ and 3-D simple sink $-1/r_1$ does not satisfy the standing wave flow near the free surface, one adds the third potential g that is harmonic in the lower half domain as shown below:

$$\phi_1 = \frac{Q}{4\pi}\left(\frac{1}{r} - \frac{1}{r_1}\right) + g\left(x, y, z; x_0, y_0, z_0\right) \tag{5.115}$$

where

$$r = \sqrt{(x - x_0)^2 + (y - y_0)^2 + (z - z_0)^2}$$
$$r_1 = \sqrt{(x - x_0)^2 + (y + y_0)^2 + (z - z_0)^2} \tag{5.116}$$

We proceed in a similar way for the 2-D pulsating source potential to determine the function g. Substitution of ϕ_1 into the free surface condition Eq. (5.110) yields the following differential equation:

$$g_y - \nu g = \frac{Q}{2\pi} \frac{\partial}{\partial y_0} \frac{1}{\sqrt{(x-x_0)^2 + y_0^2 + (z-z_0)^2}}, \quad y = 0 \qquad (5.117)$$

Since

$$\frac{\partial}{\partial y}\left(\frac{1}{r} - \frac{1}{r_1}\right) = 2\frac{\partial}{\partial y_0} \frac{-1}{\sqrt{(x-x_0)^2 + y_0^2 + (z-z_0)^2}}, \quad y = 0 \qquad (5.118)$$

and the Hankel transform of exponential function (Abramowitz and Stegun, 1972) is:

$$\int_0^\infty e^{kb} J_0(ka)dk = \frac{1}{\sqrt{a^2 + b^2}}, \quad b < 0 \qquad (5.119)$$

where $a = R$ and $b = y_0$, we may write Eq. (5.117) in the form:

$$\begin{aligned}
g_y - \nu g &= \frac{Q}{2\pi} \frac{\partial}{\partial y_0} \int_0^\infty e^{ky_0} J_0(kR)dk \\
&= \frac{Q}{2\pi} \int_0^\infty k e^{ky_0} J_0(kR)dk, \quad y = 0
\end{aligned} \qquad (5.120)$$

The harmonic function g in Eq. (5.120) may be assumed in the form:

$$g = \frac{Q}{2\pi} \int_0^\infty F(k) e^{k(y+y_0)} J_0(kR)dk \qquad (5.121)$$

with unknown $F(k)$. Substitution of Eq. (5.121) into Eq. (5.120) yields:

$$F(k) = \frac{k}{k-\nu} \qquad (5.122)$$

Hence inserting $F(k)$ in Eq. (5.121), we determine the function g:

$$g = \frac{Q}{2\pi} \text{PV} \int_0^\infty \frac{k}{k-\nu} e^{k(y+y_0)} J_0(kR)dk \qquad (5.123)$$

since

$$\frac{k}{k-\nu} = 1 + \frac{\nu}{k-\nu} \qquad (5.124)$$

we have the following result for ϕ_1:

$$\phi_1 = \frac{Q}{4\pi}\left(\frac{1}{r} - \frac{1}{r_1}\right) + \frac{Q}{2\pi}\text{PV}\int_0^\infty \frac{k}{k-v}e^{k(y+y_0)}J_0(kR)dk \qquad (5.125)$$

or

$$\phi_1 = \frac{Q}{4\pi}\left(\frac{1}{r} + \frac{1}{r_1}\right) + \frac{Qv}{2\pi}\text{PV}\int_0^\infty \frac{e^{k(y+y_0)}}{k-v}J_0(kR)dk \qquad (5.126)$$

5.3.1 Radiation condition

The remaining condition to be satisfied by both ϕ_1 and ϕ_2 is the prescribed radiation condition Eq. (5.112). Since $(1/r + 1/r_1)$ in Eq. (5.126) vanish in the far-field, the remaining task is to investigate the asymptotic expression of Cauchy PV integral that is identical to the form of the 2-D expression in Eq. (5.70). The comparison of Eq. (5.69) with the integral Eq. (5.126) suggests that the factor $\cos[k(x-x_0)]$ for large kR is expressed in terms of the Bessel function $J_0(kR)$. Therefore we consider the following integral:

$$I = \text{Re}\left[\text{PV}\int_0^\infty \frac{e^{k(y+y_0)}}{k-v}H_0^{(1)}(kR)dk\right] \qquad (5.127)$$

where, the 1st kind of Hankel function is defined in Sec. 4.3.6.2.

First we use the condition that the closed contour integral along c_1, c_2, c_3 and c_4 in complex k-domain vanishes according to Cauchy theorem as shown in Fig. 5.7. Further the integrations along c_3 and c_4 vanish for the large radius $R \to \infty$, whereas the integral along c_2 ($k = \rho e^{i\theta}$) yields residue at $k = v$, i.e.,

$$\text{Res}(k=v) = -i\pi e^{v(y+y_0)}H_0^{(1)}(vR)|_{R\to\infty} \qquad (5.128)$$

Substituting Eq. (5.128) into Eq. (5.127) and using the 1st-kind of Hankel function, we determine the PV integral I:

$$\begin{aligned}
I &= \text{Re}\left[\text{PV}\int_0^\infty \frac{e^{k(y+y_0)}}{k-v}H_0^{(1)}(kR)dk\right] \\
&= \text{Re}\left[i\pi e^{v(y+y_0)}H_0^{(1)}(kR)|_{R\to\infty}\right] \\
&= -\pi e^{v(y+y_0)}Y_0(kR)|_{R\to\infty}
\end{aligned} \qquad (5.129)$$

Thus the asymptotic expression of ϕ_1 Eq. (5.126) becomes:

$$\phi_1\big|_{R\to\infty} = -\frac{Qv}{2}e^{v(y+y_0)}Y_0(vR)\big|_{R\to\infty}$$

$$= -Q\sqrt{\frac{v}{2\pi R}}\,e^{v(y+y_0)}\sin\left(vR-\frac{\pi}{4}\right)+O\left(\frac{1}{R}\right)$$

(5.130)

According to the radiation condition Eq. (5.112), we may write the potential ϕ_2:

$$\phi_2\big|_{R\to\infty} = -\frac{1}{v}\frac{\partial\phi_1}{\partial R}$$

(5.131)

The derivative of ϕ_1 Eq. (5.130) is expressed in the form:

$$\frac{\partial\phi_1}{\partial R}\bigg|_{R\to\infty} = -Qv\sqrt{\frac{v}{2\pi R}}\,e^{v(y+y_0)}\cos\left(vR-\frac{\pi}{4}\right)+O\left(\frac{1}{R}\right)$$

(5.132)

From Eqs. (5.131) and (5.132) we have ϕ_2 in the form:

$$\phi_2\big|_{R\to\infty} = Q\sqrt{\frac{v}{2\pi R}}e^{v(y+y_0)}\cos\left(vR-\frac{\pi}{4}\right)+O\left(\frac{1}{R}\right)$$

(5.133)

Substitution of both Eqs. (5.130) and (5.133) into Eq. (5.108) gives the resultant velocity potential in the far-field $R\to\infty$:

$$\Phi\big|_{R\to\infty} = \left(\phi_1\cos\omega t+\phi_2\sin\omega t\right)\big|_{R\to\infty}$$

$$= -Q\sqrt{\frac{v}{2\pi R}}\,e^{v(y+y_0)}\sin\left(vR-\omega t-\frac{\pi}{4}\right)+O\left(\frac{1}{R}\right)$$

(5.134)

which indicates progressing wave in the far-field. Thus the radiation condition has been satisfied.

5.3.2 Resultant 3-D source potentials

Since the asymptotic expression of ϕ_2 is similar to cosine function, the ϕ_2 is expressed in terms of Bessel function $J_0(vR)$ as given in Sec. 4.3.6.1:

$$\phi_2 = \frac{Qv}{2}e^{v(y+y_0)}J_0(vR)$$

(5.135)

Hence the resultant 3-D source potential Eq. (5.108) is given in the form:

$$\Phi = \left\{ \frac{Q}{4\pi}\left(\frac{1}{r}+\frac{1}{r_1}\right) + \frac{Qv}{2\pi}\,\mathrm{PV}\int_0^\infty \frac{e^{k(y+y_0)}}{k-v}J_0(kR)dk \right\}\cos\omega t$$

$$+ \frac{Qv}{2}e^{v(y+y_0)}J_0(vR)\sin\omega t \qquad (5.136)$$

If we replace $Q/4\pi$ in Eq. (5.136) by unity, the resultant potential becomes:

$$\Phi = \left\{ \left(\frac{1}{r}+\frac{1}{r_1}\right) + 2v\mathrm{PV}\int_0^\infty \frac{e^{k(y+y_0)}}{k-v}J_0(kR)dk \right\}\cos\omega t$$

$$+ 2\pi v e^{v(y+y_0)}J_0(vR)\sin\omega t \qquad (5.137)$$

Since

$$\frac{k+v}{k-v} = 1 + \frac{2v}{k-v} \quad \text{and Hankel transform Eq. (5.119):} \quad \int_0^\infty e^{k(y+y_0)}J_0(kR)dk = \frac{1}{r_1}$$

we have another 3-D pulsating source potential in the form:

$$\Phi = \left\{ \frac{1}{r} + \mathrm{PV}\int_0^\infty \frac{k+v}{k-v}e^{k(y+y_0)}J_0(kR)dk \right\}\cos\omega t$$

$$+ 2\pi v e^{v(y+y_0)}J_0(vR)\sin\omega t \qquad (5.138)$$

Thus, altogether we have derived three types of the velocity potentials for the 3-D pulsating source.

5.3.3 3-D Green's function

The three types of pulsating source potential Eqs. (5.136), (5.137), and (5.138) also represent Green's functions if we omit the time factor. Referring to Eq. (5.137), we write:

$$\Phi = \mathrm{Re}\left[G(P;P_0)e^{-i\omega t} \right]$$

with

$$G(P;P_0) = \frac{1}{r} + g(P;P_0) \tag{5.139}$$

$$g(P;P_0) = \frac{1}{r_1} + 2\nu \mathrm{PV} \int_0^\infty \frac{e^{k(y+y_0)}}{k-\nu} J_0(kR)dk$$

$$+ i 2\pi\nu e^{\nu(y+y_0)} J_0(\nu R)$$

G ($P;P_0$) represents the Green's function and g ($P;P_0$) is a complex harmonic function in the lower domain. Numerical computation of Green's function is much involved. Green's function is a sort of reciprocating engine for determining the velocity potentials at all the points on the body surface at all frequencies as desired as will be discussed in Chap. 6. The potential due to 3-D pulsating source in the finite water is given Wehausen and Laitone (1960). An efficient method of numerical computation of the finite water Green's function is given by Telste and Noblesse (1986).

References

Journal references:

Clauss, G.F. and Bergmann, J. (1986). Gaussian Wave Packets — A New approach to Seakeeping Tests of Ocean Structures, Appl. Ocean Res., Vol. 8, No. 4, pp.190–206.

Holstein, H. (1937). Die periodische Quelle im Wasser mit freier Oberflaeche, Zeitschrift fuer angewandte Mathematik und Mechanik, Band 17, Heft 1, pp.38–47.

Kim, C.H. and Tsakonas, S. (1981). An Analysis of Heave Added Mass and Damping of a Surface-Effect Ship, J. Ship Res., Vol. 25, No.1, pp. 44–61.

Pierson, W.J., Donelan, M.A. and Hui, W.H. (1992). Linear and Nonlinear Propagation of Water Wave Groups, J. Geophys. Res., Vol. 97, No. C4, pp. 5607–5621.

Telste, J.G. and Noblessse, F. (1986). Numerical Evaluation of the Green Function of Water Wave Radiation and Diffraction, J. Ship Res., Vol. 30, No. 2, pp.69–84.

Book references:

Abramowitz, M. and Stegun, A. (1972) Handbook of Matehmatical Functions, Dover Publications, Inc., New York.

Hilderbrand, Francis B. (1959) Advanced Calculus, The MIT Press.

Kinsman, B. (1965) Wind Waves, Dover Publications, Inc., New York.

Lamb, H. (1945) Hydrodynamics, Dover Publications.

Sommerfeld, A. (1964) Partial Differential Equations in Physics, Lectures on Theoretical Physics, Vol. VI, Academic Press.

Stoker, J.J. (1957) Water Waves, The Mathematical Theory with Applications, Interscience Publishers, Inc., New York.

Wehausen, J.V. and Laitone, E.V. (1960) Surface Waves, Encyclopedia of Physics Vol. IX, Springer-Verlag, Berlin, Goettingen, Heidelberg.

Proceedings references:

Frank, W. (1967). Oscillation of Cylinders in or Below the Free Surface of Deep Fluids, NSRDC Report 2375.

Chapter 6

Linear Wave-Body Interaction

6.1 Introduction

The motion of a floating vessel is affected by the waves, winds and currents, among which the waves play the most significant role to produce the significant responses in the context.

The design technology employs the linear Volterra input-output model in Chap. 3. It applies the linear spectral method given the wave energy spectrum and RAO square for determining the response energy spectrum.

The linear hydrodynamic theory for determination of RAO of the structure to sinusoidal wave excitation lays the foundation and will be reviewed first. This chapter will be followed by the discussion of nonlinear theories in Chaps. 7 through 12.

The fundamental assumption in the linear theory is that the floating structure harmonically oscillates approximately about the center of gravity of the body. Also, the waves and body motions are small and the motion amplitude is linearly proportional to the wave amplitude.

The wave exciting force exerted on the body causes oscillatory motion, which, induces inertial oscillatory force on the structure. The oscillatory forces (Table 6.1) consist of three groups; one is the structural inertia of vessel and the other two are hydromechanical (hydrodynamic and hydrostatic) and viscous damping forces.

The fluid hydrostatic pressure generates the hydrostatic restoring force and hydrodynamic pressure creates the hydrodynamic forces consisting of motion-induced hydrodynamic inertial and hydrodynamic damping force that are proportional to the acceleration and velocity of the body, respectively, and wave-exciting force on the structure assumed to be fixed in waves. The wave-exciting force consists of the Froude-Krylov and diffraction force.

The hydrodynamic forces are usually computed by potential theory while the viscous damping force is obtained through decay test of a scaled model in the wave tank.

Table 6.1. Oscillatory forces acting on the floating body in waves (Blagoveschensky, 1962)

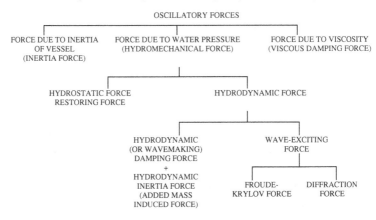

6.1.1 Linear radiation boundary condition

Let us consider a steadily oscillating body in calm water and the kinematical boundary condition to be satisfied on the body surface, i.e., the body surface moves with the same fluid particles in contact as discussed in kinematical boundary condition in Sec. 4.2.4.1:

$$\frac{DF}{Dt} = \Phi_x F_x + \Phi_y F_y + \Phi_z F_z + F_t = 0, \text{ on } F(x,y,z,t) = 0 \qquad (6.1)$$

which is nonlinear since it contains the products of the fluid velocity and body surface gradient. A perturbation procedure is applied to account for the small disturbed motion of the body and fluid from the equilibrium. Such disturbances are expressed in series with an expansion parameter ε. The first-order terms with ε are called linear. We consider a space-fixed coordinate system $oxyz$ on the calm water surface with positive z-axis upward and a body-fixed coordinate system $\overline{o}\,\overline{x}\,\overline{y}\,\overline{z}$ that is parallel to the space-fixed coordinate system with its origin $\overline{o}$ in coincidence with the center of gravity o_g of the body (Fig. 6.1). In reference to Fig. 6.1, one may write the position of a point P on the body surface using position vectors in the form when there is no oscillatory displacement:

$$\mathbf{r} = \mathbf{r}_g + \overline{\mathbf{r}} \tag{6.2}$$

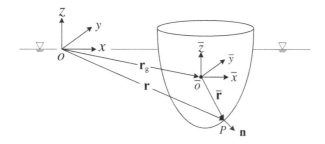

Fig. 6.1. Sketch of the space and body coordinate system.

Now assume the origins of the two coordinate systems are in coincidence and let the body undergo small translatory and rotational oscillatory motions. Then the position of a point P on the moving body is determined by the translatory displacements of the origin of the moving frame and the angular displacements of the moving frame relative to the space frame. The rigorous derivation of angular displacements using Euler angles and coordinate transforms are involved (e.g., Wehausen and Laitone, 1960). Here instead we restrict ourselves to the 1st-order. Denote the linear translation and angular displacement in vectors:

$$\varepsilon\boldsymbol{\xi}^{(1)} = \varepsilon\{\xi_1, \xi_2, \xi_3\}, \quad \varepsilon\mathbf{A}^{(1)} = \varepsilon\{\alpha_1, \alpha_2, \alpha_3\} \tag{6.3}$$

Then the displacement of a point P on the body in motion is the sum of $\varepsilon\boldsymbol{\xi}^{(1)}$ and $\varepsilon\mathbf{A}^{(1)}\times(\mathbf{r} - \mathbf{r}_g)$. Addition of these small displacements to the static position vector Eq. (6.2) gives the new position of P due to oscillatory motions:

$$\mathbf{r} = \mathbf{r_g} + \overline{\mathbf{r}} + \varepsilon\mathbf{y} \tag{6.4}$$

$$\mathbf{y} = \boldsymbol{\xi}^{(1)} + \mathbf{A}^{(1)} \times (\mathbf{r} - \mathbf{r_g}) \tag{6.5}$$

which are given in components in the form:

$$\overline{x} = x - x_g - \varepsilon\{\xi_1 + \alpha_2(z - z_g) - \alpha_3(y - y_g)\}$$
$$\overline{y} = y - y_g - \varepsilon\{\xi_2 + \alpha_3(x - x_g) - \alpha_1(z - z_g)\} \tag{6.6}$$
$$\overline{z} = z - z_g - \varepsilon\{\xi_3 + \alpha_1(y - y_g) - \alpha_2(x - x_g)\}$$

Thus, the wetted body surface changes its position during δt from the equilibrium at time t. The wetted body surface at $t + \delta t$ is expanded with respect to the equilibrium position at time t in a series:

$$F(\overline{x}, \overline{y}, \overline{z}, t + \delta t)$$
$$= F(x - x_g - \varepsilon A, y - y_g - \varepsilon B, z - z_g - \varepsilon C, t + \delta t)$$
$$= F^{(0)}(x - x_g, y - y_g, z - z_g, t) \tag{6.7}$$
$$-\varepsilon A F_x^{(0)} - \varepsilon B F_y^{(0)} - \varepsilon C F_z^{(0)} + \delta t \frac{\partial F^{(0)}}{\partial t} + O(\varepsilon^2) \equiv 0$$

where

$$A = \xi_1 + \alpha_2(z - z_g) - \alpha_3(y - y_g)$$
$$B = \xi_2 + \alpha_3(x - x_g) - \alpha_1(z - z_g) \tag{6.8}$$
$$C = \xi_3 + \alpha_1(y - y_g) - \alpha_2(x - x_g)$$

and the superfix (0) indicates 0th -order or the surface at the mean position of oscillation. Hence, the 0th-order terms are automatically zero. Thus the local derivative of the surface in Eq. (6.7) is given by:

$$\frac{\partial F(\overline{x}, \overline{y}, \overline{z}, t)}{\partial t} = -\varepsilon\dot{A}F_x^{(0)} - \varepsilon\dot{B}F_y^{(0)} - \varepsilon\dot{C}F_z^{(0)} + O(\varepsilon^2) \tag{6.9}$$

and the surface gradients are given in series referring to Eq. (6.7):

$$F_x(\overline{x},\overline{y},\overline{z},t) = F_x^{(0)}(x - x_g, y - y_g, z - z_g, t)$$
$$- \varepsilon[AF_{xx}^{(0)} + BF_{yx}^{(0)} + CF_{zx}^{(0)}] + O(\varepsilon^2)$$
$$F_y(\overline{x},\overline{y},\overline{z},t) = F_y^{(0)}(x - x_g, y - y_g, z - z_g, t)$$
$$- \varepsilon[AF_{xy}^{(0)} + BF_{yy}^{(0)} + CF_{zy}^{(0)}] + O(\varepsilon^2)$$
$$F_z(\overline{x},\overline{y},\overline{z},t) = F_z^{(0)}(x - x_g, y - y_g, z - z_g, t)$$
$$- \varepsilon[AF_{xz}^{(0)} + BF_{yz}^{(0)} + CF_{zz}^{(0)}] + O(\varepsilon^2)$$

(6.10)

The velocity gradients are similarly expanded in series, assuming the exact solutions of 0th- order are zero:

$$\Phi_x = \varepsilon\,\Phi_x^{(1)} + O(\varepsilon^2), \Phi_y = \varepsilon\,\Phi_y^{(1)} + O(\varepsilon^2), \Phi_z = \varepsilon\,\Phi_z^{(1)} + O(\varepsilon^2) \quad (6.11)$$

where the superfix (1) indicates the 1st-order (linear). Substituting surface and velocity gradients Eqs. (6.9), (6.10), and (6.11) in the kinematical boundary condition Eq. (6.1) and collecting the terms of ε, we have:

$$\nabla F^{(0)}(x - x_g, y - y_g, z - z_g, t) \cdot \nabla \Phi^{(1)}(x, y, z, t)$$
$$- \dot{A}F_x^{(0)} - \dot{B}F_y^{(0)} - \dot{C}F_z^{(0)} = 0$$

(6.12)

from which

$$\frac{\partial \Phi^{(1)}}{\partial n} = \frac{\dot{A}F_x^{(0)} + \dot{B}F_y^{(0)} + \dot{C}F_z^{(0)}}{|\nabla F^{(0)}|} = V_n, \quad \text{on } \sigma \quad (6.13)$$

where σ denotes the mean wetted surface in oscillation. $\partial \Phi^{(1)} / \partial n$ is the normal velocity of fluid on the body surface and V_n denotes the common velocity of fluid and boundary surface. From Eq. (6.13) we have:

$$\frac{\partial \Phi^{(1)}}{\partial n} = \dot{A}\,n_1 + \dot{B}\,n_2 + \dot{C}n_3, \quad \text{on } \sigma \quad (6.14)$$

$$n_1 = \frac{F_x^{(0)}}{\left|\nabla F^{(0)}\right|}, \quad n_2 = \frac{F_y^{(0)}}{\left|\nabla F^{(0)}\right|}, \quad n_3 = \frac{F_z^{(0)}}{\left|\nabla F^{(0)}\right|} \quad (6.15)$$

Thus the unit normal on σ above is into the fluid region.

Substituting Eq. (6.8) into Eq. (6.14) and dropping the first-order superfix (1), we may write the normal derivative of the linear velocity potential in the form:

$$\frac{\partial \Phi}{\partial n} = \frac{\partial \phi_1}{\partial n}\dot{\xi}_1 + \frac{\partial \phi_2}{\partial n}\dot{\xi}_2 + \frac{\partial \phi_3}{\partial n}\dot{\xi}_3$$

$$+ \frac{\partial \phi_4}{\partial n}\dot{\alpha}_1 + \frac{\partial \phi_5}{\partial n}\dot{\alpha}_2 + \frac{\partial \phi_6}{\partial n}\dot{\alpha}_3, \quad \text{on} \quad \sigma \tag{6.16}$$

with

$$\frac{\partial \phi_1}{\partial n} = n_1, \quad \frac{\partial \phi_2}{\partial n} = n_2, \quad \frac{\partial \phi_3}{\partial n} = n_3$$

$$\frac{\partial \phi_4}{\partial n} = (y - y_g)n_3 - (z - z_g)n_2$$

$$\frac{\partial \phi_5}{\partial n} = (z - z_g)n_1 - (x - x_g)n_3 \tag{6.17}$$

$$\frac{\partial \phi_6}{\partial n} = (x - x_g)n_2 - (y - y_g)n_1 \quad \text{on} \quad \sigma$$

From Eq. (6.16) one may write:

$$\Phi = \dot{\xi}_1\phi_1 + \dot{\xi}_2\phi_2 + \dot{\xi}_3\phi_3 + \dot{\alpha}_1\phi_4 + \dot{\alpha}_2\phi_5 + \dot{\alpha}_3\phi_6 \tag{6.18}$$

Equation (6.18) states that the radiation potential consists of six components, each of which is defined as the radiation potential per unit velocity amplitude. Equation (6.17) may be expressed in a compact form:

$$\frac{\partial \phi_j}{\partial n} = n_j, \quad \text{on } S, \quad j = 1, \cdots, 6 \tag{6.19}$$

with

$$n_4 = (y - y_g)n_3 - (z - z_g)n_2$$

$$n_5 = (z - z_g)n_1 - (x - x_g)n_3 \tag{6.20}$$

$$n_6 = (x - x_g)n_2 - (y - y_g)n_1$$

Equations (6.19) and (6.20) represent the linear radiation boundary condition for the radiation potential per unit velocity. It represents also the unit normal vector on the boundary surface directed into the fluid.

6.1.2 Linear diffraction boundary condition

The linear velocity potential of the fluid motion due to the motion of a freely floating body in an incident wave is represented by the sum of the incident wave and diffraction and radiation wave potentials:

$$\Phi = \Phi_I + \Phi_D + \Phi_R \qquad (6.21)$$

The linear superposition law let the potential divide into two parts, i.e., the radiation and the sum of the incident and diffraction potentials. Since the radiation potential satisfies the radiation boundary condition Eq. (6.17) at the mean position of oscillation, the sum of the incident and diffraction potential has to satisfy the kinematical boundary condition at the same mean position:

$$\frac{\partial \Phi_D}{\partial n} = -\frac{\partial \Phi_I}{\partial n}, \text{ on } \sigma \qquad (6.22)$$

Equation (6.22) is called linear diffraction boundary condition on the body surface. In other words, the diffraction normal velocity is opposite to the incident wave normal velocity. The solution to the diffraction boundary value problem will determine the wave-exciting force on the fixed body in the incident wave. We will review the formulations of the radiation and diffraction problem below.

6.1.3 Formulation of radiation problem

Assume the body oscillates in the calm water in six modes harmonically at frequency ω and the velocity potential per unit velocity amplitude Eq. (6.18) in the finite water may be given in the form:

$$\Phi_R(x, y, z, t) = \sum_{j=1}^{6} \text{Re}\left[\phi_j(x, y, z)e^{-i\omega t} \right] \qquad (6.23)$$

The radiation potential ϕ_j is complex and fulfills the continuity condition in the fluid and the boundary conditions over the surface enclosing the fluid domain as shown in Fig. 6.2. The unit normal $\mathbf{n}$ is defined outward from the fluid domain (τ), to make it identical to the definitions in Green's theorems in Sec. 4.1.5 and Fig. 4.5.

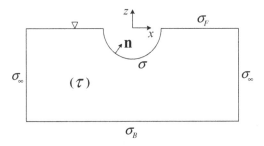

Fig. 6.2. Sketch of the boundary surfaces enclosing τ domain.

The radiation potential satisfies:

$$\nabla^2 \phi_j = 0, \text{ in } (\tau) \tag{6.24}$$

$$\frac{\partial^2 \phi_j}{\partial t^2} + g \frac{\partial \phi_j}{\partial z} = 0, \quad j = 1, \cdots, 6, \ z = 0, \text{ on } \sigma_F \tag{6.25}$$

$$\frac{\partial \phi_j}{\partial z} = 0, \quad j = 1, \cdots, 6, \text{ on } \sigma_B \tag{6.26}$$

$$\frac{\partial \phi_j}{\partial n} = n_j, \quad j = 1, \cdots, 6 \text{ on } \sigma \tag{6.27}$$

$$\lim_{R \to \infty} \sqrt{R} \left(\frac{\partial \operatorname{Re}\left[\phi_j\right]}{\partial R} + v \operatorname{Im}\left[\phi_j\right] \right) = 0,$$

$$\text{on } \sigma_\infty, R = \sqrt{x^2 + y^2}, j = 1, \cdots, 6 \tag{6.28}$$

Suppose we have determined the foregoing radiation potential per unit velocity amplitude satisfying the above conditions. Then we can determine the linear hydrodynamic pressure in the Bernoulli's equation with omit of the time factor $\exp(-i\omega t)$, in the form:

$$p_j = i\rho\omega\phi_j, \quad j = 1, \cdots, 6 \tag{6.29}$$

Since the unit normal on the body surface in Fig. 6.2 is directed outward from the fluid domain, the force and moment amplitudes due to the above hydrodynamic pressure are:

$$\mathbf{F} = \int_\sigma p\mathbf{n}d\sigma, \quad \mathbf{M} = \int_\sigma p(\mathbf{r} - \mathbf{r}_g) \times \mathbf{n}d\sigma \tag{6.30}$$

Substituting Eq. (6.29) into Eq. (6.30), and applying the definition that the velocity $n_k = \dfrac{\partial \phi_k}{\partial n}$ in Eqs. (6.19) and (6.20) is directed into the fluid region, we have the complex hydrodynamic force and moment due to unit velocity amplitude , with omission of the time factor, in the form:

$$F_{jk} = -i\rho\omega \int_\sigma \phi_j \frac{\partial \phi_k}{\partial n} d\sigma, \quad j \text{ and } k = 1, \cdots, 6 \qquad (6.31)$$

where F_{jk} represents the *jth* mode force (j = 1, 2, 3 for force; j = 4, 5, 6 for moment) due to *kth* mode motion. Let u_k be the generalized displacement amplitude of mode k that are real and positive. Then the *kth* mode velocity and acceleration amplitude are given by:

$$V_k = -i\omega u_k, \quad A_k = -\omega^2 u_k \qquad (6.32)$$

Multiplying the velocity amplitude $-i\omega u_k$ to the radiation potential per unit velocity amplitude ϕ_j in Eq. (6.31), we obtain the hydrodynamic force and moment due to the given motion u_k:

$$F_{jk} = -\rho\omega^2 u_k \int_\sigma \phi_j \frac{\partial \phi_k}{\partial n} d\sigma \qquad (6.33)$$

The hydrodynamic force F_{jk} consists of the forces proportional to the acceleration and velocity. Thus:

$$F_{jk} = -\omega^2 u_k M_{jk}'' - i\omega u_k N_{jk} \qquad (6.34)$$

Equating the foregoing two relations we have:

$$M_{jk}'' + \frac{i}{\omega} N_{jk} = \rho \int_\sigma \phi_j \frac{\partial \phi_k}{\partial n} d\sigma \qquad (6.35)$$

where the proportionality constants M_{jk}'' and N_{jk} represent added mass and wavemaking- or radiation-damping, respectively; and the suffix jk indicates the jth-mode added mass and damping due to the kth-mode motion. Since Green's theorem holds for the two harmonic functions in the domain (τ):

$$\int_\sigma \phi_j \frac{\partial \phi_k}{\partial n} d\sigma = \int_\sigma \phi_k \frac{\partial \phi_j}{\partial n} d\sigma \qquad (6.36)$$

we have the symmetry relations:

$$M_{jk}'' = M_{kj}'', \quad N_{jk} = N_{kj} \qquad (6.37)$$

Added mass and damping are called hydrodynamic inertia and damping. The values of added mass and damping of various shape body as function of frequency are found in Kochin et al. (1937); Korvin-Kroukovski (1963); Newman (1977); Sarpkaya and Issacson (1981); Lewis (1989).

6.1.4 Formulation of diffraction problem

Let the incident wave potential $\phi_0\, e^{-i\omega t}$ for deep water in oblique sea be given in the form:

$$\phi_0(x, y, z) = -i\frac{gA}{\omega}e^{vz}e^{i(vx\cos\mu + vy\sin\mu)} \tag{6.38}$$

where A and v denote the incident wave amplitude and deep water wavenumber and μ is the wave heading angle relative to longitudinal x-axis given in Sec. 3.2.4. The incident wave will be diffracted due to the body, and the wave-exciting force will be produced by the diffraction which in turn will create body motion. Thus there are both diffraction and radiation whose potentials are designated by ϕ_7, and ϕ_j ($j = 1,\cdots, 6$), respectively. All the above potentials satisfy the continuity condition, linearized free surface boundary condition and radiation condition, where no radiation condition is required for the incident wave potential ϕ_0:

$$\nabla^2\phi_j = 0, \quad j = 0,1,2,\cdots,7, \text{ in } (\tau) \tag{6.39}$$

$$\frac{\partial\phi_j}{\partial z} - v\phi_j = 0, \quad j = 0,1,\cdots 7, \ z = 0 \tag{6.40}$$

$$\lim_{R\to\infty}\sqrt{R}\left(\frac{\partial\,\mathrm{Re}\left[\phi_j\right]}{\partial R} + v\,\mathrm{Im}\left[\phi_j\right]\right) = 0 \tag{6.41}$$

$$j = 1,2,\cdots,7, \ R = \sqrt{x^2 + y^2}$$

The diffraction boundary condition on the body surface is written according to Eq. (6.22):

$$\frac{\partial\phi_7}{\partial n} = -\frac{\partial\phi_0}{\partial n}, \quad \text{on} \quad \sigma \tag{6.42}$$

Satisfying the above boundary conditions Eqs. (6.40), (6.41), and (6.42) one determines the diffraction potential and the linear hydrodynamic pressure amplitude due to the incident and diffracted wave omitting the time factor:

$$p = i\rho\,\omega(\phi_0 + \phi_7) \tag{6.43}$$

Since the unit normal on the boundary surface is directed outward from the fluid domain Fig. 6.2, and the unit normal on the body surface is directed into the fluid region as shown in Sec. 6.1.1, the force and moment amplitude are same as the radiation force given in Eq. (6.31):

$$F_j = -i\rho\omega \int_\sigma (\phi_0 + \phi_7)\frac{\partial \phi_j}{\partial n}\,d\sigma,\quad j = 1, \cdots, 6 \tag{6.44}$$

where the unit normal $\dfrac{\partial \phi_j}{\partial n} = n_j$ is defined in Eq. (6.17) and suffix j is the mode of force and moment. The wave-exciting force consists of the Froude-Krylov force

$$F_{FK_j} = -i\rho\omega\sigma \int_\sigma \phi_0 \frac{\partial \phi_j}{\partial n}\,d\sigma,\quad j = 1, 2, \cdots, 6 \tag{6.45}$$

and diffraction force

$$F_{D_j} = -i\rho\omega \int_\sigma \phi_7 \frac{\partial \phi_j}{\partial n}\,d\sigma,\quad j = 1, 2, \cdots, 6 \tag{6.46}$$

If the wave is very long compared to the characteristic length of the body, the diffraction becomes negligibly small, which is called transparent. Under the circumstances, the diffraction force becomes negligibly small.

6.1.5 Haskind-Newman relation

The Green's 2nd-identity holds for the two harmonic functions of diffraction and radiation ϕ_7 and ϕ_j in the fluid domain bounded by the entire boundary surface in Fig. 6.2.

$$\int_{\sigma_1} \phi_j \frac{\partial \phi_7}{\partial n}\,d\sigma_1 = \int_{\sigma_1} \phi_7 \frac{\partial \phi_j}{\partial n}\,d\sigma_1,\, j = 1, 2, \cdots, 6 \tag{6.47}$$

where, σ_1 indicates the entire boundary surface. However, since the integrals vanish, except on the body boundary surface, we consider the integral on the body boundary surface σ only. Substituting diffraction boundary condition Eq. (6.42) into Eq. (6.47), one obtains:

$$-\int_\sigma \phi_j \frac{\partial \phi_0}{\partial n} d\sigma = \int_\sigma \phi_7 \frac{\partial \phi_j}{\partial n} d\sigma \qquad (6.48)$$

Use of the above relation in Eq. (6.44) leads to:

$$F_j = -i\rho\omega \int_\sigma \left(\phi_0 \frac{\partial \phi_j}{\partial n} - \phi_j \frac{\partial \phi_0}{\partial n} \right) d\sigma, \; j = 1, 2, \cdots, 6 \qquad (6.49)$$

Equation (6.49) states that given the incident wave and radiation potential, one can determine the wave-exciting force called Haskind relation (1957) translated by Newman (1962). Haskind relation is also frequently called Haskind-Newman relation.

6.2 Computation of Radiation and Diffraction Potentials

The foregoing sections treated the formulations to determine the radiation as well as diffraction potential. We need to numerically compute the formulas given the mode of motion and geometry of the body in a given oblique wave. One may determine the potential employing the Green's 3rd identity or source distribution method. These methods will employ the pulsating source potential or Green's function derived in Chap. 5.

6.2.1 Green's 3rd identity method

The Green's function or the pulsating source potential is defined in the same domain as shown in Fig. 6.2. The unit normal on the boundary in the Green's 3rd identity is outward from the fluid domain as shown both in Fig. 4.5 and in Fig. 6.2. Since the surface integrals in the 3rd identity vanish on the entire boundaries except on the body surface we consider the integral on the body surface σ only.

Hence we write Green's 3rd identity for the velocity potential:

$$\phi(P) = \frac{1}{2\pi} \int_\sigma \left(\frac{\partial \phi(P_0)}{\partial n(P_0)} G(P; P_0) - \phi(P_0) \frac{\partial G(P; P_0)}{\partial n(P_0)} \right) d\sigma, P \in \sigma \qquad (6.50)$$

where

$$G(P;P_0) = \frac{1}{r(P;P_0)} + g(P;P_0) \tag{6.51}$$

Equating the normal derivative $\partial\phi(P_0)/\partial n(P_0)$ in Eq. (6.50) to the radiation condition Eq. (6.19) or diffraction condition Eq. (6.22), one can solve the boundary integral equation for the radiation or diffraction potential ϕ. The above approach for determining the velocity potential is called direct boundary integral equation method (DBIEM).

6.2.2 Source distribution method

Assume the velocity field is due to the distribution of source Q on the body surface σ in the form:

$$\phi(P) = \frac{1}{4\pi} \int_\sigma Q(P_0)G(P;P_0)d\sigma(P_0) \tag{6.52}$$

and the normal velocity of the fluid on the body surface in Eq. (6.52):

$$\frac{\partial\phi(P)}{\partial n(P)} = \frac{1}{4\pi}\int_\sigma Q(P_0)\frac{\partial}{\partial n(P)}\left(\frac{1}{r(P;P_0)} + g(P;P_0)\right)d\sigma(P_0) \tag{6.53}$$

$$P \in \sigma$$

The foregoing normal derivative of the simple source in Eq. (6.51) is derived referring to Fig. 4.5 as shown below. Drawing a small hemisphere bounded by σ_ε, on the body boundary σ in the domain (τ), we look for the normal derivative of the simple source $1/r$ in the direction opposite to r. If we take r as small as possible so that $P_0 \to P$, the integral converges to a constant $2\pi Q(P)$ at the center of the shrunken hemisphere:

$$\lim_{r\to\varepsilon}\int_{\sigma_\varepsilon} Q(P_0)\frac{\partial\left(\frac{1}{r(P;P_0)}\right)}{\partial n(P)}d\sigma_\varepsilon(P_0) \tag{6.54}$$

$$= \lim_{r\to\varepsilon, P_0\to P}\int_{\sigma_\varepsilon}\frac{1}{r^2}Q(P_0)d\sigma_\varepsilon(P_0) = 2\pi Q(P)$$

Thus, the normal velocity of fluid on the body surface Eq. (6.53) becomes:

$$\frac{\partial \phi(P)}{\partial n(P)} = \frac{1}{2}Q(P) + \frac{1}{4\pi}\int_{\sigma}Q(P_0)\frac{\partial g(P;P_0)}{\partial n(P)}\,d\sigma(P_0), \ P \in \sigma \qquad (6.55)$$

Substituting the radiation boundary condition into the LHS of Eq. (6.55), one can solve for the unknown source intensities and consequently the radiation or diffraction potential ϕ in Eq. (6.52). The above approach is called indirect boundary integral equation method (IBIEM) or more commonly called source distribution method according to Hess and Smith (1966). An example of computation study is found in Garrison (1979).

6.2.3 Discretization of IBIE by CPM

To determine the potentials, one needs to discretize Eqs. (6.52) and (6.55) by many of flat panels or constant panels. The method to use constant panels is called constant panel method (CPM). Using indices i and j for indicating points P and P_0 respectively and $\Delta \sigma_j$ for jth panel, we discretize Eqs. (6.52) and (6.55) in the forms:

$$A_{ij} = \frac{1}{4\pi}\int_{\Delta\sigma_j}G(P;P_0)d\sigma(P_0) \qquad (6.56)$$

and

$$H_{ij} = \frac{1}{2}\delta_{ij} + \frac{1}{4\pi}\int_{\substack{\Delta\sigma_j \\ i\neq j}}\left(\frac{\partial h(P;P_0)}{\partial n(P)}\right)d\sigma(P_0) \qquad (6.57)$$

where Kronecker $\delta_{ij} = 1$ for $i = j$, and 0 for $i \neq j$. Since the source intensity $Q(P)$ is assumed to be uniformly distributed on each panel $\Delta\sigma_j$, we denote it:

$$Q(P) = Q_j, \ j = 1, 2, \cdots, M \qquad (6.58)$$

Then Eqs. (6.52) and (6.55) are represented respectively by:

$$\phi_i = \sum_{j=1}^{M}A_{ij}Q_j, \ i = 1, 2, \cdots, M \qquad (6.59)$$

$$\left(\frac{\partial\phi}{\partial n}\right)_i = \sum_{j=1}^{M}H_{ij}Q_j, \ i = 1, 2, \cdots, M \qquad (6.60)$$

Substituting radiation or diffraction condition into the LHS of Eq. (6.60), one can determine the source intensities and the corresponding velocity potential on the body surface given in Eq. (6.59).

It is noted that CPM can also be used for discretization of DBIE.

6.2.4 Modified Green's 3rd identity-DBIE

The source distribution and Green's 3rd identity method are for the smooth boundary surface. If the structure has corners and edges such as those of a box shaped body or vertical truncated cylinder, we can avoid the problem by modifying the Green's 3rd identity Eq. (6.50) in the form:

$$\frac{1}{4\pi} \int_{\sigma} \left[G(P;P_0) \frac{\partial \phi(P_0)}{\partial n(P_0)} - \phi(P_0) \frac{\partial G(P,P_0)}{\partial n(P_0)} \right] d\sigma(P_0)$$

$$= C(P)\phi(P), P \in \sigma \tag{6.61}$$

where $C(P)$ is the normalized solid angle α at a point on the boundary surface being defined as $C(P) = \alpha / 4\pi$. Thus, the solid angle and coefficient on the smooth boundary surface are $\alpha = 2\pi$ and $C(P) = 1 / 2$, respectively for which the Green's 3rd identity holds. For the body having edges and corners the solid angles and coefficients are $\alpha = 3\pi$ and 3.5π and $C(P) = 3 / 4$ and $3.5 / 4$, respectively. In these cases Green's 3rd identity does not hold.

6.2.5 Quadratic higher order boundary element

Modified Green's 3rd identity will be solved by applying boundary element method (BEM). Liu et al. (1991) employs higher order boundary element method (HOBEM). In such case one may employ the quadratic isoparametric quadrilateral element as shown in Fig. 6.3, where the body surface, velocity potential and its normal derivative at eight nodal points are expressed by the same quadratic shape function N_j on each element:

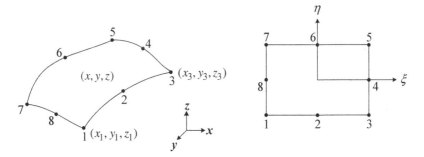

Fig. 6.3. Sketch of a quadrilateral element of a boundary surface and local coordinate system.

$$[x, y, z] = \sum_{j=1}^{NOD} N_j(\xi, \eta)\left[x_j, y_j, z_j\right] \tag{6.62}$$

$$\phi(P_0) = \sum_{j=1}^{NOD} N_j(\xi, \eta)\ \phi_j \tag{6.63}$$

$$\frac{\partial \phi(P_0)}{\partial n(P_0)} = \sum_{j=1}^{NOD} N_j(\xi, \eta)\left(\frac{\partial \phi}{\partial n}\right)_j \tag{6.64}$$

where NOD is the number of nodes and

$$\phi_j = \phi(x_j, y_j, z_j), \quad \left(\frac{\partial \phi}{\partial n}\right)_j = \frac{\partial \phi(x_j, y_j, z_j)}{\partial n(x_j, y_j, z_j)} \tag{6.65}$$

The shape function for a quadratic quadrilateral element, for instance, of eight nodes, is given for instance by (Brebbia and Walker, 1980):

$$N_j(\xi, \eta) = \begin{cases} 0.25(1 + \xi_j \xi)(1 + \eta_j \eta)(-1 + \xi_j \xi + \eta_j \eta), & j = 1, 3, 5, 7 \\ 0.5(1 + \xi_j \xi + \eta_j \eta)[1 - (\eta_j \xi)^2 - (\xi_j \eta)^2], & j = 2, 4, 6, 8 \end{cases} \tag{6.66}$$

Figure 6.3 shows a quadrilateral surface identified by eight nodes in global coordinates and the corresponding mapped points in the local ξ, η plane. For each j there is a set of ξ, η values, using which one may determine the shape functions in Eq. (6.66). This allows us to determine values of (x, y, z) on the quadrilateral surface, which is the base data for integration of the other properties including kernels G and $\partial G / \partial n$ on the body surface. The integration is indirectly carried out in the local ξ, η plane.

Using the above quadratic order boundary element one may discretize the modified Green's 3rd identity Eq. (6.61) by integrating in the local ξ, η plane in the form:

$$\sum_{k=1}^{NOD} H_{ik} \phi_k = \sum_{k=1}^{NOD} A_{ik} \left(\frac{\partial \phi}{\partial n} \right)_k, \quad i = 1, 2, \cdots, NOD \qquad (6.67)$$

where

$$H_{ik} = \sum_{e=1}^{NOE} \sum_{j=1}^{NOD} \delta_{kr} \hat{H}_{ij}^{(e)} + C_i \delta_{ik}$$

$$(6.68)$$

$$A_{ik} = \sum_{e=1}^{NOE} \sum_{j=1}^{NOD} \delta_{kr} \hat{A}_{ij}^{(e)}, \quad r = NENN(j, e)$$

$r = NENN(j, e)$, connectivity matrix, NOE number of element, C_i coefficient of solid angle, δ_{kr} Kronecker delta; which is 1 for $k = r$ and 0 for $k \neq r$

$$\hat{H}_{ij}^{(e)} = \frac{1}{4\pi} \int_{\sigma_e} \frac{\partial G(P; P_0)}{\partial n(P_0)} N_j(\xi, \eta) J(\xi, \eta) d\xi d\eta$$

$$\hat{A}_{ij}^{(e)} = \frac{1}{4\pi} \int_{\sigma_e} G(P; P_0) N_j(\xi, \eta) J(\xi, \eta) d\xi d\eta \qquad (6.69)$$

$$j = 1, 2, \cdots, NOD, \quad i = 1, 2, \cdots, NOD,$$

$$e = 1, 2, \cdots, NOE, \quad J = \text{Jacobian}$$

The connectivity matrix NENN is illustrated by taking two quadrilaterals connected side by side horizontally and denoting the nodes in the horizontal directions from left to right; 1, 2, 3, 4, 5: 6, 7, 8, 9, 10: 11, 12, 13, 14, 15, as shown in Fig. 6.4.

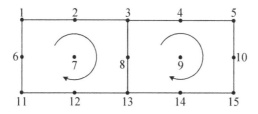

Fig. 6.4. Sketch of two adjacent quadrilateral elements and the rule to determine 9 nodes for defining each element.

Then the nodes on each quadrilateral element $e = 1$ and $e = 2$ and the connectivity matrix *NENN* are constructed by the node numbers in clockwise order for each quadrilateral element. Here, we consider 9 nodes adding one at the center of the quadrilateral element:

j	1	2	3	4	5	6	7	8	9
$e = 1$	1	2	3	8	13	12	11	6	7
$e = 2$	3	4	5	10	15	14	13	8	9

The foregoing rule will be applied for all the quadrilateral elements covering the entire mean wetted body surface.

6.2.5.1 Removal of singularity

The Green's function Eq. (6.51) consists of the basic simple source $1/r$ and harmonic wave function h.

If the field point P is on the surface of the element, P and P_0 may coincide to generate singularities $1 / r(P;P_0) \to \infty$ and $1 / r^2(P;P_0) \to \infty$. These singularities can be removed using polar coordinate transformation on the local element and locating the coordinate origin at the singular point. However, this transformation yields still another singularity $1 / \rho$ in the diagonal term H_{ii}. Here ρ is the radius of the polar coordinate (Liu et al., 1991). In order to solve $1 / \rho$ singularity, an indirect procedure is used. The unknown velocity potential ϕ in the modified Green's third identity Eq. (6.61) is replaced by a known potential ϕ_1 for the free, bottom and far-field surfaces; $\sigma_F + \sigma_B + \sigma_\infty$, and thus it is analytically integrable resulting in I:

$$I = \frac{1}{4\pi} \int_{\sigma_F + \sigma_B + \sigma_\infty} \left(G(P,P_0) \frac{\partial \phi_1(P_0)}{\partial n(P_0)} - \phi_1 \frac{\partial G(P,P_0)}{\partial n(P_0)} \right) d\sigma(P_0) \qquad (6.70)$$

The integral of ϕ_1 over the body surface σ is given by as in Eq. (6.67):

$$\sum_{k=1}^{NOD} H_{ik} \phi_{1k} = \sum_{k=1}^{NOD} A_{ik} \left(\frac{\partial \phi_1}{\partial n} \right)_k \qquad (6.71)$$

Hence the modified Green's third identity or DBIE Eq. (6.61) for the ϕ_1 over the entire boundary:

$$\sum_{k=1}^{NOD} H_{ik}\phi_{1k} = \sum_{k=1}^{NOD} A_{ik}\left(\frac{\partial\phi_1}{\partial n}\right)_k + I \tag{6.72}$$

The diagonal term H_{ii} is thus obtained from Eq. (6.72):

$$H_{ii} = \frac{1}{\phi_{1i}}\left[\sum_{k=1}^{NOD} A_{ik}\left(\frac{\partial\phi_1}{\partial n}\right)_k - \sum_{k=1}^{NOD} H_{ik}\phi_{1k} + I\right], \ i = 1,2,\cdots,NOD \tag{6.73}$$

which contains no singularity. This procedure also makes it possible to avoid direct calculation of C_i in Eq. (6.68).

We may now calculate the matrices H and A in Eq. (6.68) and solve for the velocity potential Eq. (6.67), where $(\partial\phi / \partial n)_k$ represents the normal velocity components in the radiation BVP.

6.2.5.2 Comparison of CPM and HOBEM

Since the velocity potentials have been computed using the above methods one can readily determine the added mass, radiation damping coefficient and wave-exciting force on 3-D bodies, referring to Secs. 6.1.3, 6.1.4 and 6.1.5.

Use of CPM for discretization of IBIE cannot model the corners and edges on the body surface accurately. But use of HOBEM for discretization of DBIE can precisely determine the coefficient of the solid angle by evaluating the diagonal term H_{ii} using a specifically chosen potential ϕ_1 as shown in Sec. 6.2.4.1.

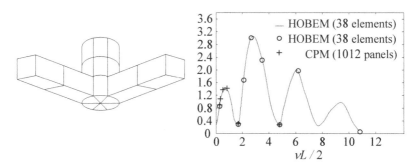

Fig. 6.5. Comparison of non-dimensional surge-exciting force as function of non-dimensional wave frequency $\nu L / 2$ of main part of ISSC TLP employing 38 quadratic higher-order elements and 1012 constant panels.

Hydrodynamic forces of variety of structures were computed by employing HOBEM and the results were compared with many typical CPM results. The ISSC TLP designed by the International Ship Structure Congress (Eatock Taylor and Jefferys, 1986) is one of the structures that had corners and edges as shown in Fig. 6.5. The surge-exciting forces by CPM with 1012 constant panels (Korsmeyer et al., 1988) and HOBEM with 38 elements per quadrant are well compared as shown in Fig. 6.5. It is noted that Fig. 6.5 represents the TLP without tendons. The convergence test is absolutely necessary in using these methods.

6.3 2-D Radiation and Diffraction

Let us consider the 2-D radiation and diffraction problems that will be applied later for strip-wise computation of the hydrodynamic loadings on slender 3-D bodies such as ships. The 2-D CPM can be applied to arbitrary shape smooth cylinder (Frank, 1967), while the hydrodynamic theories (Grim, 1960; Tasai, 1959) for Lewis form section (Lewis, 1929) are limited to the mathematical sections. Since 2-D theories will be applied for ship motion calculation using strip method later, we will set the body and space coordinate systems by the x-axis in longitudinal direction, z-axis vertically upward and y-axis to the port, taking the right-handed coordinate system.

6.3.1 2-D radiation problem

Consider the 2-D radiation problem in a similar way to 3-D radiation in Sec. 6.1.3. We assume a cylinder (ship section) is steadily swaying, heaving and rolling ($m = 2, 3, 4$), in the calm water with y-axis in the transverse direction and z-axis vertically upward. Then the 2-D radiation potential per unit velocity amplitude is given in the form as Eq. (6.23):

$$\Phi_R(y,z,t) = \sum_{m=2}^{4} \mathrm{Re}\left[\phi_m(y,z)e^{-i\omega t}\right] \tag{6.74}$$

Since the above potential is derived from the 2-D pulsating source, it satisfies already all the boundary conditions except the kinematical boundary condition on the body surface contour s. The complex radiation potential is similar to 3-D problem:

$$\phi_m = \phi_{mc} + i\phi_{ms}, \quad m = 2,3,4 \tag{6.75}$$

where suffixes c and s represent the real and imaginary components of the radiation potential, respectively.

As in the 3-D source distribution method in Sec. 6.2.2 we assume the velocity potential due to 2-D source distribution method on the section contour s in the form:

$$\phi_m(P) = \frac{1}{2\pi} \int_s Q_m(P_0) G(P; P_0) ds(P_0), \quad m = 2,3,4 \tag{6.76}$$

where G is the Green's function per unit source intensity that is given in terms of the logarithmic source and complex harmonic function g as shown in Sec. 5.2.2:

$$G(P; P_0) = \log r\left(P; P_0\right) + g(P; P_0) \tag{6.77}$$

$$g\left(P; P_0\right) = g(x, y; \xi, \eta) = -\log r_1$$

$$-2PV \int_0^\infty \frac{e^{k(z+\varsigma)}}{k - v} \cos k(y - \eta) dk \tag{6.78}$$

$$-i2\pi \, e^{v(z+\varsigma)} \cos v(y - \eta)$$

It is noted that the variables used in Chap. 5 has been changed in Eq. (6.78); $y_0 = \varsigma$ and $x_0 = \eta$. From Eq. (6.76) one can construct normal velocity of the fluid on the boundary contour s in the same manner as 3-D problem in Eq. (6.55) in the form:

$$\frac{\partial \phi_m(P)}{\partial n(P)} = Q_m(P) + \frac{1}{2\pi} \int_s Q_m(P_0) \frac{\partial g(P; P_0)}{\partial n(P)} ds(P_0), \, P \in s \tag{6.79}$$

Discretizing Eq. (6.79) with N segments, and assuming the source is uniformly distributed on each segment, one obtains:

$$\left\{ \frac{\partial \phi_m(y, z)}{\partial n(y, z)} \right\}_i = \delta_{ij}(Q_m)_i + \frac{1}{2\pi} \sum_{j=1; j \neq i}^N (Q_m)_j \int_{s_j} \left\{ \frac{\partial g}{\partial n} \right\}_{ij} ds(\eta, \varsigma)_j,$$

$$m = 2,3,4 \tag{6.80}$$

δ_{ij} = Kronecker delta: 1 and 0 when $i = j$ and $i \neq j$.

The kinematical radiation condition on the body surface is given in accordance to Eq. (6.19):

$$\left(\frac{\partial \phi_{mc}}{\partial n}\right) = n_m, \ \left(\frac{\partial \phi_{ms}}{\partial n}\right) = 0, \ m = 2,3,4 \qquad (6.81)$$

Substituting Eq. (6.81) into Eq. (6.80) and solving for the unknown complex source intensities of the radiation potential, one can determine the radiation potential Eq. (6.76). Since the radiation potential is per unit velocity amplitude, one must multiply the velocity $-i\omega u_m$ to the potential after the theoretical development. The complex hydrodynamic force thus becomes:

$$F_{mk} = -\rho\omega^2 u_k \int_s \phi_m \frac{\partial \phi_k}{\partial n} ds, \ m = k = 2,3,4 \qquad (6.82)$$

Since the motion induced hydrodynamic force consists of those proportional acceleration and velocity as given in Eq. (6.34), we can determine the hydrodynamic inertia (added mass) and wave making damping from the following equation:

$$M''_{mk} + \frac{i}{\omega} N_{mk} = \rho \int_{s(x)} \phi_m \frac{\partial \phi_k}{\partial n} ds, \ m = 2,3,4, \ k = 2,3,4 \qquad (6.83)$$

where m and k indicate the modes of motion and force, respectively. The foregoing radiation analysis is called close-fit method or source distribution method (Frank, 1967). This method encountered the problem of the irregular frequency causing divergence of added mass and damping coefficient.

6.3.2 2-D diffraction problem

We can determine the 2-D diffraction potential and wave-exciting force in similar manner as the radiation potential. However, it may not be necessary if Haskind-Newman relation is used. When it is necessary to compute the wave-exciting force directly, we may assume the incident wave progresses in the positive y-direction in the form:

$$\eta(y,t) = A \cos(ky - \omega t) \qquad (6.84)$$

The corresponding wave potential is then given by:

$$\Phi_I(y,z,t) = \frac{gA}{\omega}\frac{\cosh k(z+h)}{\cosh kh}\sin(ky-\omega t) = \mathrm{Re}[\phi_I(y,z)e^{-i\omega t}] \qquad (6.85)$$

where the complex incident wave potential is expressed:

$$
\begin{aligned}
\phi_I(y,z) &= -i\frac{gA}{\omega}\frac{\cosh k(z+h)}{\cosh kh}e^{iky} \\
&= \frac{gA}{\omega}\frac{\cosh k(z+h)}{\cosh kh}\sin ky - i\frac{gA}{\omega}\frac{\cosh k(z+h)}{\cosh kh}\cos ky
\end{aligned}
\qquad (6.86)
$$

in the finite water, and,

$$
\begin{aligned}
\phi_I(y,z) &= -i\frac{gA}{\omega}e^{\nu(z+iy)} \\
&= \frac{gA}{\omega}e^{\nu z}\sin\nu y - i\frac{gA}{\omega}e^{\nu z}\cos\nu y \equiv \phi_I^o + \phi_I^e
\end{aligned}
\qquad (6.87)
$$

in the deep water.

In Eq. (6.87), we note that the incident potential consists of odd and even function of y, which will be used for determining the odd and even diffraction potentials.

The velocity potential for diffraction is the sum of incident and diffraction potential:

$$\Phi(y,z,t) = \mathrm{Re}\left[(\phi_I + \phi_D)e^{-i\omega t}\right] \qquad (6.88)$$

And the kinematical boundary conditions for the odd and even diffraction potentials are given by:

$$\frac{\partial \phi_D^e}{\partial n} = -\frac{\partial \phi_I^e}{\partial n}, \quad \frac{\partial \phi_D^o}{\partial n} = -\frac{\partial \phi_I^o}{\partial n} \qquad (6.89)$$

Substituting Eq. (6.89) into the IBIE Eq. (6.79) or Eq. (6.80), one determines the source intensities and thus the diffraction potentials due to odd and even incident wave potentials, respectively. The odd incident and diffraction potentials are used for determining the sway-and roll-exciting force and moment of ship section, while the even potential is used for the heave-exciting force. Thus the wave-exciting forces due to odd and even functions:

$$F_m(\omega) = -i\rho\omega \int_s \left(\phi_I^o + \phi_D^o \right) \frac{\partial \phi_m}{\partial n} ds, \ m = 2,4$$

$$F_m(\omega) = -i\rho\omega \int_s \left(\phi_I^e + \phi_D^e \right) \frac{\partial \phi_m}{\partial n} ds, \ m = 3$$

(6.90)

where the normal derivative or unit normal is given in Eq. (6.17). The odd and even diffraction potentials will be used below for determining the waves in the far-field.

6.3.3 Far-field radiation and diffraction

The asymptotic expression of the radiation or diffraction potential, for $y \to \pm \infty$, is obtained from the velocity potential due to source distribution Eq. (6.76) in the form:

$$\phi_{m\pm}(y,z) = \left[\frac{1}{2\pi} \sum_{j=1}^{N} (Q_m)_j \left(\int_{S_j} G_\pm(y,z;\eta,\varsigma) d\sigma(\eta,\varsigma) \right) \right]$$

$$m = 2,3,4$$

(6.91)

where the source intensities Q have been determined by solving the kinematical boundary condition Eq. (6.80), and the asymptotic Green's function is obtained from Eq. (5.87) in Sec. 5.2.2 in the form:

$$G_\pm(y,z;\eta,\varsigma) = -ie^{\nu(z+\varsigma)\pm i\nu(y-\eta)}$$

(6.92)

Substituting Eq. (6.92) into Eq. (6.91) and integrating Eq. (6.91), we may write the asymptotic radiation and diffraction potentials in the forms:

$$\phi_{m\pm}^R(y,z) = B_{m\pm}^R e^{\nu z \pm i\nu y}, \quad \phi_{m\pm}^D(y,z) = B_{m\pm}^D e^{\nu z \pm i\nu y}$$

(6.93)

Where $B_{m\pm}^R$ and $B_{m\pm}^D$ are implicit and m for radiation are 2, 3, and 4, and for diffraction 2 and 3.

To determine the far-field wave, we consider the wave elevation:

$$\eta_\pm^R \Big|_{z=0} = \frac{i\omega}{g} \left(\frac{B_{m\pm}^R}{B_{m\pm}^D} \right) e^{\pm i\nu y}$$

(6.94)

from which one will obtain the radiated and diffracted wave amplitudes:

$$A_{m\pm}^R = \frac{i\omega}{g} B_{m\pm}^R, \quad A_{\pm}^D = \frac{i\omega}{g} B_{m\pm}^D \tag{6.95}$$

Since the radiation potential has to be multiplied by $-i\omega u_m$, one obtains the complex radiation amplitude in the form:

$$A_{m\pm}^R = vu_m B_{m\pm}^R, \quad m = 2,3,4 \tag{6.96}$$

Thus one defines the complex amplitude ratios in the far-field i.e., the ratios of both radiation and diffraction amplitude in the far-field to the incident wave amplitude A in the forms:

$$\overline{A}_{m\pm}^R = \frac{A_{m\pm}^R}{A} = \frac{vu_m}{A} B_{m\pm}^R, \quad m = 2,3,4$$

$$\overline{A}_{m\pm}^D = \frac{A_{m\pm}^D}{A} = \frac{i\omega}{gA} B_{m\pm}^D, \quad m = 2,3 \tag{6.97}$$

Hence the resultant complex amplitude ratio of the radiation and diffraction in the far-field is given by:

$$\overline{A}_{\pm}^{D+R} = \sum_{m=2}^{4} \overline{A}_{m\pm}^R + \sum_{m=2}^{3} \overline{A}_{m\pm}^D \tag{6.98}$$

The resultant complex amplitude ratio in the far-field in Eq. (6.98) will be used for determining the mean drift force of a floating body in monochromatic waves in Chap. 8.

6.3.3.1 Wave-exciting force versus radiation amplitude

Newman (1962; 1965) derived the relation between the far-field wave and wave-exciting force. A similar formula may be derived referring to Fig. 6.2 and applying Green's 2nd identity:

$$\int_{s+s_F+s_B+s_\pm} \left(\phi_I \frac{\partial \phi_m}{\partial n} - \phi_m \frac{\partial \phi_I}{\partial n} \right) ds = 0 \tag{6.99}$$

Since the integrals along the free and bottom surface vanish, there remains the integration on the body and far-field boundary surface. The wave-exciting force due to Haskind Eq. (6.49) is expressed by the integration in the far-field:

$$F_m = -i\rho\omega \int_s \left(\phi_l \frac{\partial \phi_m}{\partial n} - \phi_m \frac{\partial \phi_l}{\partial n} \right) ds$$

$$= i\rho\omega \int_{s_{+\infty}+s_{-\infty}} \left(\phi_l \frac{\partial \phi_m}{\partial n} - \phi_m \frac{\partial \phi_l}{\partial n} \right) ds \tag{6.100}$$

Consider the incident and radiation wave in the far-field $\phi_l\big|_{\pm} = -\dfrac{igA}{\omega}e^{vz}e^{\pm ivy}$

and $\phi_{m\pm}^R = \dfrac{1}{vu_m}A_{m\pm}^R e^{vz}e^{\pm ivy}$ and denote the wave-exciting force corresponding

to the far-field waves by $F_{m\pm}$:

Then, for the incident wave progressing positive y-direction:

$\phi_l = -\dfrac{igA}{\omega}e^{vz}e^{ivy}$. And the integrand in Eq. (6.100) becomes:

$$\phi_l \frac{\partial \phi_m}{\partial n} - \phi_m \frac{\partial \phi_l}{\partial n} = 0 \qquad\qquad \text{on } s_{+\infty}$$

$$= -\frac{2gAA_{m-}^R}{\omega u_m}e^{2vz} \qquad \text{on } s_{-\infty} \tag{6.101}$$

whereas for the incident wave progressing negative y-direction:

$\phi_l = -\dfrac{igA}{\omega}e^{vz}e^{-ivy}$,

$$\phi_l \frac{\partial \phi_m}{\partial n} - \phi_m \frac{\partial \phi_l}{\partial n} = -\frac{2gAA_{m+}^R}{\omega u_m}e^{2vz} \quad \text{on } s_{+\infty}$$

$$= 0 \qquad\qquad\qquad \text{on } s_{-\infty} \tag{6.102}$$

Substituting Eqs. (6.101) and (6.102) into Eq. (6.100), and performing the

integral $\displaystyle\int_{-\infty}^{0} e^{2vz}dz = \frac{1}{2v}$, one determines the wave-exciting forces in non-

dimensional form:

$$\frac{F_{m\pm}}{\rho g A B} = \left(\frac{-iA}{u_m} \right) \frac{\overline{A}_{m\mp}^R}{vB} \tag{6.103}$$

Equation (6.103) indicates that the wave-exiting force in the positive and
negative y-directions correspond to the complex amplitude ratio in the far-
field in the negative and positive y directions, respectively.

6.3.4 Radiation and radiation damping

The average power necessary to sustain the steady forced oscillation of a body with small amplitude in the calm water may be written using Eq. (6.82) in the form:

$$
\begin{aligned}
(P_{av})_{\text{body motion}} \\
&= \frac{1}{T} \int_0^T \text{Re}[F_{mk} e^{-i\omega t} dt] \text{Re}[-i\omega\, u_k e^{-i\omega t}] \\
&= \frac{1}{T} \int_0^T (\text{Re}\, F_{mk} \cos \omega t + \text{Im}\, F_{mk} \sin \omega t)(-\omega\, u_k \sin \omega t) dt \\
&= -\frac{1}{2}\omega\, u_k \,\text{Im}\, F_{mk} = -\frac{1}{2}(\omega\, u_k)^2 N_{mk}, \quad j,k = 2,3,4
\end{aligned}
\tag{6.104}
$$

The hydrodynamic damping force only contributes to consumption of power. The average power delivered to the fluid is used in transporting the radiation wave energy. For instance, in the 2-D body motion in deep water, the wave energy transported in two opposite directions is given by:

$$
(P_{av})_{\text{wave}} = 2\left(\frac{1}{2}\rho g\, |a_k^2|\, c_g \right) = \frac{1}{2}\rho g\, |a_k^2|\, c
\tag{6.105}
$$

In which, a_k is the kth mode complex radiation wave amplitude ($k = 2, 3, 4$ for sway, heave and roll and c_g and c are the group and phase velocity, respectively. Energy conservation law dictates that the average power consumed for steady oscillation is identical to the average power transported by the propagating wave:

$$
\left| (P_{av})_{\text{body motion}} \right| = (P_{av})_{\text{wave}}
\tag{6.106}
$$

from which we have the relation between the complex radiation amplitude ratio and radiation damping in the form, when $m = k$:

$$
N_{kk} = \frac{\rho g^2}{\omega^3} \left| \frac{a_k}{u_k} \right|^2_{\text{radiation}}, \quad k = 2,3,4
\tag{6.107}
$$

One designates a_k / u_k the radiation amplitude ratio, i.e., radiation amplitude per unit motion amplitude of the oscillating cylinder. It may be computed using either close-fit method or Lewis form section method. If $k = 3$, the above radiation damping is the heave-induced heave damping.

6.3.5 Hydrodynamic force of Lewis section

Lewis form section was applied earlier than the close-fit method (Frank, 1976). It is purely mathematical while the latter uses fitting approach, thus it needs frequently trial and error process. Lewis theory (1929) is still being widely employed in the practice and research.

A circle of unit radius is represented in complex ς domain:

$$\varsigma = e^{i\theta} = \cos\theta + i\sin\theta \qquad (6.108)$$

which is used for a conformal mapping of a unit circle into a contour in z-plane by the following transform:

$$z = \varsigma + \frac{a}{\varsigma} + \frac{b}{\varsigma^3}, \quad z = x + iy \qquad (6.109)$$

which is called Lewis transform (Lewis, 1929). This was used for evaluating the heaving added mass of ship hull sections in the infinitely high frequency range. Lewis form ship sections are computed by:

$$x = \cos\theta + a\cos\theta + b\cos3\theta,$$

$$y = \sin\theta - a\sin\theta - b\sin3\theta, -\frac{\pi}{2} \le \theta \le 0 \qquad (6.110)$$

where a and b are real numbers that can be determined given the half-beam-draft ratio H and fullness coefficient of the section β defined in the forms:

$$H = \frac{B}{2T}, \quad \beta = \frac{S}{BT} \qquad (6.111)$$

where S = submerged sectional area, B = beam of the section in the waterline, T = draft of the section. The a and b are given in the form (Grim, 1960):

$$a = q \; \frac{\frac{3}{4}\pi + \sqrt{\left(\frac{\pi}{4}\right)^2 - \frac{\pi}{2}p(1-q^2)}}{\pi + p\,(1-q^2)}$$

$$b = -1 + \frac{\frac{3}{4}\pi + \sqrt{\left(\frac{\pi}{4}\right)^2 - \frac{\pi}{2}p(1-q^2)}}{\pi + p\,(1-q^2)} \qquad (6.112)$$

$$p = \beta - \frac{\pi}{4} \;,\quad q = \frac{H-1}{H+1}$$

Thus, given the beam and draft of a ship section one can determine the section coordinates x and y in accordance to Eqs. (6.110) through (6.112). An example of Lewis form section of unit half-beam-draft ratio and 0.8 of fullness coefficient is shown in Fig. 6.6.

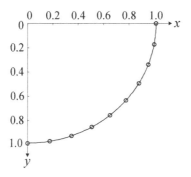

Fig. 6.6. Lewis form section for $H = 1.0$ and $\beta = 0.8$.

The 2-D added mass and damping coefficient are computed for the Lewis section families (Grim, 1960; Tasai, 1969). Figures 6.7(a) and (b) present the 2-D heave added mass coefficient $C = m'' \div \pi\rho B^2 / 8$ and radiation amplitude ratio a_k / u_k of Lewis form ship sections. As seen in Fig. 6.7(a), the added mass coefficient C of a section at high frequency asymptotes to a constant which is used for the hull vibration analysis [Fig. 6.7(a)].

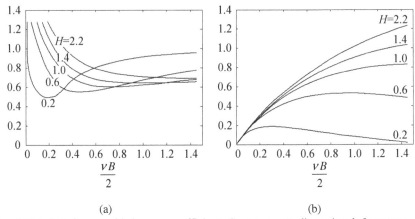

(a)   (b)

Fig. 6.7(a) 2-D heave added mass coefficient C versus non-dimensional frequency. (b) Radiation amplitude ratio a_k / u_k versus non-dimensional frequency in deep water for Lewis section of $\beta = 0.8$ for $H = 0.2$ to 2.2.

In the foregoing figures, the non-dimensional frequency for deep water is denoted as $vB/2$, where $v = \omega^2 / g$ is deepwater wave number. Using the radiation amplitude ratio in Fig. 6.7(b) one may determine the radiation damping N_{33} due to heaving motion of a Lewis form ship section according to the formula Eq. (6.107) for $k = 3$. It is to be noted that the phase angles of the amplitude ratios are not shown here.

6.4 Wavemaker Transfer Function

Various types of wavemaker have been developed; flap-type, piston-type, plunger-type and others. The LTFs of piston- and flap-type wavemaker were theoretically derived and compared with experiments by Ursell et al. (1960). The plunger-type wavemaker transfer function was computed and an extensive experimental work was conducted by Wang (1974).

6.4.1 Flap-type wavemaker

Assume an infinitely long 2-D wave tank with one end open and a flap-type wavemaker is installed at one end. Then we may write the velocity potential for the flow field consisting of progressing wave and evanescent standing wave:

$$\Phi(x, y, t) = \frac{gA}{\omega} \frac{\cosh k(y+h)}{\cosh kh} \sin(kx - \omega t)$$

$$+ \sum_{i=1}^{\infty} C_i \cos k_i(y+h) \exp(-k_i x) \cos \omega t \qquad (6.113)$$

where A is the progressing wave amplitude and C_i is arbitrary constant of the evanescent wave, which were analytically derived (Ursell, et al., 1960; Dean and Dalrymple, 1991). The above potential satisfies the linear free surface boundary condition and bottom condition. The down tank end is assumed to be open, thus the potential satisfies the radiation condition automatically. Thus this is called a simple wavemaker theory. Now we consider the linear kinematical boundary condition on the flap surface using the above velocity potential. Assume the flap surface oscillates about z-axis whose origin is on the tank bottom at frequency ω with small flap motion amplitude a (or stroke / 2) on the waterline. Thus the horizontal motion of the flap surface is given by:

$$x = a(1 + y/h) \sin \omega t \qquad (6.114)$$

The linear kinematical boundary condition is according to Eq. (6.17):

$$\frac{\partial \Phi}{\partial x} = \omega a \left(1 + \frac{y}{h}\right) \cos \omega t, \quad x = 0 \qquad (6.115)$$

Substituting the velocity potential in Eq. (6.113) into the linear kinematical boundary condition on the flap surface Eq. (6.115), one has:

$$\omega a \left(1 + \frac{y}{h}\right) = \frac{gAk}{\omega} \frac{\cosh k(y+h)}{\cosh kh} - \sum_{i=1}^{\infty} C_i k_i \cos k_i(y+h) \qquad (6.116)$$

The formula for progressing wave or radiated wave amplitude A (Ursell et al., 1956; Dean and Dalrymple, 1991) is given in the form:

$$A = \frac{-\omega a \int_{-h}^{0} \left(1 + \frac{y}{h}\right) \cosh k(y+h) dy}{\frac{gk}{\omega \cosh kh} \int_{-h}^{0} \cosh^2 k(y+h) dy} \qquad (6.117)$$

from which the LTF (ratio of wave height to stroke or ratio of wave amplitude to the half stroke) of the flap-type wavemaker is given by:

$$\frac{A}{a} = 4\frac{\sinh kh}{kh}\frac{kh\sinh kh - \cosh kh + 1}{\sin h2kh + 2kh} \qquad (6.118)$$

The simple piston-type wavemaker theory is verified by Ursell et al. (1960) with experimental data obtained in a wave flume. The average deviation between theory and experiment is of the order of 3% for wave steepness H / λ = between 0.002 and 0.03 and 10% for steepness in the range of 0.045 to 0.048.

However, these theories cannot provide the phase angles of LTFs. Hence it may only be applicable to the generation of random waves because the random waves are obtained by implementing the random phase angles in the component waves of the target spectrum. In the generation of design wave such as strongly asymmetric wave in the wave tank one needs the control of phase angles as will be discussed in Chap. 11.

The 3-D wavemaker theory is given by Dean and Dalrymple (1991).

6.4.2 Plunger-type wavemaker

The plunger-type wavemaker is a heaving body near the tank wall at one end. For instance, the heaving body is a half of wedge or half of a Lewis-form section. Wang (1974) modeled the wedge type-plunger using conformal mapping and verified the theory by experiment. The average scatter of the measured wave height about the theoretical prediction is 6.5%, which is considered within the experimental error involved in the laboratory measuring apparatus. Grim (1960) computed the LTFs of the plunger-type wavemaker of Lewis form sections as shown in Fig. 6.7(b), though no intention of its application to the plunger-type wavemaker was noted in the report. The radiation amplitude ratio of Lewis form sections are complex amplitude ratios, though no phase data are shown. The LTFs of the Lewis-form plunger-type wavemaker by Wang has both magnitudes and phase angles.

6.4.3 Reflection of waves in wave tank

The simple wavemaker theory assumes that the down tank end is open. If the tank end is closed as an ordinary wave tank, one needs to install wave energy absorbing beach to avoid or minimize the reflection of the waves as small as possible. A variety of new techniques have recently been developed in order to minimize the reflection.

The reflection coefficient is defined by the ratio of the reflected wave height to the incident wave height. The total of surface elevation due to the incident and reflected wave:

$$\eta(x,t) = \frac{1}{2}\left[H_i \cos(kx - \omega t) + H_r \cos(kx + \omega t) \right]$$
$$= \frac{1}{2} H_i \left[\cos(kx - \omega t) + K_r \cos(kx + \omega t) \right]$$

(6.119)

with

$$K_r = \frac{H_r}{H_i} = \frac{H_{max} - H_{min}}{H_{max} + H_{min}}$$

(6.120)

$$H_i = \frac{1}{2}\left(H_{max} + H_{min} \right)$$

(6.121)

The maximum and minimum wave heights are measured by traversing the wave gauge in the wave, from which the incident and reflected wave heights are found by using Eqs. (6.120) and (6.121). The reflection coefficients of random waves in a wave flume (30 m × 0.9 m × 0.8 m, horse hair beach with slope of 1 / 15) were measured at Texas A & M University according to Goda (2000), which amounts to 0.15 on the average in the range of frequency (rad / s) from 0.3 to 1.2. Outside these the reflection coefficients violently fluctuate with large magnitudes.

6.5 Hydrodynamic Loads in Oblique Seas

6.5.1 Strip-method

In order to estimate the added mass and damping of the ship floating in the oblique wave by 2-D theory in Sec. 6.3.1, we employ a strip-wise computation by slicing the entire hull length into, for instance, 10 to 20 stations, generating 9 to 19 segments of ship sections. The resultant hydrodynamic loads are the sums of the radiation and diffraction loads on these segments.

6.5.1.1 Added mass and damping

The ship in the sinusoidal oblique incident wave will experience the wave-exciting force and motion-induced or radiation force simultaneously. The radiation force will be estimated by summing the 2-D added mass and damping as given in Sec. 6.3.1 in the form:

$$M_{jk}'' + \frac{i}{\omega} N_{jk} = \rho \int_{-L/2}^{L/2} \int_{s(x)} \phi_j \frac{\partial \phi_k}{\partial n} \, ds dx,$$

$$j,k = 2,3,\cdots,6$$

$$(6.122)$$

where, in the integrand, the suffix j and k indicates jth-mode motion-induced kth-mode force on the segment. Here the mode of motion and force are extended from the sectional radiation (2, 3, 4) to pitch (5) and yaw (6), applying Eq. (6.17). The normal derivative of the radiation potential is given in Eq. (6.17). No effects of the obliqueness of the wave are to be included in the added mass and damping.

6.5.1.2 Wave-exciting force

Referring to Fig. 3.2.4, we may write sinusoidal oblique wave and potential in the deep water in the form:

$$\eta(x,y,t) = A\cos(\nu x \cos \mu + \nu y \sin \mu - \omega t)$$

$$\Phi_I(x,y,z,t) = \text{Re}\left[\phi_I e^{-i\omega t}\right]$$

$$(6.123)$$

$$\phi_I = \frac{-igA}{\omega} e^{\nu z + i(\nu x \cos \mu + \nu y \sin \mu)}$$

The foregoing incident potential is rewritten in the form:

$$\phi_I \left(y,z,\mu\right) = \frac{gA}{\omega} e^{vz} \sin\left(vx\cos\mu + vy\sin\mu\right)$$

$$- \frac{igA}{\omega} e^{vz} \cos\left(vx\cos\mu + vy\sin\mu\right) \tag{6.124}$$

$$= \phi_I^o + \phi_I^e$$

In the above ϕ_I^o and ϕ_I^e denote the odd and even function of y, respectively. One applies the same form of diffraction boundary condition Eq. (6.89), to determine the 2-D diffraction potential both in the odd and even mode in the oblique wave. In this manner one will determine the wave exciting force in oblique wave on a ship at zero speed.

$$F_m(\omega,\mu) = -i\rho\omega \int_s \left(\phi_I^o\left(\omega,\mu\right) + \phi_D^o\left(\omega,\mu\right)\right)\frac{\partial\phi_m}{\partial n} ds, \quad m = 2,4$$

$$F_m(\omega,\mu) = -i\rho\omega \int_s \left(\phi_I^e\left(\omega,\mu\right) + \phi_D^e\left(\omega,\mu\right)\right)\frac{\partial\phi_m}{\partial n} ds, \quad m = 3 \tag{6.125}$$

When the ship uniformly advances at U, in oblique seas, one has to satisfy the linearized free surface boundary condition and hydrodynamic pressure with respect to the moving ship in Sec. 4.3.2.1:

$$\left[\left(\frac{\partial}{\partial t} - U\frac{\partial}{\partial x}\right)^2 \phi + g\phi_z\right]$$

$$\times\left[\phi_I\left(x,y,z,;v,\mu;t\right) + \phi_D\left(x,y,z,;v,\mu;t\right)\right] = 0, z = 0 \tag{6.126}$$

and

$$p = \rho\left(i\omega + U\frac{\partial}{\partial x}\right)\left(\phi_I\left(x,y,z;v,\mu;t\right) + \phi_D\left(x,y,z;v,\mu;t\right)\right) \tag{6.127}$$

For the details of ship motion at constant speed, refer to Salvesen et al. (1970) and Kim et al. (1980).

6.6 Motion of Structure

The floating offshore structures are usually moored in various ways using a variety of structural configurations and materials. Thus one has to take into account the restoring forces due to both hydrostatics and mooring structures. The response motion of such structures to the incident waves are computed by solving the linear coupled 6 degree freedom motion equations in the frequency domain. It is usual that the fluid loadings are computed assuming ideal fluid, while the viscous damping is separately determined from decay test of a scaled model.

6.6.1 Linear equation of motion in frequency domain

Newton's law is applied to the basic equation of motion of a structure due to the sinusoidal wave:

$$-M_{jk}\omega^2 u_k = M''_{jk}\omega^2 u_k + iN_{jk}\omega u_k - B_{jk}u_k + F_j \qquad (6.128)$$

where LHS is the inertial force of the structure and RHS is the sum of reactionary hydrodynamic forces according to Eq. (6.34), hydrostatic restoring force, and wave-exciting force F_j as derived in Eq. (6.44).

We consider a moored structure being connected to mooring lines oscillates about the mean offset position. By taking into consideration of the effects of the mooring structures attached to the main structure, we modify the coupled equation Eq. (6.128) in the form:

$$\sum_{k=1}^{6}\begin{bmatrix} -\omega^2(M_{sjk}) - i\omega(N_{sjk} + N_{fjk}) \\ +(B_{sjk} + B_{fjk}) \end{bmatrix} u_k = F_j, \ j = 1, 2, \cdots, 6 \qquad (6.129)$$

where $u_k = (\xi_1, \xi_2, \xi_3, \alpha_1, \alpha_2, \alpha_3)$ and F_j indicates jth mode first-order wave exciting forces and moments. M_{sjk}, N_{sjk} and B_{sjk} represent the inertial mass, damping and restoring force of the structure, whereas, $M''_{fjk} (= M''_{jk})$, N_{fjk} and B_{fjk} are the added mass, radiation damping including viscous damping if necessary and restoring force coefficient due to fluid. The above equation is the linear coupled six degree of freedom motion of the structure in the sinusoidal wave. We define here the virtual mass matrix = structure mass matrix in the air + added mass matrix. The added mass matrix has been given already in Sec. 6.1.3. The structural mass matrix is given below (Newman, 1977).

$$
M_{sjk} = \begin{pmatrix}
M & 0 & 0 & 0 & Mz_g & -My_g \\
0 & M & 0 & -Mz_g & 0 & Mx_g \\
0 & 0 & M & My_g & -Mx_g & 0 \\
0 & -Mz_g & My_g & I_{11} & I_{12} & I_{13} \\
Mz_g & 0 & -Mx_g & I_{21} & I_{22} & I_{23} \\
-My_g & Mx_g & 0 & I_{31} & I_{32} & I_{33}
\end{pmatrix}
\tag{6.130}
$$

where M_{sjk} is the structure mass matrix and (x_g, y_g, z_g) is the vector position of the center of gravity. I_{ii} denotes moment of inertia and I_{ij} product of inertia. If the moments are estimated from the axes through the center of gravity, the terms with (x_g, y_g, z_g) will vanish. In the case of a ship, there remains only the diagonal terms.

6.6.1.1 Hydrostatic restoring force and moment

Consider a ship that is symmetric about a principal vertical plane. The hydrostatic pressure below the waterline gives the buoyancy that balances the weight of the structure in equilibrium. The hydrostatic pressure force acting on the small layer over the waterplane area gives the restoring force and moment to restore the small vertical displacement and small inclinations. Small heave, roll and pitch of a structure in waves create heave restoring force, roll and pitch restoring moments (Lewis, 1989):

$$
\begin{aligned}
B_{33} &= \rho g A_{WP} \\
B_{44} &= \rho g \nabla \overline{GM}_T \\
B_{55} &= \rho g \nabla \overline{GM}_L
\end{aligned}
\tag{6.131}
$$

$$
\overline{GM}_T = \frac{I_T}{\nabla} + \overline{KB} - \overline{KG}, \quad \overline{GM}_L = \frac{I_L}{\nabla} + \overline{KB} - \overline{KG}
\tag{6.132}
$$

A_{WP}	: waterplane area
∇	: volume of displacement
$\overline{GM}_T$	: transverse metacentric height
$\overline{GM}_L$	: longitudinal metacentric height

It should be noted that the transverse and longitudinal moment of inertia I_T and I_L of the waterplane area are about the longitudinal and transverse axes passing the center of floatation (centroid of the water plane area), respectively. In the above, B_{ii} ($i = 3,4,5$) represents the heave restoring force per unit heave displacement, roll restoring moment per unit roll, and pitch restoring moment per unit pitch, respectively. A more general form of B_{jk} is discussed in Newman (1977).

6.6.1.2 Viscous damping

The viscous damping due to roll in the calm water and drift damping in waves as given in Chap. 8 are estimated from decay tests in the wave tank. The basis of it may rest on the equation of weakly damped oscillation with viscous damping force $c\dot{x}$ on the body:

$$m\ddot{x} + c\dot{x} + kx = F\left(t\right) \tag{6.133}$$

The solution of the weakly damped oscillation is given by:

$$x = Xe^{-\varsigma\omega_n t} \sin\left(\sqrt{1-\varsigma^2}\,\omega_n t + \phi\right) \tag{6.134}$$

where X is the initial maximum amplitude, $\omega_n \sqrt{k/m}$ natural frequency (rad / s), $\varsigma = c/c_c$, damping ratio, and $c_c = 2m\omega_n$ critical damping. The logarithmic decrement is:

$$\delta = \ln\frac{x_1}{x_2} = \frac{2\pi\varsigma}{\sqrt{1-\varsigma^2}} \tag{6.135}$$

when ς is small, $\delta \approx 2\pi\varsigma$.

The decay test gives the logarithmic decrement δ, and thus the experimental damping coefficient c is determined.

6.6.1.3 RAO of SDOF motion

Consider a body that steadily oscillates in the wave train of $\eta = A \cos$ $(vy-\omega t)$. One considers only a single degree of freedom (SDOF) motion, such as heave, sway or roll assuming each of which is uncoupled with other motions:

$$\left[-\omega^2 (M_S + M''_{mm}) - i\omega N_{mm} + B_{mm} \right] u_m = F_m \qquad (6.136)$$

The RAO for the uncoupled motion is given by:

$$\frac{u_m}{A} = \frac{F_m / A}{B_{mm} - \omega^2 (M_S + M''_{mm}) - i\omega N_{mm}} \qquad (6.137)$$

The heave or sway RAO (heave amplitude/wave amplitude; sway amplitude/wave amplitude) approaches to unity in the infinitely long wave train, while they become negligibly small at the very short wave. The non-dimensional roll RAO (roll amplitude / maximum wave slope) also reaches unity at the long period. The peak motion of heave or roll occurs at resonance frequency, while surge or sway does not have such peak because it has no restoring force and thus no resonance. The heave resonance frequency may be derived by taking the denominator as zero in Eq. (6.137). And if one assumes that the damping is negligible compared to other terms and the ship's mass is approximately equal to its heaving added mass, one may have an approximate heave natural frequency:

$$\omega_3 \approx \sqrt{\frac{B_{33}}{M_s + M''_{33}}} \approx \sqrt{\frac{\rho g A_{WP}}{2M_s}} \qquad (6.138)$$

The semi-submersible and spar platform have very long heave resonance periods compared to the ship because of the smaller waterplane areas than the ship. These platforms are designed to avoid the influence of the high wave energy of the spectrum. The surge or sway mode of lateral motion of a soft moored compliant structure has a very low resonance frequency due to the soft mooring stiffness of the mooring cables (structural restoring force).

6.6.2 Deck-wetting and slamming of ship

The strip motion theory was first applied by Korvin-Kroukovsky (1955) and Korvin-Kroukovsky and Jacobs (1957) to the prediction of heaving and pitching motions of a ship running in longitudinal regular waves. Following their investigations in the above two works, a number of modifications were made to the theory, but these offered no discernable distinction. Nevertheless, a number of small differences resulted from formulation in choice of 2-D theory, range of wave heading and ship speed, etc. (Kim et al., 1980).

Ships running fast in head seas usually undergo large vertical relative motions creating the situation for severe deck wetting and slamming on the ship bottom. And the effect of 3-D body shape such as flare and waves are largely involved. Thus 2-D strip theory is limited in handling with such deck-wetting. In such case 3-D theory is locally applied. Though development of 3-D technique has been quite some time, it still gives inconsistent result to the ship motion in general. Thus it is common practice to employ the strip theory and 3-D theory together to obtain an educated judgment for design.

It has been known that there fundamentally unsolved problems have been seated in the 3-D theory for long time. Basic studies about the problems have recently been initiated by Noblesse and Yang (2005), Noblesse et al. (2005) and others.

6.6.2.1 Vertical relative motion

Takaishi et al. (1973) and others conducted analyses of vertical relative motion of ships in oblique seas. The vertical relative motion is defined by the difference between the vertical motion of a ship section and dynamic swell at the section $(p_I + p_D + p_R)/\rho g$. Kim et al. (1980) computed the vertical relative motion similar to Takaishi et al., which were generally in good agreement. However, the theoretical computation appeared to be very large compared to the experimental data according to the comment by Bales in the discussion of the paper (Kim et al., 1980).

It was also found that the linear theory overestimated the experimental data of vertical relative motion (Cummins, 1973) as shown in Fig. 6.8. The

RAOs for the vertical relative motion of a ship were measured at the variety of sea severity by cross-spectral analysis (Dalzell, 1962). It was found that RAOs significantly varied depending on the degree of sea severity.

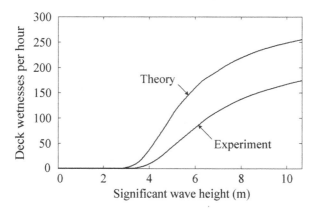

Fig. 6.8. Predicted frequency of deckwetness for destroyer DE 1052 using theoretical and experimental RAO.

It seems certain that the large discrepancy between the linear theory and experimental data in higher seas is due to the nonlinearity of the waves and ship's nonlinear motion.

6.6.2.2 Hydrodynamic pressure on the hull surface

Kim (1982) computed the hydrodynamic pressure distributions of the seagoing ship model. The hydrodynamic pressure was investigated by many researchers in Japan (Tasai and Takaki, 1969; Goda, 1968; Tasai, 1968; Fukuda et al., 1971; Nakamura et al., 1973; Sugai et al., 1973; Matsuyama, 1975).

The computations (Kim, 1982) were in good agreement with the other analytical and experimental results. The prediction technique of the hydrodynamic pressures on the hull surface of sea-going vessels is one of the most important research tasks concerning the safety of ships in the high seas.

6.6.2.3 Ship's slamming

The slamming pressure or the impact pressure on the hull bottom is due to the large vertical relative motion velocity. For instance it is given by Ochi and Motter (1973) in the form:

$$p = \frac{1}{2}\rho k \left(\frac{dr}{dt}\right)^2 \qquad (6.139)$$

where k is the coefficient derived from experimental data, ρ density of sea water, and dr / dt is the vertical relative velocity amplitude. Since the slamming load is random, one employs the most probable extreme pressure to occur in N-observations as in Sec. 1.2.6.5. The observed data provides the method of estimation of extreme slamming pressure for design application. This vertical slam is different from the horizontal impact (slam) on a surface-piercing fixed structure which will be discussed in Chap. 11.

6.7 Hydrodynamic Interaction Between Two Bodies

Cargo transfer from one ship to another in a sea environment is a crucial problem for the safety of ship handling operation. Kim (1972) and Ohkusu (1976) developed theories for the vertical and lateral relative motions between the two adjacent vessels in two-dimensional ways. Kodan (1984) used Ohkusu (1976) for strip-wise computation of relative motions between two floating ships in oblique waves and conducted an extensive model test to verify the theory. Kim and Fang (1985) and Fang and Kim (1986) compared their strip theory with the experimental data by Korsmeyer (1983) and Kodan (1984). The 3-D computations of the relative motions are due to Van Oortmerssen (1979) and Loken (1981), which employed CPM.

6.7.1 Relative heaving motions of two cylinders in beam seas

We will review the 2-D method for computing the hydrodynamic interaction between two cylinders in the beam seas (Kim, 1972). The basic idea is identical regardless of the bodies being 2-D and 3-D. We may consider two arbitrary shape cylindrical bodies a and b oscillate independently in arbitrary modes $k = 2, 3, 4$, in beam seas. But here we review a simplest case of hydrodynamically coupled linear heaving motions and the relative heaving motion in beam seas.

6.7.1.1 Linear hydrodynamic coupling

The hydrodynamic force on each single body of two heaving bodies is a function of the heaving motions of two bodies ζ_a and ζ_b in the calm water.

The assumption of small motions allows the hydrodynamic forces be expanded in Taylor series about the equilibrium position of oscillations:

$$F_a(\zeta_a, \zeta_b) = \zeta_a \frac{\partial F_a(0,0)}{\partial \zeta_a} + \zeta_b \frac{\partial F_a(0,0)}{\partial \zeta_b}$$

$$F_b(\zeta_a, \zeta_b) = \zeta_a \frac{\partial F_b(0,0)}{\partial \zeta_a} + \zeta_b \frac{\partial F_b(0,0)}{\partial \zeta_b} \tag{6.140}$$

The meaning of the partial derivative with respect to a is that cylinder a moves while b is fixed Eq. (6.140). The same rule applies to the rest of the partial derivatives. These motion-induced force derivatives (force per unit displacement) are represented in terms of the coupled added mass and damping coefficient:

$$\frac{\partial F_a}{\partial \zeta_a} = \omega^2 m''_{aa} + i\omega N_{aa}, \frac{\partial F_a}{\partial \zeta_b} = \omega^2 m''_{ba} + i\omega N_{ba} \tag{6.141}$$

$$\frac{\partial F_b}{\partial \zeta_a} = \omega^2 m''_{ab} + i\omega N_{ab}, \frac{\partial F_b}{\partial \zeta_b} = \omega^2 m''_{bb} + i\omega N_{bb} \tag{6.142}$$

Equation (6.141) represents the hydrodynamic force on cylinder a due to the partial motion of body a and b, respectively. The m''_{aa} represents the inertial force on a due to motion of cylinder a, while m''_{ba} denotes the inertial force on cylinder a due to the motion of cylinder b. The first subscript indicates the moving body, and the second indicates the body on which the hydrodynamic force is exerted. The same rule is applied to the coupled damping forces. Equation (6.142) represent the force derivatives on b due to the motion of cylinders a and b, respectively.

6.7.1.2 The coupled equations of heaving motions

Employing the foregoing linear hydrodynamic coupling terms, one writes the coupled heaving equations:

$$
\left[-\omega^2 (m_a + m_{aa}'') - i\omega N_{aa} + B_a \right] \zeta_a
$$
$$
+ \left[-\omega^2 m_{ba}'' - i\omega N_{ba} \right] \zeta_b = F_{H_a}
$$
$$
\left[-\omega^2 m_{ab}'' - i\omega N_{ab} \right] \zeta_a \tag{6.143}
$$
$$
+ \left[-\omega^2 (m_b + m_{bb}'') - i\omega N_{bb} + B_b \right] \zeta_b = F_{H_b}
$$

where m_a, and m_b. are the masses of the two structures, respectively; B_a and B_b represent the restoring force coefficient of cylinder a and b respectively, and F_{Ha} and F_{Hb} represent the heave-exciting forces on bodies on cylinders a and b, respectively. The motion-induced hydrodynamic forces and wave-exciting forces are determined by satisfying the radiation and diffraction boundary conditions as shown below.

6.7.2 Radiation problem

One assumes the heave mode radiation potential of bodies a and b together floating in the water in the form referring to Sec. 6.3.1:

$$
\phi_3(y,z) = \frac{1}{2\pi} \int_{s_a} Q_{3a}(\eta,\zeta) G(y,z;\eta,\zeta)\, ds(\eta,\zeta)
$$
$$
+ \frac{1}{2\pi} \int_{s_b} Q_{3b}(\eta,\zeta) G(y,z;\eta,\zeta)\, ds(\eta,\zeta) \tag{6.144}
$$

where Q_{3a} and Q_{3b} are heave mode complex sources on a and b and G is the 2-D Green's function given in Eq. (6.77). In accordance to the rule of the partial derivative in deriving the linear hydrodynamic forces in Sec. 6.7.1.1, one writes the radiation boundary conditions on a, while b is fixed:

$$
\frac{\partial \phi_3^a}{\partial n} = n_3^a \quad \text{on} \quad a
$$
$$
\frac{\partial \phi_3^b}{\partial n} = 0 \quad \text{on} \quad b \tag{6.145}
$$

Thus the heave-induced forces are written according to Eq. (6.83) in the forms:

$$\rho \int_{S_a} \phi_3^a \frac{\partial \phi_3^a}{\partial n} \, ds = m_{aa}'' + \frac{i}{\omega} N_{aa}$$

$$\rho \int_{S_b} \phi_3^b \frac{\partial \phi_3^a}{\partial n} \, ds = m_{ab}'' + \frac{i}{\omega} N_{ab} \qquad (6.146)$$

Another radiation condition when a is fixed while b is in motion:

$$\frac{\partial \phi_3^a}{\partial n} = 0 \quad \text{on} \quad a$$

$$\frac{\partial \phi_3^b}{\partial n} = n_3^b \quad \text{on} \quad b \qquad (6.147)$$

from which:

$$\rho \int_{S_a} \phi_3^a \frac{\partial \phi_3^b}{\partial n} \, ds = m_{ba}'' + \frac{i}{\omega} N_{ba}$$

$$\rho \int_{S_b} \phi_3^b \frac{\partial \phi_3^b}{\partial n} \, ds = m_{bb}'' + \frac{i}{\omega} N_{bb} \qquad (6.148)$$

6.7.3 Diffraction problem

We assume the diffraction potential is identical to the radiation potential in Eq. (6.144), and apply the same diffraction boundary condition on each body surface:

$$\frac{\partial \phi_D}{\partial n} = -\frac{\partial \phi_I}{\partial n} \quad \text{on} \quad a \, , \quad \frac{\partial \phi_D}{\partial n} = -\frac{\partial \phi_I}{\partial n} \quad \text{on} \quad b \qquad (6.149)$$

where

$$\phi_I(y,z) = -i \frac{gA}{\omega} e^{kz+iky} = \phi_I^o + i\phi_I^e \qquad (6.150)$$

which consists of the odd and even potentials of y for determining the swaying- and heaving-forces and rolling-exciting moment. Here we are interested in determining the have-exciting forces on body a and b according to Eq. (6.90):

$$F_{H_a} = -i\rho\omega \int_{S_a} \left(\phi_I^e + \phi_D^e \right) \frac{\partial \phi_3}{\partial n} ds$$

$$F_{H_b} = -i\rho\omega \int_{S_b} \left(\phi_I^e + \phi_D^e \right) \frac{\partial \phi_3}{\partial n} ds$$

(6.151)

In the above we have described how to determine the motion-induced forces and wave-exciting forces. Thus, we can solve the coupled motion equation Eq. (6.143) for determining the unknown motions ζ_a and ζ_b of bodies a and b, respectively. With these motions, one may evaluate the relative heaving motion between two bodies. Fang and Kim (1986) computed the relative motions of the models tested by Korsmeyer (1983) and Kodan (1984) applying the strip theory (Kim et al., 1980), and they compared the heave RAOs with other experiment and analysis as shown in Fig. 6.9.

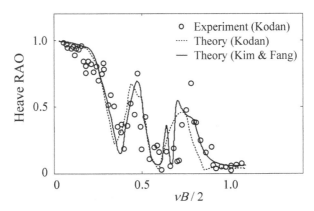

Fig. 6.9. Comparison of heave RAO of ship in parallel position to same size barge floating in quartering seas $\mu = 45°$ (Kodan, 1984).

6.7.4 Loads on Catamaran ship

Thus far we have considered two separate bodies that are in arbitrary motions independently. When bodies a and b are identical and rigidly connected to each other the structure becomes a twin body such as a catamaran ship, semi-submersible and TLP. Haskind relation Eq. (6.103) gives the resultant force due to the resultant motion of the entire body, thus the individual forces on each body cannot be obtained by the formula. The individual loads on a and b are necessary in order to compute the hydrodynamic loads on the bridging deck section such as shearing and bending loads, which are determined by the motion-induced forces and wave-exciting forces and moments on the one sided hull. The bending moment was called the spreader bending moment of the deck (Kim, 1972; Kim, 1976). It is involved in computing the dynamic spreader bending moment. Here we demonstrate the need of computation of each wave-exciting force on the body a and b separately in order to determine the bending moment. Figure 6.10 is the result of a 2-D calculation of sway-exciting forces on a catamaran ship. It is evident in the figure that half of the total force is not equal to each sway-exciting force.

The sway-exciting force on each body is the vectorial sum of odd and even exciting forces. The meaning of the odd and even wave potential is now different from each independent bodies a and b as explained in Eq. (6.151).

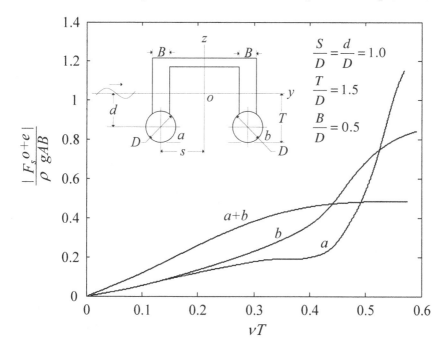

Fig. 6.10. The sway-exciting forces on cylinders a, b and $a + b$; and $v = \omega^2 / g$ is the deep water wave number. A is the incident wave amplitude.

References

Journal references:

Eatock Taylor, R. and Jefferys, E.R. (1986). Validity of Hydrodynamic Load Predictions for a Tension Leg Platform, Ocean Eng., Vol. 13, No. 5, pp. 449–490.

Fang, M.C. and Kim, C.H. (1986). Hydrodynamically Coupled Motions of Two Ships Advancing in Oblique Seas, J. Ship Res., Vol. 30, No. 3, pp.159–171.

Fukuda, J., Nagamoto, M., Konuma, M. and Takahashi, M. (1971). Theoretical Calculations on the Motions, Hull Surface Pressures and Transverse Strength of a Ship in Waves, J. Soc. Naval Architects of Japan, Vol. 129.

Goda, K. (1968). Hydrodynamic Pressure on a Midship in Waves, J. Soc. Naval Architects of West Japan, No. 35 (in Japanese).

Hess, J.L. and Smith, A.M.O. (1966). Calculation of Potential Flow about Arbitrary Bodies, Progress in Aeronautical Science, Vol. 8.

Kim, C.H. (1976). Motions and Loads of a Catamaran Ship of Arbitrary Shape in a Seaway, J. Hydronautics, AIAA, pp. 106–115.

Kim, C.H., Chou, F. S. and Tien, D (1980). Motions and Hydrodynamic Loads of a Ship Advancing in Oblique Waves, Society of Naval Architects and Marine Engineers Trans., Vol. 88, pp. 225–256.

Korvin-Kroukovsky, B.V. (1955). Investigation of Ship Motions in Regular Waves, SNAME Trans., Vol. 63.

Korvin-Kroukovsky, B.V. and Jacobs, W.B. (1957). Pitching and Heaving Motions of a Ship in Regular waves, SNAME Trans., Vol. 65.

Lewis, F.M. (1929). The Inertia of the Water Surrounding a Vibrating Ship, SNAME Trans., Vol. 37, pp. 1–20.

Liu, Y.H., Kim, C.H. and Lu, S.X. (1991). Comparison of Higher-Order Boundary Element and Constant Panel Method for Hydrodynamic Loadings, Int. J. Offshore and Polar Eng., Vol. 1, No.1, pp. 8–17.

Nakamura, et al. (1973). Hydrodynamic Pressures on a Restrained Ship in Oblique Waves, J. Soc. Naval Architects of Japan (in Japanese).

Ochi. M.K and and Motter, L.E. (1973). Prediction of Slamming Characteristics and Hull Responses for Ship Design, SNAME Trans., Vol. 81.

Salvesen, N., Tuck, E.O., and Faltinsen, O. (1970). Ship Motions and Sea Loads, SNAME Trans., Vol. 78.

Sugai, K., Goda, K., Kitagawa, H., Takei, Y., Kan, M., Miyamoto, T., Oomatsu, S. and Okamoto, M. (1973). Model Tests on Hydrodynamic Pressures Acting on the Hull of an Ore-Carrier in Oblique Waves, J. Soc. Naval Architects of Japan, Vol. 133 (in Japanese).

Takaishi, Y., Yoshimoto, T., Ganno, M. (1973). On the Relative Wave Elevations at the Ship's Side in Oblique Seas, J. Soc Naval Architects of Japan (in Japanese).

Tasai, F. (1959). On the damping Force and Added Mass of Ships Heaving and Pitching, J. of Zosen Kyokai, Japan.

Tasai, F. (1968). Pressure Fluctuation on the Ship Hull Oscillating in Beam Seas, J. Soc. Naval Architects of Japan, No. 35 (in Japanese).

Ursell, F., Dean, R.G. and Yu, Y.S. (1960). Forced small-amplitude water waves: a comparison of theory and experiment, J. Fluid Mech. Vol. 7, pp. 33-52.

Wang, S. (1974). Plunger-Type Wavemakers: Theory and Experiment, J. Hydronautics Res., Vol. 12, No. 3, pp. 357-405.

Yoshimoto, T. and Ganno, M. (1973). On the Relative Wave Elevations at the Ship's Side in Oblique Seas, J. Soc. Naval Architects of Japan, Vol. 132 (in Japanese).

Book references:

Blagoveshchensky, S.N. (1962) Theory of Ship Motion, Vol. 2, Dover Publication.

Brebbia, C.A. and Walker, S. (1980) Boundary Element Techniques in Engineering, Butterworth Ltd.

Dean, R.G. and Dalrymple, R.A. (1991) Water Wave Mechanics for Engineers and Scientists, Advanced Series in Ocean Engineering, Vol. 2, World Scientific Publishing Co.

Garrison, C.J. (1979) Hydrodynamic Loading of Large Volume Offshore Structures: Three Dimensional Source Distribution Methods, Numerical Method in Offshore, English edition by O.C. Zienkiewicz, et al., Wiley, London, UK

Goda, Y. (2000) Random Seas and Design of Maritime Structures, Advanced Series on Ocean Engineering, Vol. 15, World Scientific Publishing Co.

Kochin, N.E., Kibel, I.A. and Rose, N.V. (1964) Theoretical Hydromechanics, English translation of 5th Russian edition, Wiley, New York.

Korvin-Kroukovsky, B.V. (1963) Theory of Seakeeping, The Society of Naval Architects and Marine Engineers, Jersey City, N.J.

Lewis, E.V. (1989) Motions in Waves and Controllability, Principles of Naval Architecture, Vol. 3, The Society of Naval Architects and Marine Engineers, Jersey City, N.J.

Newman, J.N. (1977) Marine Hydrodynamics, MIT Press.

Sarpkaya T. and Isaacson, M. (1981) Mechanics of Wave Forces on Offshore Structures, Van Nostrand Reinhold Co, New York.

Vossers, G. (1962) Behavior of Ships in Waves, Resistance, Propulsion and Steering of Ships, The Technical Publishing Co. H. Stam, N.V., Haarlem, The Netherlands.

Wehausen, J.V. and Laitone, E.V. (1960) Surface Waves, in Encyclopedia of Physics Vol. IX, Springer-Verlag, Berlin, Goettingen, Heidelberg.

Proceedings references:

Bales, N.K. (1979). Minimum Freeboard Requirements for Dry Foredecks. A Design Procedure, SNAME Spring Meeting, Houston, Texas.

Cummins, W.E. (1973). Pathologies of the Transfer Functions, The Seakeeping Symposium Commemorating the 20th Anniversay of the St. Denis-Pierson Paper.

Dalzell, J.F. (1962). Some Further Experiments on the Application of Linear Superposition Techniques to the Response of a Destroyer Model in Extreme Irregular Long-Crested Head Seas, Report 918, Stevens Institute of Technology, Hoboken, New Jersey.

Frank, W. (1967). Oscillation of Cylinders in or below the Free Surface of Deep Fluids, NSRDC Report 2375.

Grim, O. (1960). Eine Methode fuer eine genauere Berechnung der Tauch-und Stampfbewegungen in glattem Wasser und in Wellen, Berecht, Nr. 1217 Hamburgische Schiffbau Versuchs Anstaldt.

Grim, O. und Schenzle, P. (1968). Berechnung der Torsionsbelastung eines Schiffes im Seegang, Forschungscentrum des Deutschen Schiffbaus, Bericht Nr. 5.

Haskind, M.D. (1957). The Exciting Forces and Wetting of Ships in Waves, DTMB 307 Translation by J.N. Newman in November 1962.

Inoue, Y. and Ali, M.T. (2003). A Numerical Investigation on the Behaviour of Multiple Floating Bodies of Arbitrary Arrangements in Regular Waves, Proc 13th Int. Offshore and Polar Eng. Conf. Vol. 3, pp. 558–565.

Kim, C.H. (1972). The Hydrodynamic Interaction Between Two Cylindrical Bodies Floating in Beam Seas, Report SIT-OE-72-10, Stevens Institute of Technology, Hoboken, New Jersey.

Kim, C.H. and Mercier, J.A. (1972). Analysis of Multiple Float-Supported Platforms in Waves, Ninth Symposium on Naval Hydrodynamics, Vol. 1, Unconventional Ships Ocean Engineering, pp.793–956.

Kim, C.H. (1982). Hydrodynamic Loads on the Hull Surface of Seagoing Vessel, SNAME Spring Meeting /STAR Symposium, Honolulu, Hawaii.

Kim, C.H. and Fang, M.C. (1985). Vertical Relative Motion Between Two Adjacent Platforms in Oblique Waves, Proc. 4th ASME Int. Offshore Mechanics and Arctic Eng. Symp., Vol. 1, pp. 114–124.

Kodan, N. (1984). The Motions of Adjacent Floating Structures in Oblique Waves, Proc. 3rd Offshore Mechanics and Arctic Engineering, New Orleans, Vol. 1, pp. 206–213.

Korsmeyer, F.T. (1983). T-ACS Offloading Mission Model Tests, Report SIT-DL-83-9-2333, Stevens Institute of Technology, Hoboken, N.J.

Korsmeyer, F.T., Lee, C.H., Newman, J.N. and Sclavounous, P.D. (1988). The Analysis of Wave Effects on Tension-Leg Platform, Proc. Offshore Mech. Arctic Eng., Vol. 2, pp. 1–15.

Loken, A.E. (1981). Hydrodynamic Interaction between Several Floating Bodies of Arbitrary Form in Waves, Proc. Int. Symp. on Hydrodynamics in Ocean Engineering, NIT, Trondheim, Vol. 1, pp. 284-306.

Newman, J.N. (1962). Translation of The Exciting Forces and Wetting of Ships in Waves by Haskind , M.D. (1957), DTMB Translation No. 307 .

Newman, J.N. (1962). The Exciting Forces on Fixed Bodies in Waves, J. Ship Res., Vol. 6, No. 3.

Newman, J.N. (1965). The Exciting Forces on Moving Body in Waves, J. Ship Res., Vol. 9, No. 3.

Noblesse, F. and Yang, C. (2005). Wave Potential in Representation of 3D Flow About a Ship Advancing Through Regular Waves in Deep Water, Proc. 15th Int. Offshore and Polar Eng. Conf., Vol. 1, pp. 54–59.

Noblesse, F., Espinosa, R., Yang, C. and Loener, R. (2005). Dispersion Curves for Diffraction-Radiation by a Ship Advancing Through Regular Waves in Finite Water Depth, Proc 15th Int. Offshore and Polar Eng. Conf., Vol. 1, pp. 61–67.

Ohkusu, M. (1976). Ship Motions in Vicinity of Structures, Proc. Int. Conf. Behaviour of Off-Shore Structures, Vol. 1, pp. 284–306.

Tasai, F. and Takaki, M. (1969). Theory and Calculation of Ship Responses in Regular Waves, Symposium on Seaworthiness of Ships, Society of Naval Architects. of Japan.

Van Oortmerssen, G. (1979). Hydrodynamic Interaction between Two Structures Floating in Waves, Proc. Int. Conf. Behavior of Offshore Structures, London, pp. 339–356.

Division III

Volterra Quadratic Model

Chapter 7

Volterra Quadratic Model and
Cross-Bi-Spectrum

7.1 Volterra Quadratic Model

The lowest order Volterra model in Chap. 3 deals with the Gaussian input-output of a linear system that can predict ship's response to the Gaussian seas (St. Denis and Pierson, 1953). The response energy density spectrum is estimated using linear system characteristics called RAO (or LTF).

The Volterra model can be applied both steady and unsteady problems stemming from the randomly varying wave excitation.

The quadratic model has been found very useful for the nonlinear analysis of ships in weakly nonlinear random seas. A further extension to cubic model will be briefly reviewed in Chap. 10.

Bendat (1990) explained the Volterra quadratic model as of an uncorrelated linear and quadratic system as shown in Fig. 7.1.

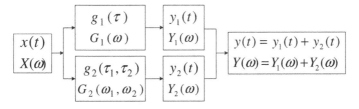

Fig. 7.1. Single input/single output model with parallel linear and quadratic system where the output are uncorrelated.

251

Barret (1963) gives the general input output theory using infinite functional series that expresses the weakly nonlinear response. Hasselmann (1966) proposed the quadratic model for statistical analysis of 2nd-order nonlinear ship motion in the 2nd-order nonlinear irregular waves. The formula has the same form as the Volterra quadratic model, though the explanation about its origin is unclear. Vassilopoulos (1967) presented the application of the Volterra quadratic model for ship motion in random seas. Dalzell (1974; 1976) developed the Volterra quadratic model for practical application by conducting a model test for added resistance in head seas in the wave tank. The Volterra quadratic model's basic assumption is that the input is the zero-mean Gaussian. Since the real input is non-Gaussian, Kim and Powers (1988) proposed recently non-Gaussian input Volterra quadratic model.

7.1.1 Volterra quadratic model in time domain

The Volterra quadratic model represents a parallel linear and quadratic system given by:

$$
\begin{aligned}
y(t) = &\int_{-\infty}^{\infty} g_1(\tau_1) x(t - \tau_1) d\tau \\
&+ \int_{-\infty}^{\infty} \int_{-\infty}^{\infty} g_2(\tau_1, \tau_2) x(t - \tau_1) x(t - \tau_2) d\tau_1 d\tau_2
\end{aligned}
\tag{7.1}
$$

Although the real input $x(t)$ cannot precisely be Gaussian, it is assumed Gaussian and zero mean, which is an important point of argument as will be discussed later in the non-Gaussian input method. The system characteristics in the time domain are given by the linear and quadratic impulse responses (LIR and QIR), $g_1(\tau_1)$ and $g_2(\tau_1, \tau_2)$, respectively. The LIR $g_1(\tau_1)$ was reviewed in Sec. 3.1.5. These kernels are time invariant because they are the functions of the lags τ_1 and τ_2. The first integral represents LIR convolved with a single input and QIR is doubly convolved with two inputs. The linear wave excites the linear system $g_1(\tau_1)$ and (linear wave) $\times$ (linear wave) excites the quadratic system $g_2(\tau_1, \tau_2)$.

The Volterra quadratic model Eq. (7.1) in terms of impulse response functions can be transformed to the response in terms of frequency response functions by applying the Fourier transforms of convention A, in Sec. 2.1.5. In Chap. 3, we derived the linear Fourier transform pairs, as shown below:

$$g_1(\tau) = \frac{1}{2\pi} \int_{-\infty}^{\infty} G_1(\omega) e^{+i\omega\tau} d\omega \tag{7.2}$$

$$G_1(\omega) = \int_{-\infty}^{\infty} g_1(\tau) e^{-i\omega\tau} d\tau \tag{7.3}$$

Similarly we derive the Fourier transform pair between the QIRF $g_2(\tau_1, \tau_2)$ and QFRF (or QTF) G_2 (quadratic frequency response function):

$$g_2(\tau_1, \tau_2) = \frac{1}{(2\pi)^2} \int_{-\infty}^{\infty} \int_{-\infty}^{\infty} G_2(\omega_1, \omega_2) e^{i\omega_1\tau_1 + i\omega_2\tau_2} d\omega_1 d\omega_2 \tag{7.4}$$

$$G_2(\omega_1, \omega_2) = \int_{-\infty}^{\infty} \int_{-\infty}^{\infty} g_2(\tau_1, \tau_2) e^{-i\omega_1\tau_1 - i\omega_2\tau_2} d\tau_1 d\tau_2 \tag{7.5}$$

where $g_2(t_1, t_2)$ is assumed symmetrical in its arguments,

$$g_2(\tau_1, \tau_2) = g_2(\tau_2, \tau_1) \tag{7.6}$$

Thus from the Fourier transform pairs, symmetry relations of G_2 hold in the forms:

$$\begin{aligned} G_2(\omega_1, \omega_2) &= G_2(\omega_2, \omega_1) \\ G_2^*(\omega_1, \omega_2) &= G_2(-\omega_1, -\omega_2) = G_2(-\omega_2, -\omega_1) \end{aligned} \tag{7.7}$$

where $*$ denotes the complex conjugate. The above symmetry relations are illustrated in bi-frequency domain as shown in Fig. 7.2.

7.1.2 Volterra quadratic model in frequency responses

The Volterra quadratic model Eq. (7.1) is given in terms of impulse response functions with assumption of Gaussian input. The same can be expressed in terms of frequency response functions after some manipulations using Eqs. (7.1) through (7.7), as shown by Hasselmann (1966) and Dalzell (1976). Consider a bichromatic wave (or dual wave), i.e., sum of two mono-chromatic waves:

$$x(t) = \text{Re}\left[\sum_{j=1}^{2} A_j e^{i(\omega_j t - \varepsilon_j)} \right] \tag{7.8}$$

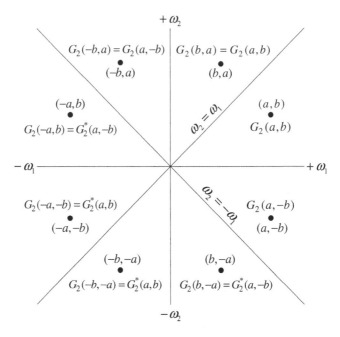

Fig. 7.2. Distribution of QTFs in bi-frequency domain showing the symmetries of QTFs.

where $A_j = A(\omega_j)$ and ε_j are the amplitude and random phase of a monochromatic wave, respectively. Eq. (7.8) may also be written in the form:

$$x(t) = \sum_{j=1}^{2} A_j \cos(\omega_j t - \varepsilon_j) = \frac{1}{2} \sum_{j=1}^{2} A_j \left\{ e^{i(\omega_j t - \varepsilon_j)} + e^{-i(\omega_j t - \varepsilon_j)} \right\} \qquad (7.9)$$

Substituting Eq. (7.9) into Eq. (7.1), and applying Fourier transforms Eqs. (7.3) and (7.5) with symmetry relations Eq. (7.7), we obtain the output in terms of LFRFs and QFRFs. Using electrical engineering terminology, the response functions are frequently called linear and quadratic transfer functions LTFs (G_1) and QTFs (G_2):

$$y(t) = \mathrm{Re}\left\{ A_1 G_1(\omega_1)e^{i(\omega_1 t - \varepsilon_1)} + A_2 G_1(\omega_2)e^{i(\omega_2 t - \varepsilon_2)} \right\}$$

$$+ \frac{1}{2}\left\{ A_1^2 G_2(\omega_1, -\omega_1) + A_2^2 G_2(\omega_2, -\omega_2) \right\}$$

$$+ \frac{1}{2}\mathrm{Re}\left\{ A_1^2 G_2(\omega_1, \omega_1)e^{i(2\omega_1 t - 2\varepsilon_1)} + A_2^2 G_2(\omega_2, \omega_2)e^{i(2\omega_2 t - 2\varepsilon_2)} \right\}$$

$$+ \mathrm{Re}\left\{ A_1 A_2 G_2(\omega_1, \omega_2)e^{i[(\omega_1 + \omega_2)t - (\varepsilon_1 + \varepsilon_2)]} \right\}$$

$$+ \mathrm{Re}\left\{ A_1 A_2 G_2(\omega_1, -\omega_2)e^{i[(\omega_1 - \omega_2)t - (\varepsilon_1 - \varepsilon_2)]} \right\} \tag{7.10}$$

The response Eq. (7.10) may be given in a compact form, for a practical application:

$$y(t) = y_1(t) + y_2(t) = \mathrm{Re}\sum_{j=1}^{2} A_j G_1(\omega_j)e^{i(\omega_j t - \varepsilon_j)}$$

$$+ \mathrm{Re}\frac{1}{2}\sum_{j=1}^{2}\sum_{k=1}^{2} A_j A_k \left\{ \begin{array}{l} G_2^+(\omega_j, \omega_k)e^{i[(\omega_j + \omega_k)t - (\varepsilon_j + \varepsilon_k)]} \\ + G_2^-(\omega_j, -\omega_k)e^{i[(\omega_j - \omega_k)t - (\varepsilon_j - \varepsilon_k)]} \end{array} \right\}$$

or $\hspace{10cm}$ (7.11)

$$y(t) = y_1(t) + y_2(t) = \mathrm{Re}\sum_{j=1}^{2} A_j G_1(\omega_j)e^{i(\omega_j t - \varepsilon_j)}$$

$$+ \mathrm{Re}\frac{1}{2}\sum_{j=1}^{2}\sum_{k=1}^{2} A_j A_k G_{2jk}^{\pm} e^{i[(\omega_j \pm \omega_k)t - (\varepsilon_j \pm \varepsilon_k)]}$$

The QTFs in Eq. (7.11) are represented in terms of sum and difference frequency groups. The terminologies for sum and difference frequency have frequently been used in two ways. In the first way, the sum frequency group consists of the double and sum frequency terms, while the difference frequency group consists of the mean and difference frequency terms. The second way has the sum frequency without double frequency terms and difference frequency without mean terms.

It is noted that the symmetry relations make the response Eq. (7.10) in the closed form. The output has a total of eight frequency responses to the bichromatic (dual) wave excitation. The first two terms are the linear responses, while the rest of six QTFs consist of: two mean terms as functions of each wave frequency, two double frequency terms, and the sum and difference frequency terms. Elimination of the redundant terms due to the symmetry of G_2 in Fig. 7.2 leads to the distribution in the two octants as shown in Fig. 7.3. The octant of bi-frequency plane above the positive ω_1-axis corresponds to the portion of the quadratic frequency response that defines sum-frequency interactions, and the octant below the positive ω_1-axis corresponds to the portion of the function that defines difference-frequency interactions. We may add in the figure the distribution of LTFs along the monochromatic-frequency ω_1-axis. The mean and double frequency terms are distributed along the limiting lines: $\omega_1 = -\omega_2$ and $\omega_1 = \omega_2$. The terms on these two limiting lines are created when the bichromatic waves become identical. Since QTFs is a continuous function of bi-frequency, the low frequency response tends toward the approximate model derived by Newman (1974).

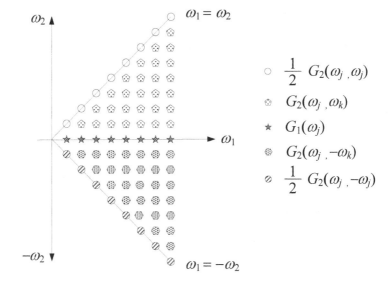

Fig. 7.3. Schematic diagram of LTFs and QTFs distributed in bi-frequency domain.

7.1.3 Difference and sum frequency coordinate system

We introduce the frequency coordinate transformation from ω_1, ω_2-plane into a difference- and sum-frequency Ω_1, Ω_2-plane, in accordance with:

$$\Omega_1 = \omega_1 - \omega_2, \quad \Omega_2 = \omega_1 + \omega_2 \qquad (7.12)$$

the inverse of which are given by:

$$\omega_1 = \frac{\Omega_1 + \Omega_2}{2}, \quad \omega_2 = \frac{\Omega_2 - \Omega_1}{2} \qquad (7.13)$$

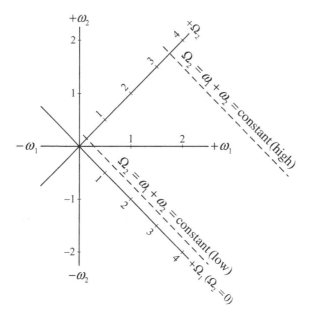

Fig. 7.4. Schematic diagram of difference- and sum-frequency plane with output frequency Ω_2.

Ω_1-and Ω_2-axis are coincident with the symmetry lines of the function $G_2(\omega_1, \omega_2)$. The quadrant of Ω_1, Ω_2-plane, in which both difference- and sum-frequencies are positive, corresponds to the quadrant selected earlier in Fig. 7.3 in interpreting the response to the bichromatic wave excitation.

In accordance to the discussion of bichromatic wave excitation, the sum-frequency Ω_2 is identical to output frequency of the quadratic system. Any combination of input frequencies $\omega_1 + |\omega_2|$ on the line $\Omega_2 = \omega_1 + \omega_2 =$ constant will generate a component at the frequency. The output frequency of QTFs along Ω_1-axis is zero. The QTFs distributed in the domain bounded by the dotted line and Ω_1-axis will produce dominantly a slowly-varying response if the output frequency $\Omega_2 =$ constant is low. The above case appears in the slow drift response of offshore structures. If the output frequency $\Omega_2 =$ constant is high, the response force in the high frequency region above $\Omega_2 =$ constant may be applied to the analysis of high frequency response such as springing and ringing of TLP.

7.1.4 Slowly-varying surge drift motion of a barge

The slowly-varying surge motion of a soft moored barge (LBT = 75.8 m, 29.6 m, 8 m) was measured in a bichromatic wave train, and the eight frequency wave excitations and eight frequency responses were observed in the data (Krafft and Kim, 1991).

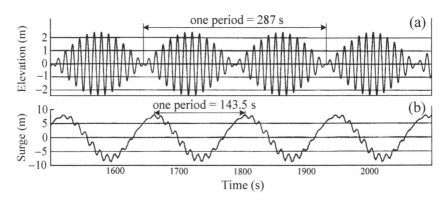

Fig. 7.5(a) Bichromatic wave. (b) Surging motion in bichromatic wave.

It has used a model of length scale 1/100 in the water depth of 0.8 m of wave tank. The measured responses are clearly identified by the Volterra quadratic model Eq. (7.10). The bichromatic wave consists of two waves of amplitude $A_1 = 1.2$ m at frequency $\omega_1 = 0.572$ rad/s and amplitude $A_2 = 1.2$ m at frequency $\omega_2 = 0.528$ rad/s, respectively. It was so designed to make the quadratic difference frequency $\Omega_1 = 0.044$ rad / s (period = 143.5 s) be equal to the resonance frequency of the moored barge. As shown in Figs. 7.5(a) and (b), the bichromatic wave produces the carrier wave of period 2π / $[(\omega_1+\omega_2)/2] = 11.4$ s and the amplitude modulated wave of period 4π / $(\omega_1-\omega_2) = 287$ s. Since the carrier wave frequency is far less than the resonance frequency, we can be certain that the slowly varying surge is not due to the linear carrier wave. In Fig. 7.5(b) we observe that the period of the large surge is identical to the resonance period 143.5 s while the modulated linear wave period is 287 s. Thus, we can be certain that the large surge is not due to the modulated wave but the 2nd-order difference frequency wave of the dual wave. The rapidly varying small amplitude (1st-order) is riding over the large surge (2nd-order). Thus, it is noted that these linear carrier and modulated waves cannot contribute to the large slowly-varying surging motion.

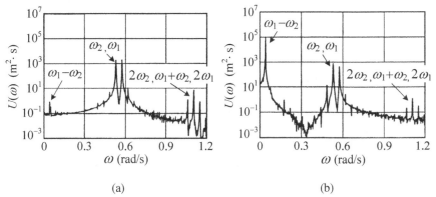

(a)   (b)

Fig. 7.6(a) Wave energy spectrum $S(\omega)$ is one-sided. (b) Surging energy spectrum $S(\omega)$ is one-sided.

Now we observe the energy density spectra of the wave and surging motion in Figs. 7.6(a) and (b). In each figure, there are two linear wave terms and six quadratic terms, i.e., two double frequency, sum and difference frequency and two zero frequency responses. It is noted that the zero frequency responses of the waves are invisible in Fig. 7.6(a) while the surge motion energy spectrum at the zero frequency is visible in Fig. 7.6(b). All of these responses are precisely identical to those quadratic terms in Eq. (7.10).

We can readily see that the difference frequency term plays a role to produce large drift motion. Though the difference frequency 2nd-order wave is negligibly small, the surge amplitude is very large because the difference frequency coincides with the resonance frequency.

We emphasize that the large surge oscillation is not due to the linear large modulated wave but by the 2nd-order difference frequency wave. And the peak of slow-drift motion is not at the peak of the amplitude modulated wave.

As we have observed in the above, the quadratic nonlinear interaction of the bichromatic wave must have produced the 2nd-order difference frequency wave-exciting force for the surge at the resonance frequency. It is clear that the quadratic difference frequency surge-exciting force, though was not measured, had affected the difference frequency surge motion. Basically the 2nd-order wave, 2nd-order wave force and 2nd-order motion must all be present simultaneously. This contradicts to the assumption of the Gaussian input in the Volterra quadratic model. However, if the sea state is low, the assumption would be acceptable because the low sea state behaves like a linear Gaussian random wave.

7.1.5 Longitudinal drift force of a tanker model

Here we observe a different kind of 2nd-order low frequency response from the foregoing observation of soft moored surging motion. The large surging motion was due to the difference frequency being identical to the resonance frequency. We observe here the low frequency 2nd-order surge-drift force of a moored tanker in the regular wave group as shown in Figs. 7.7(a) and (b), respectively. The peak surge-drift force occurs at the peak wave group. Further, it is to be noted that the peak surging motion of the barge does not occur at the peak of liner regular wave, but at the point where the difference frequency coincides with the resonance frequency.

Special attention has to be given in measuring the low frequency 2nd-order force (Pinkster, 1980). When measuring the low frequency 2nd-order force as the model undergoes motions with frequencies that coincide with the frequencies of the 2nd-order forces, the constraint of designed system must allow the model to move freely at wave frequencies while at the same time 2nd-order low frequency motions corresponding to the low frequency wave drift force must be fully suppressed.

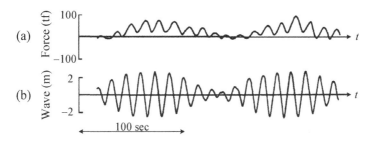

Fig. 7.7(a) Longitudinal drift force of a tanker in regular wave group. (b) Regular wave group.

It is certain that the rapidly varying force riding over the slowly varying force is due to the linear carrier wave. The drift force evidently consists of the mean and fluctuating component in the time series in Fig. 7.7(a).

It is of primary interest here to find the reason why the largest drift force occurs in the largest wave group. The same behavior will be found in the simulation of slowly-varying drift force of a ship in random seas in Chap. 8. It is postulated that the peak of 2nd-order force should occur at the peak of the 2nd-order wave which occurs at the peak of the linear wave group.

7.2 Statistical Estimates of Second-Order Response

The statistical properties of the 2nd-order response of the Volterra quadratic model to the Gaussian excitation will be derived. The input is Gaussian zero-mean and Fourier transforms are in accordance to convention A as defined in Sec. 2.1.5.

7.2.1 One- and two-sided energy density spectra

Assuming that the excitation $x(t)$ is a stationary, Gaussian, zero mean process, we form the autocorrelation of the process as given in Sec. 1.2.5:

$$R_{xx}(\tau) = E[x(t)x(t-\tau)] \tag{7.14}$$

Using the Fourier transform of convention A, we have:

$$S_{xx}(\omega) = \int_{-\infty}^{\infty} R_{xx}(\tau)e^{-i\omega\tau}d\tau \qquad (7.15)$$

$$R_{xx}(\tau) = \frac{1}{2\pi}\int_{-\infty}^{\infty} S_{xx}(\omega)e^{i\omega\tau}d\omega \qquad (7.16)$$

The energy density spectrum is an even function of ω:

$$R_{xx}(0) = \frac{1}{\pi}\int_{0}^{\infty} S_{xx}(\omega)d\omega \qquad (7.17)$$

and the variance in Eq. (7.17) is also determined by integration of one-sided energy density spectrum U_{xx}:

$$R_{xx}(0) = \int_{0}^{\infty} U_{xx}(\omega)d\omega \qquad (7.18)$$

Hence, one has:

$$S_{xx}(\omega) = \pi U_{xx}(\omega) \qquad (7.19)$$

When we apply the Fourier transform convention A in the Volterra quadratic model, it is necessary to scale the two-sided auto-energy density spectrum to a one-sided one in accordance to Eq. (7.19). If we apply convention B, one-sided spectrum will be the double of two-sided spectrum as shown in Chap. 1.

7.2.2 Parseval's formula

The Parseval's formula for a product of two functions of one variable was employed to determine the cross-spectrum in Chap. 2. The n-dimensional form of Parseval's formula will be employed here to manipulate higher order random processes according to Barrett (1963):

$$\int_{-\infty}^{\infty}\int_{-\infty}^{\infty}\cdots\int_{-\infty}^{\infty} f_1(\tau_1,\tau_2,\cdots,\tau_n)f_2(\tau_1,\tau_2,\cdots,\tau_n)d\tau_1 d\tau_2 \cdots d\tau_n$$

$$= \frac{1}{(2\pi)^n}\int_{-\infty}^{\infty}\int_{-\infty}^{\infty}\cdots\int_{-\infty}^{\infty} F_1^*(\omega_1,\omega_2,\cdots,\omega_n) \qquad (7.20)$$

$$\cdot F_2(\omega_1,\omega_2,\cdots,\omega_n)d\omega_1 d\omega_2 \cdots d\omega_n$$

where the (*) denotes complex conjugate and $f_i\,(\tau_1,\cdots)$ and $F_i\,(\omega_1,\cdots)$ are pairs of Fourier transform of convention A as defined in Sec. 2.1.5:

$$F_j(\omega_1,\omega_2,\cdots,\omega_n) = \int_{-\infty}^{\infty}\int_{-\infty}^{\infty}\cdots\int_{-\infty}^{\infty} f_j(\tau_1,\tau_2,\cdots,\tau_n)e^{\left(-i\sum_{r=1}^{n}\omega_r\tau_r\right)}d\tau_1 d\tau_2\cdots d\tau_n$$

$$f_j(\tau_1,\tau_2,\cdots,\tau_n) = \frac{1}{(2\pi)^n}\int_{-\infty}^{\infty}\int_{-\infty}^{\infty}\cdots\int_{-\infty}^{\infty} F_j(\omega_1,\omega_2,\cdots,\omega_n)e^{\left(i\sum_{r=1}^{n}\omega_r\tau_r\right)}d\omega_1 d\omega_2\cdots d\omega_n$$

$$(7.21)$$

7.2.3 Expected value of quadratic response

The expected value of the total output $y(t)$ in Eq. (7.1) may be written:

$$E[y(t)] = \int_{-\infty}^{\infty} g_1(\tau_1)E[x(t-\tau_1)]d\tau_1$$
$$+ \int_{-\infty}^{\infty}\int_{-\infty}^{\infty} g_2(\tau_1,\tau_2)E[x(t-\tau_1)x(t-\tau_2)]d\tau_1 d\tau_2$$

$$(7.22)$$

The first term in Eq. (7.22) is zero because the random input wave is zero mean Gaussian. Hence, the expected value of the total response comes from the quadratic response alone:

$$E[y(t)] = \int_{-\infty}^{\infty}\int_{-\infty}^{\infty} g_2(\tau_1,\tau_2)R_{xx}(\tau_1-\tau_2)d\tau_1 d\tau_2 \qquad (7.23)$$

To apply Parseval's formula to Eq. (7.23), we let:

$$f_1(\tau_1,\tau_2) = g_2(\tau_1,\tau_2), \ \ f_2(\tau_1,\tau_2) = R_{xx}(\tau_1-\tau_2) \qquad (7.24)$$

Then one obtains:

$$F_1(\omega_1,\omega_2) = G_2(\omega_1,\omega_2) \qquad (7.25)$$

And

$$F_2(\omega_1,\omega_2) = \int_{-\infty}^{\infty}\int_{-\infty}^{\infty} R_{xx}(\tau_1-\tau_2)e^{-i\omega_1\tau_1-i\omega_2\tau_2}d\tau_1 d\tau_2 \qquad (7.26)$$

Multiplying $\exp[i\omega_1\tau_2 - i\omega_1\tau_2]$ to Eq. (7.26) and applying Fourier transform of convention A to the autocorrelation function R_{xx}, we obtain energy density spectrum multiplied by an expression that is given in terms of Dirac delta function defined in Sec. 2.1.6.2, as shown:

$$F_2(\omega_1,\omega_2) = S_{xx}(\omega_1) \int_{-\infty}^{\infty} e^{-i\tau_2(\omega_1+\omega_2)} d\tau_2 = S_{xx}(\omega_1)(2\pi)\delta(\omega_1+\omega_2) \qquad (7.27)$$

Substituting Eqs. (7.24), (7.25) and (7.27) into Eq. (7.20), we find that Eq. (7.23) takes the following form:

$$E[y(t)] = \frac{1}{2\pi} \int_{-\infty}^{\infty} \int_{-\infty}^{\infty} G_2^*(\omega_1,\omega_2) S_{xx}(\omega_1)\delta(\omega_1+\omega_2) d\omega_1 d\omega_2 \qquad (7.28)$$

Integrating with respect to ω_2 and scaling the two-sided to one-sided spectrum according to Eq. (7.19), we have:

$$\begin{aligned} E[y(t)] &= \frac{1}{2\pi} \int_{-\infty}^{\infty} G_2(\omega,-\omega) S_{xx}(\omega) d\omega \\ &= \frac{1}{2} \int_{-\infty}^{\infty} G_2(\omega,-\omega) U_{xx}(|\omega|) d\omega \end{aligned} \qquad (7.29)$$

and since the integrand is even function of ω:

$$E[y(t)] = \int_0^{\infty} G_2(\omega,-\omega) U_{xx}(\omega) d\omega \qquad (7.30)$$

Equation (7.30) is the basic formula for estimation of the expected value of the response of quadratic system to Gaussian random seas that was given by Maruo (1957). It is noted that the expected value is the mean of a quantity sampled over a long time as defined in Sec. 1.2.1. If the estimated mean is from a finite sample as usual, the expected value should vary according to the sample length.

7.2.4 Energy density spectrum of quadratic response

The energy density spectrum of the total output is determined by applying Fourier transform of convention A to the autocorrelation of the output $y(t)$. Applying the Parseval's formula as given in the preceding section, we obtain the 2nd moment or autocorrelation of the output about zero-mean in the form:

$$M_{yy}(\tau) = E\left[\{y(t) - E[y(t)]\}\{y(t+\tau) - E[y(t+\tau)]\}\right]$$

$$= \frac{1}{2\pi}\int_{-\infty}^{\infty}|G_1(\omega)|^2 S_{xx}(\omega)e^{i\omega\tau}d\omega \qquad (7.31)$$

$$+\frac{2}{(2\pi)^2}\int_{-\infty}^{\infty}\int_{-\infty}^{\infty}|G_2(\omega_1,\omega_2)|^2 S_{xx}(\omega_1)S_{xx}(\omega_2)e^{i(\omega_1+\omega_2)\tau}d\omega_1 d\omega_2$$

Letting:

$\omega_1 = \omega - \xi$; $\omega_2 = \xi$, in the second term, we have:

$$M_{yy}(\tau) = \frac{1}{2\pi}\int_{-\infty}^{\infty}|G_1(\omega)|^2 S_{xx}(\omega)e^{i\omega\tau}d\omega$$

$$+\frac{1}{2\pi^2}\int_{-\infty}^{\infty}\int_{-\infty}^{\infty}|G_2(\omega-\xi,\xi)|^2 S_{xx}(\omega-\xi)S_{xx}(\xi)d\xi e^{i\omega\tau}d\omega \qquad (7.32)$$

Performing the Fourier transform of convention A on each term of Eq. (7.32), we obtain the two-sided output energy spectrum:

$$S_{yy}(\omega) = |G_1(\omega)|^2 S_{xx}(\omega)$$

$$+\frac{1}{\pi}\int_{-\infty}^{\infty}|G_2(\omega-\xi,\xi)|^2 S_{xx}(\omega-\xi)S_{xx}(\xi)d\xi \qquad (7.33)$$

Scaling the two-sided spectrum to one-sided one according to Eq. (7.19), and using transform of variables $\xi = (\omega + u)/2$ we obtain:

$$U_{yy}(\omega) = |G_1(\omega)|^2 U_{xx}(\omega)$$

$$+\int_0^{\infty}\left|G_2\left\{\left(\frac{\omega-u}{2},\frac{\omega+u}{2}\right)\right\}\right|^2 U_{xx}\left(\frac{|\omega-u|}{2}\right)U_{xx}\left(\frac{|\omega+u|}{2}\right)du \qquad (7.34)$$

And since $du = 2d\xi$:

$$U_{yy}(\omega) = |G_1(\omega)|^2 U_{xx}(\omega)$$

$$+2\int_0^{\infty}|G_2(\omega-\xi,\xi)|^2 U_{xx}(|\omega-\xi|)U_{xx}(|\xi|)d\xi \qquad (7.35)$$

Equation (7.35) represents one-sided response energy spectrum to Gaussian excitation.

7.2.5 Cross-bi-spectrum

The cross correlation between input and output or the third moment is formed as:

$$M_{xxy}(\tau_1, \tau_2) = E[x(t+\tau_2)x(t+\tau_1)(y(t) - E[y(t)])] \tag{7.36}$$

The double Fourier transform of the above cross correlation is called the cross-bi-spectrum (CBS) (Tick, 1961):

$$CBS(\omega_1, \omega_2) = \frac{1}{(2\pi)^2} \int_{-\infty}^{\infty} \int_{-\infty}^{\infty} e^{-i(\tau_1\omega_1 + \tau_2\omega_2)} M_{xxy}(\tau_1, \tau_2) d\tau_1 d\tau_2 \tag{7.37}$$

Equations (7.36) and (7.37) are the FT pair based on convention B. We modify the third moment in the following form and apply FT of convention A (Dalzell, 1976):

$$M_{xxy}(\tau_1, \tau_2) = E\left[x(t+\tau_1)x(t-\tau_1)\{y(t-\tau_2) - E[y(t-\tau_2)]\}\right] \tag{7.38}$$

Since we assume the Gaussian input, in the development of formulation, the average of single wave input and that of triple wave input are zero (Newland, 1986). Thus, Eq. (7.38) results in the form:

$$M_{xxy}(\tau_1, \tau_2) = \frac{2}{(2\pi)^2} \int_{-\infty}^{\infty} \int_{-\infty}^{\infty} G_2^*(\omega_1, \omega_2) S_{xx}(\omega_1) S_{xx}(\omega_2)$$
$$\times e^{\{i\tau_1(\omega_1-\omega_2)+i\tau_2(\omega_1+\omega_2)\}} d\omega_1 d\omega_2 \tag{7.39}$$

Let the bi-frequency ω_1 and ω_2 in the above be transformed to the difference and sum frequency according to Eqs. (7.12) and (7.13):

$$M_{xxy}(\tau_1, \tau_2) = \frac{1}{(2\pi)^2} \int_{-\infty}^{\infty} \int_{-\infty}^{\infty} G_2^*\left(\frac{\Omega_1 + \Omega_2}{2}, \frac{\Omega_1 - \Omega_2}{2}\right)$$
$$\times S_{xx}\left(\frac{\Omega_1 + \Omega_2}{2}\right) S_{xx}\left(\frac{\Omega_1 - \Omega_2}{2}\right) e^{i(\tau_1\Omega_1 + \tau_2\Omega_2)} d\Omega_1 d\Omega_2 \tag{7.40}$$

where the Jacobian $J = 1/2$ has been applied in the coordinate transform.

Thus, the modified CBS is the double Fourier transform of the third moment Eq. (7.40):

$$C(\Omega_1,\Omega_2) = \int_{-\infty}^{\infty}\int_{-\infty}^{\infty} M_{xxy}(\tau_1,\tau_2)e^{-i(\tau_1\Omega_1+\tau_2\Omega_2)}d\tau_1 d\tau_2$$

$$= G_2^*(\omega_1,\omega_2)S_{xx}(\omega_1)S_{xx}(\omega_2) \tag{7.41}$$

Hence, the QTF is obtained in the form:

$$G_2(\omega_1,\omega_2) = \frac{C^*(\omega_1-\omega_2,\omega_1+\omega_2)}{S_{xx}(\omega_1)S_{xx}(\omega_2)}, \quad -\infty \le (\omega_1,\omega_2) \le \infty \tag{7.42}$$

provided the denominator is not zero. This simple fact becomes a serious problem in estimating the CBS. It happens because the energy spectrum has very small values in the tails. Scaling the two-sided energy spectrum to one-sided one according to Eq. (7.19):

$$G_2(\omega_1,\omega_2) = \frac{C^*(\omega_1-\omega_2,\omega_1+\omega_2)}{\pi^2 U_{xx}(|\omega_1|)U_{xx}(|\omega_2|)}, \quad 0 \le (\omega_1,|\omega_2|) \le \infty \tag{7.43}$$

7.2.6 Algorithm for cross-bi-spectrum

The algorithm for estimation of the foregoing cross-bi-spectrum is similar to the one-dimensional Blackman-Tukey method for cross-and auto-spectrum. The Blackman-Tukey method is extended to 2-D form applying the Hamming smoothing in two-dimensional way. Dalzell (1974) referred to Shaman (1963) applying the Fourier transform convention A. The final result:

$$\hat{\hat{C}}(\Omega_1,\Omega_2) = \left(\frac{\Delta t}{e_1+e_2}\right)^2 \cdot$$

$$\sum_{j=0}^{m}\left\{\begin{array}{l} r(j)(e_1+e_2\cos\pi j/m)\cos(\pi P_1 j/m)\cdot \\ \left[\sum_{k=-n}^{n}\begin{array}{l}(e_1+e_2\cos\pi k/n)\cdot \\ [\cos(\pi P_2 k/n)+i\sin(\pi P_2 k/n)]\cdot \\ \dfrac{1}{N_S}\sum_N x(N+j)x(N-j)y(N+k)\end{array}\right]\end{array}\right\} \tag{7.44}$$

where
$\hat{C}$ = estimate of cross-bi-spectrum
$\Omega_1 = \pi P_1 / m\Delta t$, $\Omega_2 = \pi P_2 / n\Delta t$, m and n are the maximum lags in the difference frequency (Ω_1) and in the sum frequency direction (Ω_2), respectively. The maximum lag is the same as the one that was used in Blackman-Tukey Fourier transform for energy spectral densities in Chap. 2. e_1 = 0.54 and e_2 = 0.46 with a discrete version of the scalar spectrum lag window (Hamming window);
$r(j) = 1$ for $j = 0$ and $r(j) = 2$ for otherwise;
$x(N)$ = the input time series corrected to zero sample mean;
$y(N)$ = the output time series;
N_s = the number of possible products summed;
Δt = sampling interval used

7.3 Volterra Quadratic Model with Non-Gaussian Input

The laboratory random wave is in fact nonlinear and non-Gaussian disregarding the level of sea severity. It may be more inclined to be Gaussian if the severity is very low. The degree of nonlinearity depends on the significant wave height H_s or the sea severity. The laboratory random waves may be regarded approximately Gaussian if the H_s is less than 4 m, and approximately 2nd-order when the sea severity H_s is less than 9 m (Kumar and Kim, 2002). If H_s is higher than 9 m, the nonlinearity may be higher than 2nd-order, as discussed in Chap. 9. Thus the Volterra quadratic model with assumption of Gaussian input is approximately correct when the sea severity is low.

In view of the above remarks, Kim and Powers (1988) proposed the Volterra quadratic model with non-Gaussian input and validated the model using a prescribed highly non-Gaussian mathematical input and output. The LTFs and QTFs of the Volterra quadratic model are derived by Fourier transform of the Volterra quadratic model Eq. (7.1) assuming non-Gaussian input:

$$Y(f_m) = G_1(f_m)X(f_m) + \sum_{i+j=m} G_2(f_i, f_j)X(f_i)X(f_j) \qquad (7.45)$$

where $X(f_m)$ and $Y(f_m)$ are Fourier transforms of input and output and $G_1(f_m)$ and $G_2(f_i, f_j)$ are LTFs and QTFs, respectively. Multiplying X and XX to Eq. (7.45):

$$\begin{bmatrix} E[X^2] & E[X^*XX] \\ E[X^*X^*X] & E[X^*X^*XX] \end{bmatrix}\begin{bmatrix} [G_1] \\ [G_2] \end{bmatrix} = \begin{bmatrix} E[X^*Y] \\ E[X^*X^*Y] \end{bmatrix} \tag{7.46}$$

$E[\ ]$ stands for the expected value ensemble average over the quantities inside the brackets (Bendat,1990), from which LTFs and QTFs are derived in the form:

$$\begin{bmatrix} [G_1] \\ [G_2] \end{bmatrix} = \begin{bmatrix} E[X^2] & E[X^*XX] \\ E[X^*X^*X] & E[X^*X^*XX] \end{bmatrix}^{-1}\begin{bmatrix} E[X^*Y] \\ E[X^*X^*Y] \end{bmatrix} \tag{7.47}$$

If the input is Gaussian the off-diagonal terms will disappear. Hence the solution for Gaussian input assumption becomes:

$$G_1 = \frac{E[X^*Y]}{E[X^2]}, \quad G_2 = \frac{E[X^*X^*Y]}{E[X^*X^*XX]} \tag{7.48}$$

which is identical to the LTFs and QTFs in Gaussian input model Eq. (7.43). If the input is assumed to be non-Gaussian, the off-diagonal terms will remain.

It should be noted that the method does not provide the mean terms along the axis $f_i = -f_j$. The details of the formulas are referred to Kim and Powers (1988) and Kim and Powers (1991; 1995).

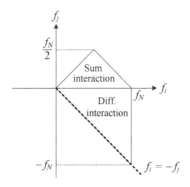

Fig. 7.8. Sketch of distribution of difference and sum frequency in the bi-frequency domain, f_N = Nyquist frequency. (Kim and Powers, 1988).

7.3.1 Gaussian and non-Gaussian method

We repeat the definitions of the Gaussian and non-Gaussian method that will be frequently referred to in the following discussions. That is, the Gaussian method assumes Gaussian input and employs the Blackman Tukey method for the spectral estimates, whereas, the non-Gaussian method assumes non-Gaussian input and employs the ensemble average method (or direct method) for the spectral estimates.

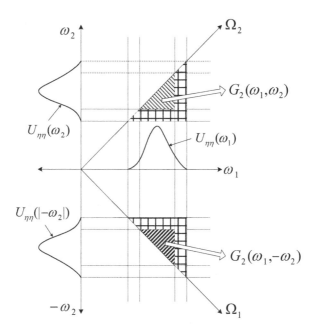

Fig. 7.9. Schematic diagram of dividing the CBS by wave energy spetra with small tails causing infinitely large magnitude.

As found in Eqs. (7.43), (7.47), and (7.48), both Gaussian and non-Gaussian methods have the same difficulty in estimating the CBS because the denominator in the equations has very small values in the tails of energy density spectra producing infinitely large values of QTFs. This situation is illustrated in Fig. 7.9. In order to avoid the uncertainty in the estimation, Kim and Kim (2004) proposed a nominal rule to take the spectral densities at the tails to be 10% of the peak energy spectral density. This truncation will lose a formidable amount of data so that the magnitudes of estimation will be less than the measured data, which is a fundamental deficit that may not be easily avoided.

7.3.2 Routines for analyses of quadratic estimates

We estimate the LTFs and QTFs of a system using the Gaussian and non-Gaussian methods. These estimates called the system identification, are used to determine the response energy density spectra and coherency functions. The estimates are also used for reconstruction of the response in the time domain. The comparison of the reconstruction with the measured response will demonstrate the quality of the estimation. Further, one must eventually investigate the effects of the sample length and sea severity on the estimate responses. Dalzell (1974) recommended increasing the time for the quadratic response 10–12 times the linear sample length.

It is always necessary to examine the Gaussian and non-Gaussianness of the processes by comparing the peak value distribution with Rayleigh. One will encounter uncertainty when examining the result of the experimental analysis. In such cases, the result of hydrodynamic theory will be used for comparison with the experimental system identification.

7.3.2.1 Coherency test

We conduct the coherency between the measurement and estimation of the linear and quadratic response (system identification) in the frequency domain. The linear and quadratic coherence functions are defined by:

$$\gamma^2_{xy_1}(\omega) = \frac{U_{y_1 y_1}(\omega)}{U_{yy}(\omega)} \qquad q^2_{xy_2}(\omega) = \frac{U_{y_2 y_2}(\omega)}{U_{yy}(\omega)} \qquad (7.49)$$

where the suffixes y_1, y_2 represent the reconstructed linear and quadratic output, respectively, while y is the measured response. γ^2 and q^2 represent the linear and quadratic coherence functions respectively. $U\, y_1 y_1$, etc. indicate one-sided response spectra. The sum of these coherence functions must lie in between 0 and 1. A goodness-of-fit measure for the validity of the quadratic nonlinear model can be found by seeing how close the sum of these linear and quadratic coherence functions is to unity (Bendat, 1990).

One may reconstruct the response by multiplying the estimated LTFs and QTFs to the Fourier frequency amplitude spectrum of the measured input waves as shown in Fig. 7.10.

7.3.2.2 Reconstruction test

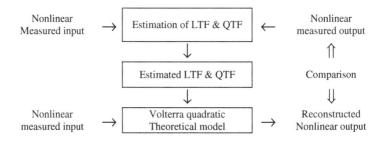

Fig. 7.10. Schematic diagram for reconstruction procedure.

One compares the reconstruction using Eq. (7.10) and measured output time series as shown in Fig. 7.10 to see the difference between them. The foregoing is used for the evaluation of errors, in terms of the normalized mean square error (NMSE) defined in the form:

$$\text{NMSE} = \sqrt{\frac{\dfrac{1}{N}\displaystyle\sum_{t=1}^{N}\left\{[y_{\text{recon}}(t)-\overline{y}_{\text{rec}}(t)]-[y_{\text{exp}}(t)-\overline{y}_{\text{exp}}(t)]\right\}^{2}}{\dfrac{1}{N}\displaystyle\sum_{t=1}^{N}[y_{\text{exp}}(t)-\overline{y}_{\text{exp}}(t)]^{2}}} \qquad (7.50)$$

where y and N denote the output and number of data, and suffixes *recon* and *exp* denote the reconstructed and measured data, respectively.

7.3.2.3 Convergence of mean slow drift response

The cumulative mean of slow drift response will indicate the convergence of the response as function of time:

$$\overline{y}_{\text{cum}}(t_{N})=\frac{1}{N}\sum_{i=1}^{N}y(t_{i}) \qquad (7.51)$$

N indicates the number of data in the response $y(t)$. Consecutively measured data of several pieces such as A, B, C, and so forth are given. Then cumulative means are made for data A and for data A + B and for data A + B + C, etc. If each data has length 1.0 hrs, the cumulative mean for A + B will be 2 hrs corresponding to the data points.

7.4 Case Studies of Quadratic Volterra Model

We will review experimental studies of quadratic system employing Volterra quadratic model with Gaussian and non-Gaussian input. In general we assume the input and output are weakly nonlinear. It is quite possible that the quality of experimental input and output from the model test is inadequate. We may check the quality of the measured wave data by simply comparing the energy density spectrum estimated from the measured data with the target spectrum and by comparing the variances of these spectra. If the difference of the variances is greater than about 10% we may question the quality of the measured data. The standard sample length used in the commercial laboratories for 2nd-order responses is three hrs, though there is strong evidence that the three hrs may be too short to guarantee the statistical convergence of slow varying responses. In this case, one may need to increase the time for the quadratic response 10–12 times the linear sample length, as recommended by Dalzell (1974). And the effects of sea severity have to be taken into consideration. Keeping these problems in mind, we will review experimental studies in laboratories.

7.4.1 Added resistance and slow drift force

Added wave resistance and slow drift responses are in the same category because they are both found in the low difference frequency domain as discussed in Sec. 7.1. The study of added resistance by Dalzell (1976) is the first ever experimental investigation that applied the Volterra quadratic model shown in Eqs. (7.1) and (7.10).

7.4.1.1 Added resistance by Gaussian method

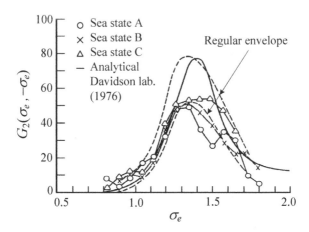

Fig. 7.11. Comparison of analytical and experimental mean added resistance operators (Dalzell, 1976).

The added resistance of a series 60 ship model running in head seas were measured with varying sea severities A, B and C corresponding to H_s of 2.7 m, 5.5 m and 10.1 m, for the ship of 152.4 m long (Dalzell, 1976; wave tank: 91.96 m × 3.66 m × 1.83 m). This work uses the Gaussian method for the estimation of QTFs and emphasizes that the mean added resistance operator is the most reliable statistical data among the many QTFs. Since the QTF is in dimension of force divided by length squared, the mean added resistance operator is defined as the normalized mean added resistance along Ω_1-axis by multiplying L^2/Δ, where L is the ship length; Δ ship's displacement.

Figure 7.11 indicates the mean resistance operator derived from the estimates of cross-bi-spectra for the given seas and the regular wave tests covering the upper and lower limits. An analytically predicted result is also added. We see here that the regular wave tests results are in good agreement with the random wave test data and the analytical result is within the regular envelopes. Thus the cross-bi-spectral estimates made by Dalzell (1976) including the analytical work to be discussed further in Chap. 8 are all validated. The random wave test is more expensive than the regular wave test. However, it is hoped that the random wave test may be used as a new tool for the research of 2nd-order responses.

We find in Fig. 7.11 that the mean added resistance operator depends on the sea severity. As the sea severity increases, the mean added resistance slightly increases, though there is some anomaly for the sea state A in the region of higher encounter frequency σ_e. The common notion of the system characteristics is that QTFs shall not change with the variation of the sea state. However, in reality the estimate of QTFs depends on the sea severity. The foregoing study does not specify the sample length in the research. However, Dalzell (1974) suggested a need of sample length 10–12 times longer than those required in linear seakeeping problems in order to have comparable accuracy. The entire result of experimental added resistance in the bi-frequency domain will be further compared with the hydrodynamic calculation by Dalzell and Kim (1979) in Chap. 8.

7.4.1.2 Slow drift by non-Gaussian method

Kim and Powers (1988) proposed non-Gaussian method for the nonlinear system identification of the slow drift responses, and demonstrated the validity of the method analyzing the prescribed mathematical input and output that is highly nonlinear. The algorithm was further improved in the foregoing method using orthogonal function approach (Kim and Powers, 1991; 1995). The goodness of fit was shown in the coherency functions. Birkelund (2002) further improved the above approach, especially in the algorithm, by applying principal component analysis with multi-tail filtering.

7.4.1.3 Deterministic method for slow drift analysis

The ultimate objective of the quadratic response research is to develop an improved technique for prediction of slow drift response due to random sea excitations. Given the 2nd-order response in random seas is sensitive to the nonlinear interaction of many combinations of wave frequencies, it was considered important to conduct a deterministic experimental investigation of the QTFs for slow drift motion of an idealized vessel/mooring system in a series of bichromatic waves. Krafft and Kim (1991) employed a linearly moored, shallow drafted barge vessel. The model was tested in the same way as described in Sec. 7.1.4 by controlling the combination of bichromatic waves to produce the resonance frequency. A variety of bichromatic waves were generated by varying the difference frequency for each fixed sum frequency, an example of which is showed in Fig. 7.12. The test data can be used to refine the cross-bi-spectral analysis. Given the QTFs one can simulate the drift motion using QIRF presented in Eq. (7.1).

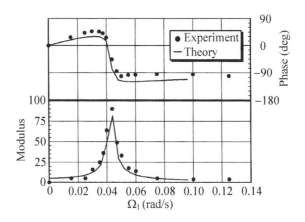

Fig. 7.12. Comparison of experimental QTFs with theory.

Similar dual wave tests were carried out by Murray et al. (1987) for the study of the effects of wave grouping on the slow drift motion of a barge model and comparison to a numerical model based on Newman's approximation (1974).

7.4.1.4 Interpretation of 2nd-order experiment

Stansberg (1997; 2001) proposed a new signal processing technique for the nonlinear hydrodynamic testing. It is considered especially important when nonlinear and complex processes in extreme sea states are investigated, where standard numerical tools often need to be checked or calibrated to experiments.

7.4.1.5 Mini TLP analysis by non-Gaussian method

Birkelund et al. (2002) analyzed the surge drift of Mini TLP model tested at Offshore Technology Research Center at Texas A & M University (Niedzwecki et al., 2001). The target spectrum was a JONSWAP significant wave height H_s = 3.9 m and modal period T_m = 16 s with peakedness factor of 2 representing 100 year return storm sea in West Africa. The analysis showed that the low frequency surge energy density distribution was negligibly small. However, the mean values and surge resonance period were not analyzed.

7.4.2 2nd-order surge-exciting force on barge

Kim and Kim (2003; 2004) experimentally investigated 2nd-order surge-exciting force on a fixed barge (LBT = 75.8 m, 29.6 m, 8m) in random head seas. The purpose of study was to examine the behavior of QTFs for surge-exciting force for a blunt shaped and fixed structure. It was thought to be remarkably different from the behaviors of response motion of conventional compliant structures to the seas. The study was also aimed at finding the effects of Gaussian and non-Gaussian method, sample length, and sea severity on the responses.

The random seas of H_s = 3 m to 9 m of P-M spectra were measured for 3 hrs to 40 hrs long and cross-bi-spectral analyses were conducted using Gaussian and non-Gaussian method.

7.4.2.1 Estimate of LTFs and QTFs of surge force

The estimate of LTFs was practically invariant with the variations of sea severity and record length. Fig. 7.13 shows the effects of low and high seas H_s = 3 m and H_s = 9 m on QTFs and Gaussian and non-Gaussian method with fixed sample length of 3 hrs.

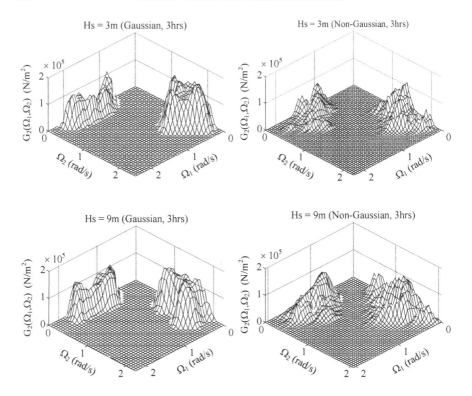

Fig. 7.13. QTFs for surge force at the seas of H_s = 3 m and 9 m, for 3 hrs sample applying Gaussian and Non-Gaussian method.

The Gaussian method gives larger QTFs than non-Gaussian method. The large sum frequency QTFs appear larger in general compared to the difference frequency QTFs. The foregoing behavior is regarded as one of the features of 2nd-oder wave force on a fixed blunt shaped body compared to the familiar conventional slow drift response of a compliant structure.

7.4.2.2 Surge force energy density

Figure 7.14 shows the surge force energy density spectrum given in semi-log scale. The low frequency energy density is distributed in the region less than 0.2 rad/s and high frequency density is mostly distributed in the region greater than 1.2 rad/s. Although the energy densities of second-order forces are small, the effects of these forces on the motion will be significant if the body is moored and thus the natural frequency falls in the ranges of low or high frequency.

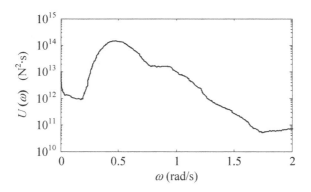

Fig. 7.14. Surge force energy density spectrum for $H_s = 9$ m.

7.4.2.3 Coherency test of surge force

The coherency functions Eq. (7.49) were evaluated and illustrated in Fig. 7.15.

At the low sea state, linear coherency function is about 1.0 in the range of frequency 0.2–1.2 rad/s. The quadratic order coherency function amounts to about 0.3 in the low frequency region 0–0.2 rad/s, while about 0.3 in the high frequency region 1.2–1.4 rad/s. The foregoing behavior implies that the 2nd-order force signal is so weak that the 2nd-order coherency is negligibly small at the low sea state.

However, at the severer sea of $H_s = 9$ m, the linear coherency function holds in the range of frequency 0.2–0.7 rad/s, and the quadratic coherency function is about 0.8 in the low frequency range of 0–0.2 rad/s and about 0.3 in the range of 0.8–1.4 rad/s. In other words, in the high seas, the coherency of 2nd-order force holds in the low frequency region, while the same is not true in the high frequency region.

The effect of sample length on the coherency function is not shown. However, the 40 hrs sample with high sea combined, gives the coherency close to unity in applying both methods.

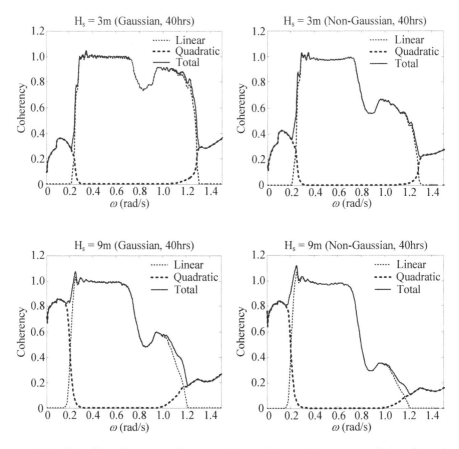

Fig. 7.15. The effect of sea severities H_s of 3 m and 9 m with same sample length on the coherency function for the surge force applying Gaussian and non-Gaussian method.

7.4.2.4 Reconstruction of surge force

Reconstructions of surge forces of 40 hrs sample in the low and high seas are obtained employing the estimated LTFs and QTFs, and the Gaussian and non-Gaussian method. As shown in Figs. 7.16(a) and (b), the reconstructions using both Gaussian and non-Gaussian methods give good correlations with the measured responses when the sea severity is low. However, the degree of agreement is slightly reduced when the sea severity is higher.

We quantize the differences by taking the normalized mean square errors. It is apparent that the errors in the high sea are higher than those at the low sea. The normalized mean square errors are affected by the sea severity rather than by sample length, while the effects of Gaussian and non-Gaussian method on the error are negligible. The estimated values of the response through the reconstruction employs the estimation of the QTFs that have to be truncated to avoid the unrealistically high values due to the tail of small energy density spectra as explained in Sec. 7.3.1.

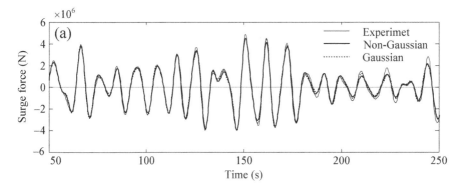

Fig. 7.16(a) Comparisons of reconstructed time series 40 hrs at $H_s = 3$ m.

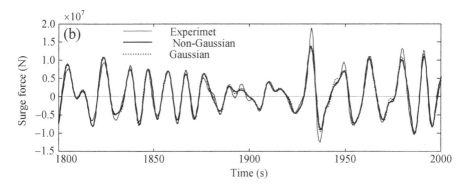

Fig. 7.16(b) Comparisons of reconstructed time series 40 hrs at $H_s = 9$ m.

7.4.2.5 Effects of sea severity on the mean surge force

The effects of sea severity and sample length on the mean surge forces are shown in Fig. 7.17. The mean QTFs slightly increase with the sea severity, which is similar to the mean added resistance as shown in Fig. 7.12.

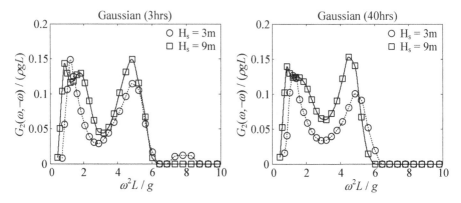

Fig. 7.17. Comparison of the effects of sea severity on the estimated mean force QTFs and effects of sample length 3 hrs and 40 hrs.

At higher seas, the mean force slightly increases. Likewise the mean force slightly increases as the sample gets longer.

7.4.2.6 Cumulative mean of surge force

In Figs. 7.18(a) and (b), we observe both cumulative mean forces take about 30 hrs to converge, respectively. The mean surging forces of 3 hrs samples in both sea severities are higher than those of 30 hrs. The effect of the Gaussian and non-Gaussian methods are negligible, which may be attributed to the relatively low sea state.

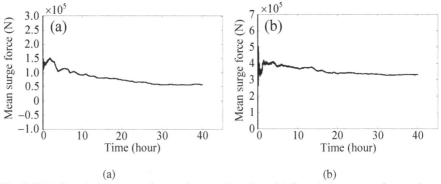

Fig. 7.18(a) Cumulative mean of surge force at $H_s = 3$ m. (b) Cumulative mean of surge force at $H_s = 9$ m.

7.4.2.7 The effects of nonlinearity of surge force

To find the nonlinearity of the response we compare the distribution of peak values of crest heights of the measured and reconstructed surge force in Fig. 7.19. The peak values of crest heights generally exceed the Rayleigh distribution, which means nonlinear. The exceedence of crest height is small at low sea while it is high at high sea.

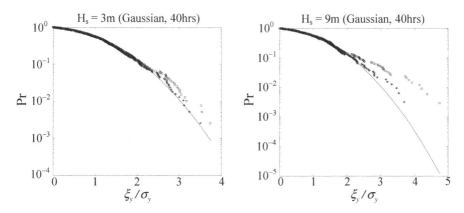

Fig. 7.19. Comparison of peak values of the crest heights of measured ($\circ$) and reconstructed surge forces ($*$) for $H_s = 3$ m and 9 m.

7.4.3 Surge drift of Mini TLP

The Mini TLP (Col. Diam. 8.64 m, draft 28.51 m) that was analyzed by Birkelund et al. (2002) in Sec. 7.4.1.3.did not include the discussion of the mean surge and surge resonance motion. Kim and Kim (2005) reanalyzed the Mini TLP response using Gaussian and non-Gaussian method. It was aimed to investigate the characteristic behavior of the slow drift of the Mini TLP including resonance frequency.

7.4.3.1 QTFs and energy density spectrum of Mini TLP

QTFs of the surge response of Mini TLP were estimated using Gaussian and non-Gaussian method as shown in Figs. 7.20(a) and (b).

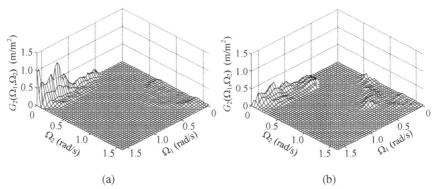

(a)   (b)

Fig. 7.20(a) QTFs of surge of Mini TLP with Gaussian method. (b) QTFs of surge of Mini TLP with non-Gaussian method.

As shown in Figs.7.20(a) and (b), the QTFs are distributed mainly in the low difference frequency domain and the output frequency ranges between 0.0–0.15 rad/s, which indicates that the 2nd-order responses are mainly at low output frequency. It appears that the non-Gaussian QTFs are much smaller than the Gaussian.

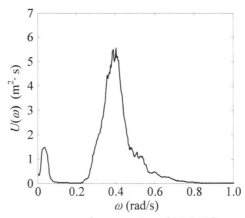

Fig. 7.21. One-sided energy spectrum of surge motion of Mini TLP.

The energy density spectrum of the surge response was determined as shown in Fig. 7.21. The low frequency energy density distribution in the band 0–0.15 rad/s is due to the 2nd-order drift response.

7.4.3.2 Coherency test of Mini TLP

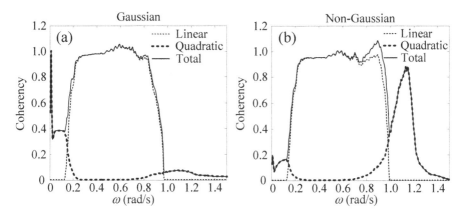

Fig. 7.22(a) Coherency functions of Mini TLP by Gaussian method. (b) Coherency functions of Mini TLP by non-Gaussian method.

Gaussian method gives weak coherencies in the low and high frequency regions. In contrast to this, non-Gaussian method gives weaker and stronger coherencies in the low and high frequency region. The coherency functions in the low and high frequency regions appear to be very small compared to unity in the wave frequency region indicating that the system is dominantly linear. Thus, it is concluded that the second-order drift is negligible compared to the linear response. It is interesting to note that the coherency function at zero frequency in Gaussian method becomes unity, while it is 0.2 in the non-Gaussian method. This is attributed to that the Gaussian method contains the mean at zero frequency while non-Gaussian method does not.

7.4.3.3 Reconstruction of response of Mini TLP

The surge drift motion is reconstructed using the LTFs and QTFs estimated in Sec. 7.4.3.1 and compared with the segmented 500 s long experimental data as shown in Fig. 7.23. The reconstructions are overall in good agreement with the experiment. However, it is difficult to distinguish the effects of the two methods by seeing the figures.

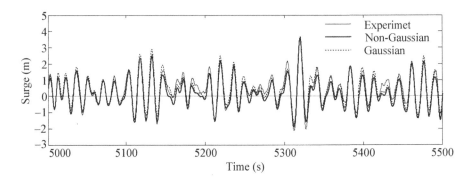

Fig. 7.23. Comparison of reconstructed surges of Mini TLP with experiment.

7.4.3.4 Detection of surge resonance of Mini TLP

Since it was difficult to recognize the resonance period by observing the experimental surge data, we have made a decision to use low-pass filter in the frequency band 0.03 rad/s and 0.05 rad/s, which represents the part of slow drift energy density spectrum Fig. 7.21. The low-pass filter was applied to the amplitude spectrum of the measured wave data. From these low-pass time series one could obtain the average periods of the selected large amplitude slow drift motions as shown in Table 7.1. It was found that the average periods are very close to the resonance period of 166 s which was measured in the model test.

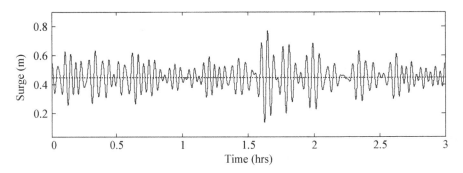

Fig. 7.24. The surge motion of Mini TLP through the low-pass filter (0.03–0.05 rad/s).

Table 7.1 Average surge periods of Mini TLP estimated from the low pass time series in Fig. 7.24.

	Gaussian method	Non-Gaussian method	Measured
Average of surge period (s)	162.8	163.3	165.9

The average of large amplitude motion period in Fig. 7.24 is in the range of 163 s–166 s, which is approximate to the resonance period, the measured resonance period 165.9 s was in this range. The low pass filter could extract the small 2nd-order slowly-varying motion as shown in Fig. 7.25.

Figure 7.25 demonstrates that the 2nd-orer drift response is negligibly small compared to the linear response. In other words, if the linear and 2nd-order are superimposed, the resultant response is expected to have approximately the same linear response. This is attributed to that the sea severity is so low that the linear wave and force are dominant compared to the 2nd-order drift motion.

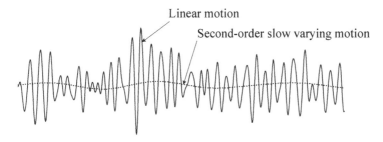

Fig. 7.25. Comparison of the linear and 2nd-order surge drift of Mimi TLP.

7.4.3.5 Cumulative mean surge of Mini TLP

The effect of sample length cannot be investigated from the experimental data because the Mini TLP test was performed for only 1500 s. The cumulative mean value shows an increasing trend indicating the need to record a longer length to reach the statistical convergence (Fig. 7.26).

The surge was estimated using the QTFs in the frequently range of 0.0–0.25 rad/s and the complex amplitude spectrum of the measured wave. The foregoing surge was taken to evaluate the mean cumulative according to

Eq. (7.51). The results of these are compared with the experimental data in Table 7.2, which shows that the Gaussian method over-estimates while the non-Gaussian method significantly underestimates. It is difficult to explain.

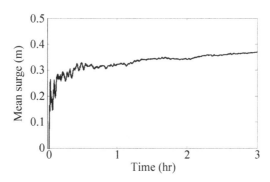

Fig. 7.26. Cumulative mean of surge motion of Mini TLP.

Table 7.2 Comparison of estimations and experimental data
for the mean surge of 3 hrs record.

Measured	Gaussian method	Non-Gaussian method
0.3698 m	0.449 m	0.193 m

Neal (1974) gave a theoretical method to determine the characteristic behavior of the low and high frequency QTFs for the hydrodynamic forces by bi-spectral analysis given a prescribed mathematical distribution of QTFs. Naess (1985) derived a mathematical formula based on the Volterra quadratic model for determining the extreme value of the 2nd-order responses.

References

Journal references:

Barrett, N. (1963). The Use of Functionals in the Analysis of Non-Linear Physical Systems, J. Electronics and Control, Vol. 15, No. 6, pp. 567–615.

Dalzell, J.F. (1974). Cross-Bispectral Analysis: Application to Ship Resistance in Waves, J. Ship Res., Vol. 18, No. 1, pp. 62–72.

Dalzell, J.F. and Kim, C.H. (1979). An Analysis of the Quadratic Frequency Response for Added Resistance, J. Ship Res., Vol. 23, No. 3, pp. 198–208.

Hasselmann, K. (1966). On Nonlinear Ship Motions in Irregular Waves, J. Ship Res., Vol. 10, No. 1, pp. 64–68.

Kim, K.I. and Powers, E.J. (1988). A Digital Method of Modeling Quadratically Nonlinear Systems with a General Random Input, IEEE Trans. Acoustics Speech Signal Processing., Vol. 36, pp. 1758–1769.

Krafft, M.J. and Kim, C.H. (1991). Experimental Investigation of Drift Damping and Slow Drift Motion in Bi-Frequency Domain, Int. J. Offshore and Polar Eng., Vol. 1, No. 3, pp. 235–238.

Maruo, H. (1957). The Excess Resistance of a Ship in Rough Seas, International Shipbuilding Progress, Vol. 4, No. 35, pp. 337–345.

Murray, J.J., Wilkie, B.P. and Muggeridge, D.B. (1987). An Experimental and Analytical Study of the Effects of Wave Grouping on the Slowly Varying Drift Oscillation of a Floating Rectangular Barge, Ocean Eng., Vol. 14, No. 4, pp. 255–274.

Naess, A. (1985). Statistical Analysis of Second-Order Response of Marine Structures, J. Ship Res., Vol. 29, No. 4, pp. 270–284.

Pinkster, J.A. (1980). Low-Frequency Second-Order Wave Exciting Forces on Floating Structures, MARIN Publication No. 600, Wageningen, The Netherlands.

Shaman, P. (1964). Bispectral Analysis of Stationary Time Series, Scientific Paper #18, Statistical Laboratory, School of Engineering and Science, N.Y.U.

St. Denis, M. and Pierson, W. J. (1953). On the motions of ships in confused seas, SNAME Trans. Vol. 61, pp. 280-357.

Tick, L.J. (1961). The Estimation of Transfer Functions of Quadratic System, Technometrics., Vol. 3, pp. 563–567.

Vassilopoulos, L.A. (1967). The Application of Statistical Theory of Non-Linear Systems to Ship Motion Performance in Random Seas, International Shipbuilding Progress, Vol. 14, No. 150, pp. 54–65.

Book references:

Bendat, J.S. (1990) Nonlinear System Analysis, Wiley Interscience.

Kim, S.B. and Powers, E.J. (1995) Chap. 7, Estimation of Volterra Kernels via higher-order statistical signal processing, Higher-Order Statistical Signal Processing, Wiley, New York, NY, pp 213–239.

Newland, D. E. (1986) An Introduction to Random Vibrations and Spectral Analysis, Longman Inc., New York.

Volterra, V. (1959) Theory of Functionals and Integral and Integro-Differential Equations, Dover Publications Inc., New York. (Original; 1930).

Proceedings references:

Birkelund, Y., Powers, E.J. and Hanssen, A. (2002). On the Estimation of Nonlinear Volterra Models in Offshore Engineering, Proc. 12th Int. Offshore and Polar Eng. Conf., Vol. 3, pp. 180–187.

Dalzell, J.F. (1972). Application of the Cross-Bi-Spectral Analysis to Ship Resistance in Waves, Report SIT-DL-72-1606, Davidson Laboratory, Stevens Institute of Technology.

Dalzell, J.F. (1975). The Applicability of the Functional Polynomial Input-Output Model to Ship Resistance in Waves, Report SIT-DL-75-1794, Davidson Laboratory, Stevens Institute of Technology.

Dalzell, J.F. and Kim, C.H. (1976). Analytical Investigation of the Quadratic Frequency Response for Added Resistance, Report SIT-DL-76-1878, Davidson Laboratory, Stevens Institute of Technology.

Dalzell, J.F. (1976). Application of the Functional Polynomial Model to the Ship Added Resistance Problem, 11th Symp. Naval Hydrodynamics, London, UK, pp. III.19-III-39.

Kim, S.B. and Power, E.J. (1991). Higher-Order Spectral Estimation of Drift Force QFRF's for Random Waves, Proc. 23rd Ann. Offshore Technology Conf., Houston, Texas, OTC 6595, pp. 313–320

Kim, N.S. and Kim, C.H. (2003). The Effect of Sea Severity on the Cross-Bi-Spectral Estimate of Quadratic Response Function for Surge Exciting Forces, Proc. 13th Int. Offshore and Polar Eng. Conf., Vol. 3, pp. 413–420.

Kim, N.S. and Kim, C.H. (2004). Gaussian and Non-Gaussian-Input Method for Extraction of QTFs from Test Data of Offshore Structures, Proc. 14th Int. Offshore and Polar Eng. Conf., Vol. 3, pp. 416–422.

Kim, N.S. and Kim, C.H. (2005). Surge Motion of Mini TLP in Random Seas-Comparison between Experiment and Theory, Proc 15th Int. Offshore and Polar Eng Conf., Vol. 1, pp. 244–251.

Neal, E. (1974). Second-Order Hydrodynamic Forces Due to Stochastic Excitation, Proc. 10th ONR Symp., Cambridge, Massachusetts, pp. 517–539.

Newman, J.N. (1974). Second-Order, Slowly-Varying Forces on Vessels in Irregular Waves, International Symposium on Marine Vehicles and Structures in Waves, London, pp. 193–197.

Niedzwecki, J.M, Liagre, P.F., Roesset, J.M. Kim, M.H. and Teigen, P. (2001). An Experimental Research Study of a Mini-TLP, Proc. 11th Int. Offshore and Polar Eng. Conf., Vol. 4, pp. 631-633.

Stansberg, C.T. (1997). Linear and Nonlinear Systemidentification in Model Testing, Int. Conf. Nonlinear Aspects of Physical Model Tests, OTRC, TAMU, USA.

Stansberg, C.T. (2001). Data Interpretation and System Identification in Hydrodynamic Model Testing, Proc. 12th Int. Offshore and Polar Eng. Conf., Vol. 3, pp. 1–9.

Yamanouchi, Y. (1974). Paper No 18: Ship's Behaviour on Ocean Waves as a Stochastic Process, Symp., Marine Vehicles Dynamics, London.

Chapter 8

Second-Order Response in Linear Wave

8.1 Introduction

In the unsteady irrotational flow we have Bernoulli's equation containing the linear and 2nd-order pressure. These pressures produce the 1st- and 2nd-order forces on the floating structure. We are interested in the mean value of the integral of the 2nd-order pressures over the body surface. The foregoing approach to determine the mean 2nd-order force is called pressure integration or near-field method.

We may also determine the same mean 2nd-order force applying the momentum principle given the far-field wave elevation, which is called momentum or far-field method.

Pressure integration method will be applied to determine slowly-varying added resistance and slowly-varying lateral drift force on ships in random seas. We will review the foregoing slowly-varying forces.

The body motion due to the mean force produces wave drift damping in the viscous fluid. The longitudinal mean drift forces of two moored vessels in tandem positions are important in handling the ships.

8.1.1 Momentum method for mean drift force on cylinder

The mean drift force due to the monochromatic wave is determined by applying the linear momentum principle as Maruo (1960).

An experimental scheme to determine the mean drift force of a floating ship in monochromatic wave is illustrated in Fig. 8.1. We assume that the body has been displaced certain distance to the right from the initial rest position due to the mean drift force, which is balanced with the component of the weight; $D = W \sin\theta$. At the balanced position the ship is assumed to oscillate due to the monochromatic wave. The foregoing scheme is well known to engineers.

We want to determine the mean drift force D by applying the linear momentum principle. Consider the incident wave propagates crossing the control surface $S_{-\infty}$ and entering the control volume, then passing the body and further crossing the other control surface $S_{+\infty}$. In the mean time, it is assumed that the body drifts to the position until the restraint balances with the mean drift force while oscillating. The oscillation in monochromatic wave will generate diffraction and radiation waves. Applying the momentum theorem, Maruo (1960) found the mean drift force on a floating body in monochromatic wave in the form:

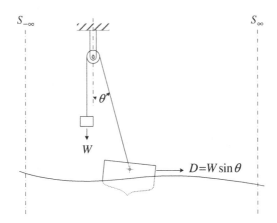

Fig. 8.1. Schematic diagram of the mean offset of a ship due to the mean drift force (Maruo, 1960).

$$\overline{F}_y = \frac{1}{2}\rho g A^2 \left| \overline{A}_{-}^{R+D} \right|^2 \qquad (8.1)$$

where $\overline{A}_{-}^{R+D} = \sum_{m=2}^{4} \overline{A}_{m-}^{R} + \sum_{m=2}^{3} \overline{A}_{m-}^{D}$ is the ratio of the resultant complex amplitude of the diffraction and radiation at the control surface $S_{-\infty}$ to the incident wave amplitude. m indicates the mode of motion (2, 3, 4; sway, heave and roll). The method for computing the resultant complex amplitude ration $\overline{A}_-^{R+D}$ is given in Sec. 6.3.3.

Kim and Chou (1973) employed Maruo's formula Eq. (8.1) and computed the mean swaying drift force and yawing drift moment of a ship in the oblique monochromatic waves using strip method as shown in Sec. 6.5.1. Computation used the Frank's close-fit method given in Sec. 6.3.1. The foregoing mean drift forces and moment are expressed in non-dimensional form:

$$\frac{\overline{F}_y}{1/2\rho g A^2 L} = \frac{1}{L}\int_{-L/2}^{L/2}|A_-|^2 dx \tag{8.2}$$

$$\frac{\overline{M}_z}{1/2\rho g A^2 L^2} = \frac{1}{L^2}\int_{-L/2}^{L/2}|A_-|^2 x dx \tag{8.3}$$

where A is the wave amplitude, and L is the ship length. The non-dimensional expression is prepared to compare with experimental data by Ogawa (1967) and Lalangas (1963).

It is noted that the procedure by Kim and Chou (1973) determines first the body motions and then satisfies the radiation boundary condition.

8.1.2 Reflection and transmission coefficient

We consider at this point the wave energy fluxes across the control surfaces of the control volume in Fig. 8.1. Average wave energy flux (Chap. 4) entering into the control volume: $\frac{1}{2}\rho g A^2 c_g$,while average energy leaving outward from the control volume to both sides:

$$\frac{1}{2}\rho g \mid \overline{A}_-^{R+D}\mid^2 c_g + \frac{1}{2}\rho g \mid \overline{A}_+^{R+D} + A\mid^2 c_g \tag{8.4}$$

where $\overline{A}_{\pm}^{R+D}$ denotes the sum of radiation and diffraction complex amplitude in the far-field from the body ($\pm \mid y \mid$) as explained in Sec. 6.3.3. Since we consider the ideal fluid motion, we can apply energy conservation law between the energy entering and leaving from the domain:

$$\frac{|\overline{A}_-^{R+D}|^2}{A^2} + \frac{|\overline{A}_+^{R+D} + A|^2}{A^2} = C_R^2 + C_T^2 = 1 \qquad (8.5)$$

where C_R and C_T are the reflection and transmission coefficient, respectively as illustrated in Fig. 8.2. We observe in the figure the reflection is low and transmission is high at the low frequency region, while the reflection reaches highest level and transmission reaches lowest level at very high frequency region. These coefficients vary depending on the geometry of the structures, water depth and wave heading. Thus for instance, the effectiveness of breakwater geometry may be investigated by computing the magnitude of reflection coefficient at various wave heading. If the reflection is large the breakwater is effective.

Comparison of Eq. (8.1) with Eq. (8.5) leads to:

$$\frac{\overline{F}_y}{A^2} = \frac{1}{2}\rho g C_R^2 \qquad (8.6)$$

Thus, the mean drift force per unit incident wave amplitude square is proportional to the square of the reflection coefficient C_R. Since the high reflection occurs at high wave frequency, large drift force is expected to occur in the waves of high frequency or in shorter waves. Theoretically we will have an asymptotic value of unity as the wave frequency approaches infinity as shown in Fig. 8.2.

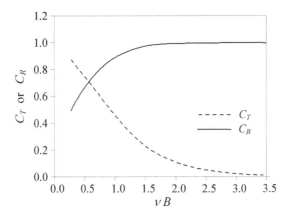

Fig. 8.2. Reflection (solid circle) and transmission coefficients (circle with triangle) of a Lewis form section, v is deep water wave number B represents the beam of the cylinder.

In the opposite situation, the mean drift force coefficient or reflection coefficient will be reduced to a negligibly small value as the frequency reaches zero. In such infinitely long waves, no reflection will occur, and thus there will be no mean drift force.

8.1.3 3-D momentum equation

The following is the general momentum equation by Newman (1967). The mean lateral drift force and moment of arbitrary shaped body in arbitrary wave heading are determined using the far-field information in Eq. (8.7):

$$\begin{Bmatrix} \bar{F}_x \\ \bar{F}_y \end{Bmatrix} = -\int_{S_\infty} \left[p \begin{Bmatrix} \cos(n,x) \\ \cos(n,y) \end{Bmatrix} + \rho \begin{Bmatrix} \dfrac{\partial \phi}{\partial x} \\ \dfrac{\partial \phi}{\partial y} \end{Bmatrix} \dfrac{\partial \phi}{\partial n} \right] dS$$

(8.7)

$$\bar{M}_z = -\int_{S_\infty} \left[p(\mathbf{r} \times \mathbf{n})_z + \rho(\mathbf{r} \times \mathbf{V})_z V_n \right] dS$$

Newman (1967) computed the mean drift force of series 60 ship model applying the slender body theory and velocity potential in the far-field.

8.1.4 A review of various methods for mean drift force

Faltinsen and Loken (1978) reviewed a number of methods of computing the mean drift force. Majority of analytical works were on the mean drift force and moment of a ship, free or fixed, in oblique waves. Experimental works that can be compared with the analytical results were limited to data from Ogawa (1967). Figures 8.3(a) and (b) show the drift force and moment on series 60 ships of $C_B = 0.6$, with an exception of $C_B = 0.7$, in 60° heading, for the fixed and free ship model. The swaying drift force coefficient asymptotes to a constant as the wavelength decreases infinitesimally short. This may be inferred from the asymptotic reflection coefficient at infinitely high frequency as discussed in Sec. 8.1.2. It is noted that Salvesen (1974) used pressure integral method, while the others used far-field method.

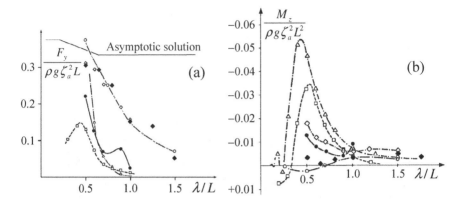

Fig. 8.3(a) Swaying mean drift force on a series 60 ship, $C_B = 0.6$, heading angle 60°. (b) Yawing mean drift moment on a series 60 ship, heading angle 60°.

O	Kim and Chou's formula (restrained), $C_B = 0.6$, 1973
□	Salvelsen's formula Mariner Hull (free), $C_B = 0.6$, 1974
△	Newman's long wavelength formula (free), $C_B = 0.6$, 1974
●	Newman-Helmholtz formula (free), $C_B = 0.6$, 1974
◆	Ogawa's experimental values (restrained), $C_B = 0.7$, 1967
◊	Newman-Helmholtz formula (restrained), $C_B = 0.6$, 1974

The forgoing review is on the force referring to Fig. 8.3(a). According to Faltinsen and Loken (1978), "Newman-Helmholtz formula and Kim and Chou's formula are in good agreement with Ogawa's experimental results". Also "Salvesen's result for Mariner hull with block coefficient 0.61 is significantly smaller than the others. This is due to the wrong assumption of weak scatter formula," This will be discussed in the next section.

According to Faltinsen and Loken (1978), "In reference to Fig. 8.3(b): Newman-Helmholtz formula and Kim and Chou's method (1973) agree with experimental data to the same degree". "Both results by Newman's long wavelength formula and Salvesen's formula disagree significantly with the experimental values."

8.2 Added Resistance

The slowly-varying 2nd-order force is of practical importance in a variety of problems including added resistance and course variations of ships, horizontal oscillations of moored vessels, and vertical oscillations of vessels with small waterplane area. The slowly-varying added resistance (or drift force) are due to the 2nd-order low frequency force. Since the QTFs in Chap. 7 are continuous function of the bi-frequency, one may approach taking the limit to the mean value where bi-frequency QTFs become the mean drift force as function of monochromatic wave. Given the mean drift force Newman (1974) proposed an approximate formula to determine the slowly-varying 2nd-order drift force. The slowly-varying added resistance or drift force of a ship in random seas was measured through the cross-bi-spectral analysis and the results were compared with the hydrodynamic computation (Dalzell and Kim, 1976; 1979).

8.2.1 Formula of added wave resistance

Dalzell and Kim (1976, 1979) conducted hydrodynamic analysis of the bi-frequency QTFs for the added resistance of series 60 model ship running in head seas in the wave tank to compare with the cross-bi-spectral analysis between the random seas and added resistance.

The added resistance was formulated based on the weak scatter assumption (Salvesen, 1974). The hydrodynamic computation employed the strip ship motion theory developed at Davidson Laboratory (Kim, 1974; Kim et al., 1980), which is different from others in the estimation of the diffraction force of a ship running in the seas. (Wave tank: 91.46 m × 3.66 m × 1.83 m). The foregoing method computes the diffraction potential directly, whereas the others use radiation potential for determining the wave-exciting force following Haskind-Newman relation as given in Sec. 6.1.5.

The hydrodynamic force on a body in the free surface:

$$\mathbf{F} = -\rho \frac{d}{dt} \iint_{S_B+S_F} \Phi \mathbf{n} dS - \rho \iint_{S_B} \left(\frac{\partial \Phi}{\partial n} \nabla\Phi - \frac{1}{2} |\nabla\Phi|^2 \mathbf{n} \right) dS \qquad (8.8)$$

where $\mathbf{F}$ denotes the hydrodynamic force, Φ total velocity potential, $\mathbf{n}$ outward unit normal vector from the fluid domain, ρ mass density, S_B submerged surface of body, S_F portion of free surface inside a far-field control surface S_∞.

No approximations were made in deriving Eq. (8.8). The expression is taken as being exact within potential-flow theory. Since it is assumed the potential Φ is composed of a steady-state wave resistance potential, Φ_S, and a time-dependent potential Φ_T due to waves, the above equation was changed by eliminating the steady part and the linear contribution of time-dependent terms, and retaining only integrals that contribute to quadratic nonlinearity even under the assumption of linear potentials. The cardinal assumption in (Salvesen, 1974) is that the dominant contribution to the quadratic or 2nd-order nonlinearities comes from the combination of linear or first-order potentials and not from nonlinearities of the 2nd-order potentials themselves. The quadratic part of the added resistance or drifting force is thus taken:

$$\mathbf{R} = -\rho \iint_{S_B} \left(\frac{\partial \Phi_T}{\partial n} \nabla \Phi_T - \frac{1}{2} |\nabla \Phi_T|^2 \, \mathbf{n} \right) dS \qquad (8.9)$$

where Φ_T is the first-order time-dependent potential. This expression is essentially the same as that taken by Salvesen (1974) in formulating the mean drifting force. Writing the time-dependent potential as the sum of an incident wave potential, Φ_I, and a disturbance potential Φ_B, i.e., sum of radiation and diffraction potential:

$$\Phi_T = \Phi_B + \Phi_I \qquad (8.10)$$

we have:

$$\mathbf{R} = \rho \iint_{S_\infty} \left(\Phi_B \frac{\partial}{\partial n} - \frac{\partial \Phi_B}{\partial n} \right) \left(\nabla \Phi_I + \frac{1}{2} \nabla \Phi_B \right) dS \qquad (8.11)$$

Salvesen (1974) made the assumption that the disturbance potential is a week scatter so that $\Phi_B \ll \Phi_I$. This assumption is considered to be "questionable for oblique ship-wave headings" as found in the survey of the lateral drift force in Sec. 8.1.4 and Figs. 8.3(a) and 8.3(b). However, the assumption is considered quite reasonable for head seas and Salvesen's calculations bear this out. Accordingly, Eq. (8.11) becomes:

$$\mathbf{R} = \rho \iint_{S_\infty} \left(\Phi_B \frac{\partial}{\partial n} - \frac{\partial \Phi_B}{\partial n} \right) \nabla \Phi_I \, dS \tag{8.12}$$

This equation corresponds to equation (32) of Salvesen (1974). Application of Green's theorem changes the integration in Eq. (8.12) to one over the body surface:

$$\mathbf{R} = \rho \iint_{S_B} \left(\Phi_B \frac{\partial}{\partial n} - \frac{\partial \Phi_B}{\partial n} \right) \nabla \Phi_I \, dS \tag{8.13}$$

8.2.1.1 Similarity between Volterra quadratic model and hydrodynamic QTFs

In Eq. (8.13), the disturbance and incident potential are the sums of potentials of frequency components respectively, in the forms:

$$\Phi_B = \mathrm{Re}\left[\phi_B(\omega_i) e^{-i\omega_i t} + \phi_B(\omega_j) e^{-i\omega_j t} \right]$$
$$\Phi_I = \mathrm{Re}\left[\phi_I(\omega_i) e^{-i\omega_i t} + \phi_I(\omega_j) e^{-i\omega_j t} \right] \tag{8.14}$$

Making use of Eq. (8.14) in Eq. (8.13), one obtains quadratic part of the added resistance in the form:

$$R_2 = \frac{1}{2}\left[A_i^2 H_2(\omega_i, -\omega_i) + A_j^2 H_2(\omega_j, -\omega_j) \right]$$
$$+ \frac{1}{2}\mathrm{Re}\left[A_i^2 H_2(\omega_i, \omega_i) e^{-i2\omega_i t} \right]$$
$$+ \frac{1}{2}\mathrm{Re}\left[A_j^2 H_2(\omega_j, \omega_j) e^{-i2\omega_j t} \right] \tag{8.15}$$
$$+ \mathrm{Re}\left[A_i A_j H_2(\omega_i, \omega_j) e^{-i2(\omega_i + \omega_j)t} \right]$$
$$+ \mathrm{Re}\left[A_i A_j H_2(\omega_i, -\omega_j) e^{-i2(\omega_i - \omega_j)t} \right]$$

where $H_2(\omega_i, \omega_j)$ is the hydrodynamically determined QTFs. Thus the 2nd-order added resistance induced in the bichromatic wave trains has been brought into essentially the same form as the six terms of the Volterra quadratic model Eq. (7.10). In other words, the QTFs of Volterra quadratic model for added resistance are identical to the hydrodynamic QTFs. This may be attributed to the same assumption of the linearized free surface boundary condition in the hydrodynamic theory and Volterra quadratic model.

We will find in Chap. 9 there exit the similarity between QTFs of the Volterra quadratic model and the Stokes 2nd-order wave theory as well as the similarity between the Volterra quadratic model and the 2nd-order force theory. The hydrodynamically computed QTFs are not identical in the form to that of the Volterra quadratic model in these cases. This may be attributed to the fact that the Volterra quadratic model satisfies the linear free surface boundary condition, whereas the Stokes 2nd-order wave and force theories satisfy 2nd-order free surface boundary conditions. The differences of QTFs between Volterra quadratic model and those of the wave and force theories in Chap. 9 are found to be in the factors of ¼ and 2, respectively. It is noted that the 2nd-order free surface conditions of the wave and force are also not identical because of the inclusion of diffraction potential in the 2nd-order free surface boundary condition.

8.2.1.2 Computation of added resistance

Equation (8.13) is computed applying the strip theory. Each of the second-order terms in Eq. (8.13) is a product of two different linear responses to the bichromatic wave. After some manipulation, one obtains the added resistance which is identical in the form to the Volterra quadratic model with omission of LTFs as given in Eq. (7.10).

In order to compare the above theoretical computation with the experimental data by cross-bi-spectrum, length is non-dimensionalized with ship length L and force is non-dimensionalized with displacement Δ. Since the QTF for added resistance is in dimension of force/length^2, it is nondimensionalized by multiplying with L^2/Δ: We define mean added resistance operator in the form

$$\hat{G}_2\left(\sigma_e, -\sigma_e\right) = G_2\left(\sigma_e, -\sigma_e\right) L^2 / \Delta \qquad (8.16)$$

where the non-dimensional encounter frequency is expressed by $\sigma_e = \omega_e / (2\pi g / L)^{1/2}$, with the encounter frequency ω_e.

8.2.1.2.1 Comparison of analytical and experimental mean added resistance

As noted before, the mean added resistance operator is the value of the QTFs along the difference frequency axis Ω_1, and it is the special section of the function for which there is a substantial confirming information. The model used in Dalzell (1976) was of the ship, series 60, 0.60 C_B parent, of 152.4 m long ship. This model was also one of several used in experiments reported by Strom-Tejsen et al. (1973).

Figure 8.4(a) shows results from Dalzell-Kim's analysis plotted on what is effectively Fig. 8(a) of Strom-Tejsen et al. (1973). The foregoing work on the series 60, 0.60 block model had involved two speeds (Froude numbers 0.266 and 0.283). Dalzell-Kim's method was evaluated by using the mean of the two speeds, $F_n = 0.274$, and the results were non-dimensionalized in accordance with the conventions employed in Strom-Tejsen et al. (1973).

Dalzell-Kim's results clearly are closer to experiment compared to the results of analytical methods of Strom-Tejsen et al. (1973). In addition, at high encounter frequencies Dalzell-Kim's method indicates the leveling-out trend often observed in experimental mean added resistance operators.

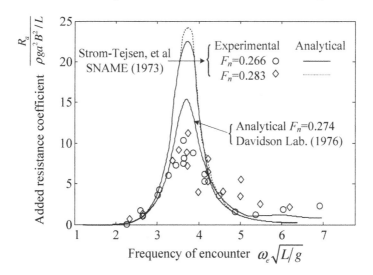

Fig. 8.4(a) Comparison of mean added resistance of series 60 model; Circles present $F_n = 0.266$, and diamonds $F_n = 0.283$.

Let us recall the cross-bi-spectral analyses of the experimental work (Dalzell, 1976) as shown in Fig. 7.11. It had also involved two speeds, F_n = 0.15 and 0.20. The lower of the two speeds was selected and the QTFs were evaluated. For this case Fig. 7.11 indicates the comparison of experimental mean added resistance operator with computational mean added resistance operator of Dalzell-Kim's analytical method. Except for the addition of analytical result, Fig. 7.11 was the same as that in Dalzell (1976).

Two kinds of experimental data are included in Fig. 7.11. The regular wave envelope indicates the magnitude and scatter of the results of regular wave experiments. The lines broken by various symbols are the results of cross-bi-spectral analyses of data obtained in three irregular seas. In Fig. 7.11, the agreement with experiment of the analytical estimates of the mean added resistance operator appears to be somewhat better than that shown in Fig. 8.4(a). In this case most of the analytical results are within the regular wave envelope.

Figures 8.4(a) and 7.11 clearly demonstrate the validity of the Dalzell-Kim's analytical method relative to experimental data obtained both at Davidson Laboratory and elsewhere. For the particular ship form and speeds shown, the differences between Dalzell-Kim's method and the mean experimental result appear to be of the same order as the scatter in the experiments (which are noted for their difficulty). Thus Dalzell-Kim's method, which involves a different approach to the computation of exciting forces has promise for refinement of the estimating methods for the mean added resistance operator.

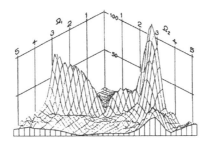

Fig. 8.4(b) Isometric view of moduli of analytically estimated QTFs for added resistance of series 60 model in head seas.

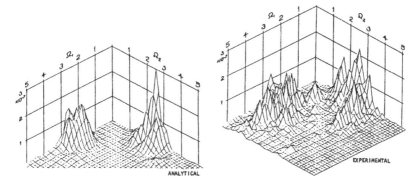

Fig. 8.4(c) Analytically predicted and experimentally derived cross-bispectral moduli: Froude number 0.15, Sea state A ($H_s = 2.74$ m).

8.2.1.2.2 Comparison of analytical and cross-bi-spectral method

Isometric views of moduli of complete analytical QTFs for added resistance are shown in Fig. 8.4(b). The analytical results for the series 60, 0.60 block model at a Froude number of 0.15 were non-dimensionalized in accordance with the experimental convention noted in the previous section. In the figure the plotted domain of the bifrequency plane appears triangular. This is the result of computing the function to a maximum absolute input frequency of $\omega_1 = 2.62$ rad/s. The right-to-left section at the front of the plot is shown in bolder lines with vertical lines defining the position of computed points. This section represents the modulus of the function along the line $\omega_1 = 2.62$ rad/s.

Comparison of the magnitude of analytical and experimental estimates might best be carried out indirectly comparing CBS instead of QTFs. The experimental estimates of QTFs were made from the ratio of two cross-bi-spectral estimates in accordance with Eq. (7.42)

$$\hat{G}_2\left(\omega_1,\omega_2\right)=\hat{G}_2\left(\frac{\Omega_1+\Omega_2}{2},\frac{\Omega_2-\Omega_1}{2}\right)=\frac{C_D*\left(\Omega_1,\Omega_2\right)}{C_\eta\left(\Omega_1,\Omega_2\right)} \qquad (8.17)$$

where

$C_D*\left(\Omega_1,\Omega_2\right)=$ cross-bi-spectrum of wave, wave and resistance

$C_\eta\left(\Omega_1,\Omega_2\right)$ = cross-bi-spectrum of wave, wave and (wave)^2

$C_\eta\left(\Omega_1,\Omega_2\right)$ = $S_{\eta\eta}\left(\omega_1\right)S_{\eta\eta}\left(\omega_2\right),\ S\left(\omega\right)=$ two-sided energy spectrum

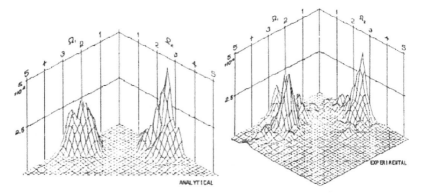

Fig. 8.4(d) Analytically predicted and experimentally derived cross-bispectral moduli: Froude number 0.15, Sea state B (H_s = 5.5 m).

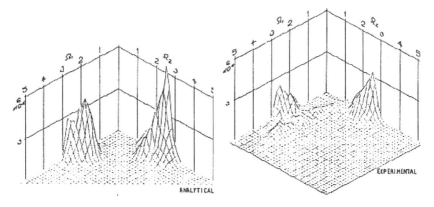

Fig. 8.4(e) Analytically predicted and experimentally derived cross-bispectral moduli: Froude number 0.15, Sea state C (H_s = 10.1 m).

If we multiply cross-bi-spectrum of wave, wave and (wave)^2 to the analytical estimate of QTFs we will have the cross-bi-spectrum of wave, wave and resistance. We will compare the cross-bi-spectrum of wave, wave and resistance (CBS) of the analytical and experimental results for a variety of sea state.

Varying the sea state and using the analytical QTFs in Fig. 8.4(b), one can determine the analytical CBS according to Eq. (7.42) as shown in Figs. 8.4(c)–(e). These analytical CBS are compared with the experimental CBS. On the whole, in Figs. 8.4(c)–(e), the maximum peaks which lie near the axes correlate fairly well. The largest such peaks are located at approximately the same sum or difference frequencies in both analytical and experimental results. The difference in magnitude is about 30 percent for sea states A and B and for the peak located in the Ω_1-axis in sea state C. The peaks along the sum frequency axis for sea state C are different by about a factor of 2. These margins of difference are roughly the same as that shown for the mean resistance operator in Fig. 7.11.

A close comparison of the experimental results in Figs. 8.4(c)–(e) discloses a range of humps lying along the line $\Omega_2 = 1$. The magnitude decreases as the sea severity increases. This characteristic is not present in the analytically derived CBS. There is in addition a significant hump along the line $\Omega_2 \approx 3.2$ rad/s in Fig. 8.4(c) for sea state A. This last corresponds to the contributions of a noise frequency which as indicated in Dalzell (1976), has little to do with the added resistance. The results of detailed comparison between analytically and experimentally determined QTFs are given in Dalzell and Kim (1976).

The above comparison illustrates that the hydrodynamic theory is partially useful for the interpretation or prediction of the fluctuating added resistance in the head random seas.

Similar works are given in Pinkster (1980).

8.2.1.3 Simulation of slowly surging drift in large wave group

Kim and Breslin (1976) simulated the slow surge drift force and motion of a moored ship in head seas, using the low pass filter added resistance at zero speed which is extracted from the computed QTFs in the low output frequency region as shown in Fig. 8.5 by Dalzell and Kim (1976).

The QTFs that contribute to the slowly-varying drift force are distributed in the low output frequency region as shown in Fig. 8.5. Thus, we use low pass filter QTFs in the low frequency region in employing the slowly-varying drift force $D(t)$. The slowly-varying drift force is due to the low-difference frequency as given in the form:

$$D(t) = \mathrm{Re} \sum_{j=1}^{N} \sum_{k=1}^{N} \frac{1}{2} A_j A_k \left\{ G_2(\omega_j, -\omega_k) e^{i(\omega_j - \omega_k - \varepsilon_j + \varepsilon_k)t} \right\} \qquad (8.18)$$

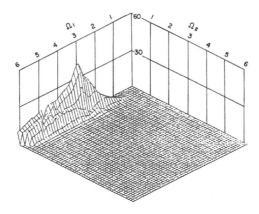

Fig. 8.5. Low pass filtered QTFs for the drift force of series 60 model ship in head sea (Sea state B, H_s = 5.5 m).

where A_j and A_k represent the random wave amplitudes and ε_j and ε_k are random phase angles. The surge drift motion equation is then written:

$$M_v \ddot{\xi}_1(t) + N\dot{\xi}_1(t) + K\xi_1(t) = D(t) \qquad (8.19)$$

where M_v and N denote the virtual mass (infinite frequency) and radiation damping of the series 60 model ship in head seas and K is the stiffness of the mooring lines. Computing Eqs. (8.19) and (8.18) simultaneously, one obtains the slowly-varying surge drift force and motion as shown in Figs. 8.6(b) and (c). Runge-Kutta numerical scheme is employed in the simulation of Eq. (8.19). The slowly-varying drift force consists of the mean and fluctuation parts.

Interestingly the large drift forces occur when the ship is in the large wave groups as shown in Fig. 8.6(a). Similar behavior was observed in the model test data of slow drift force of a tanker measured in regular wave group (Pinkster, 1980) in Sec. 7.1.5. The difference here is that we have simulated the slowly-varying surge drift force of a ship in random head seas, while the previous case is the model test in a regular wave group. The slowly-varying drift force was measured restraining the 2nd-order slow-drift oscillation and allowing the linear motion due to linear waves in Sec. 7.1.5. It is the same here in that the slowly-varying motion was restrained by taking zero speed of advance and the linear motions were allowed in the computation of drift force. Both experiment and simulation produces the same behavior as expected. It is of our interest to know the reason for the large drift force occurring in the large wave groups. It is postulated that the 2nd-order wave group will be generated in the large linear regular wave group. If the 2nd-order wave group is large the 2nd-order force will be large according to Eq. (8.18). We will search for the 2nd-order wave group that may become large in the large linear wave group in Chap. 9.

It is to be noted the low frequency surging drift motion in Fig. 8.6(c) represents a resonance motion. The large surging motion occurs when the drift force frequency coincides with the resonance frequency of the mooring system.

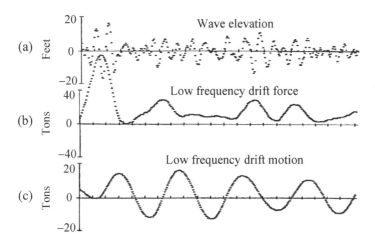

Fig. 8.6(a) The random wave time series. (b) The low pass filtered slowly-varying drift force of series 60 ship in the random head sea $H_s = 5.45$ m. (c) The slowly-varying drift motion of series 60 ship in the random head sea $H_s = 5.45$ m.

8.3 Lateral Drift Force and Moment in Oblique Seas

Applying the 2nd-order transverse drift force of a ship in beam seas (Soeding, 1978), Kim and Dalzell (1980) analyzed the transverse drift force and moment of a ship in oblique seas at zero-speed of advance in bi-frequency domain.

8.3.1 Lateral drift force in beam seas

We assumed that in the calculation of added resistance the contribution to the 2nd-order nonlinearities comes dominantly from the combination of 1st-order potentials and not from nonlinearities of the 2nd-order potentials themselves. The same assumption is applied to the computation of the 2nd-order transverse force on a ship in waves.

Consider first a freely floating body in a monochromatic wave steadily oscillating with restrained drift motion as shown in Fig. 8.1. We determine the 2nd-order force on a ship section in beam seas including the mean force coefficient and its asymptotic behavior at infinite frequency. Since the vertical relative motion gives a large effect on the 2nd-order transverse force, a detailed analysis was conducted. We have then moved to the computation of QTFs for the 2nd-order transverse force on a floating cylinder in the bichromatic waves in the beam sea as that of added resistance in head seas. The foregoing foundations had let us to simulate the slowly-varying lateral drift force of a ship in oblique random seas.

8.3.1.1 The 2nd-order mean transverse force

Referring to Fig. 8.7, let o-yz and O-YZ be the body- and space-fixed frames, respectively. The y-axis indicates the waterline and z-axis indicates the vertical centerline of the cylinder. The Y-axis rests on the calm water surface and the Z-axis points vertically upward. The two frames were

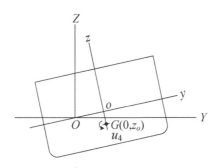

Fig. 8.7. Schematic diagram of a body deviated instantaneously from the equilibrium.

initially identical at the equilibrium condition. After the body is set in motion and reached a steady state the body-fixed frame will take a deviated position as shown. If we treat the problem in linear fashion, we will need to consider the boundary conditions at the mean position of oscillation (Chap. 6). These linear solutions will still be employed here.

Now we consider the effects of the instantaneously deviated position of the body on the changes of the velocity potential, pressure and force from those at equilibrium. The instantaneously displaced hull configuration can be identified in reference to the body-fixed frame using some known data.

We denote the displacement amplitudes, omitting the time factor $e^{-i\omega t}$, by u_m (m = 2, 3, 4; sway, heave, and roll). The roll-axis is assumed to pass through the center of the gravity G ($y = 0$, $z = z_0$).The coordinates of a point $A(Y, Z)$ on the body surface at an instantaneously deviated position are expressed by the sum of the translation amplitude u_m (m = 2, 3) and the linear displacement of the point A due to roll (u_4) of the body about the center of gravity:

$$Y = u_2 + y\cos u_4 - (z - z_0)\sin u_4$$
$$Z = u_3 + y\sin u_4 + (z - z_0)\cos u_4 \qquad (8.20)$$

We find nonlinear function of roll in the above equation. Expanding cos u_4 and sin u_4 up to 2nd-order we may express the coordinates of point A in terms of the linear, and 2nd-order motion displacements, and the body coordinates:

$$Y = u_2 + y - (z - z_0)u_4 - \frac{1}{2}yu_4^2$$
$$Z = u_3 + (z - z_0) + yu_4 - \frac{1}{2}u_4^2(z - z_0) \qquad (8.21)$$

The deviation of point A is given by $Y-y$ and $Z-(z-z_0)$. Hence the corresponding deviation of the velocity potential is approximately estimated by Taylor expansion of the potential at the equilibrium position or mean position of oscillation:

$$\phi(Y,Z,t) = \phi(y,z,t) + (Y - y)\phi_y(y,z,t)$$
$$+ \left[Z - (z - z_0)\right]\phi_z(y,z,t) \qquad (8.22)$$

Thus Bernoulli's pressure at a point $A(Y, Z)$:

$$p(Y,Z,t) = -\rho g Z - \rho \phi_t(y,z,t) - \rho(Y-y)\phi_{yt}(y,z,t)$$
$$- \rho[Z-(z-z_0)]\phi_{zt}(y,z,t) - \frac{1}{2}\rho[\nabla\phi(y,z,t)]^2 \qquad (8.23)$$

The integration of the foregoing pressure on the instantaneous contour of the body gives the lateral force in the form:

$$f_y = \int_c p(Y,Z,t)(-dZ) + \int_{c_1} p(Y,Z,t)(-dZ) \qquad (8.24)$$

where $dZ = \sin\alpha\, dS$, dS indicates the segment and α is the angle of the segment to the Y-axis, c is the mean wetted contour and c_1 is the deviated instantaneous wetted contour from the mean waterline. The dS is taken positive in counterclockwise movement along the contour. Using Eqs. (8.21) through (8.24) we have:

$$f_y = \rho g B u_3 u_4 + u_4 \int_c \rho \phi_t dy$$
$$+ \rho \int_c \left\{ [u_2 - (z-z_0)u_4]\phi_{yt} + [u_3 + yu_4]\phi_{zt} + \frac{1}{2}(\nabla\phi)^2 \right\} dz \qquad (8.25)$$
$$+ \rho \int_{c_1} (gZ + \phi_t)dZ$$

B = beam of the cylinder, $dy = \cos\alpha\, ds$, $dz = \sin\alpha\, ds$, ds segment on the contour c, α is the angle of the segment with respect to the y-axis. And ds is positive in counterclockwise movement along the contour.

The sum of the first two terms in Eq. (8.25) is identical to $\omega^2 m_0 u_3 u_4$, where m_0 is the mass of the cylinder. This is because we can write the heaving equation of the cylinder in the form:

$$-\omega^2 m_0 u_3 = -\rho g B u_3 - \rho \int_c \phi_t dy \qquad (8.26)$$

where the ϕ is the sum of the velocity potential of radiation due to sway, heave and roll, and the incident and diffraction potentials. However, the contribution to the integral in Eq. (8.26) comes from the heave-mode radiation, incident and diffraction potentials in Eq. (8.26), because the integral gives the heaving force only because of the expression of heave force may be determined by Σ (pressure $\times$ Δy).

The contour c_1 of the last integral in Eq. (8.25) consists of the wetted contours on the right and left sides of the hull measured from the mean waterline. Since $Z + (1/g)$ ϕ_t is the relative displacement of the hull to the wave elevation, introducing a variable Z_r for the relative wave elevation to the hull displacement, we have the integral Eq. (8.25) in the form:

$$\rho \int_{c_1} (gZ + \phi_t) dZ = -\rho g \int_{c_2} Z_r dZ_r \tag{8.27}$$

where c_2 is the wetted contour for the relative displacement on the right and left sides of the hull, and $dZ_r = dZ = \sin \alpha dS$. Hence:

$$-\rho g \int_{c_2} Z_r dZ_r = -\frac{1}{2} \rho g (r_+^2 - r_-^2) \tag{8.28}$$

where

$$r_{\pm} = e_{\pm} - (u_3 \pm \frac{B}{2} u_4)$$

$$e_{\pm} = -\frac{1}{g} \phi_t \left[\pm \frac{B}{2}, 0, t \right] \tag{8.29}$$

Substitution of Eqs. (8.26) and (8.28) into Eq. (8.25) yields the 2nd-order lateral force of cylinder per unit length in the form:

$$\begin{aligned} f_y (\omega) = &\ \omega^2 m_0 u_3 u_4 \\ &+ \rho \int_c \{ [u_2 - (z - z_g) u_4] \phi_{yt} + [u_3 + y u_4] \phi_{zt} \} dz \\ &+ \frac{1}{2} \rho \int_c (\nabla \phi)^2 dz + \frac{1}{2} \rho g (-r_+^2 + r_-^2) \end{aligned} \tag{8.30}$$

Equation (8.30) states that the transverse force f_y consists of the gyroscopically coupled heave inertia force and roll motion, the product of linear motion and linear force, Bernoulli's quadratic pressure force, and the vertical relative-wave-elevation-induced-force.

The first term of f_y in Eq. (8.30) may be averaged in the form:

$$\begin{aligned} \overline{f}_1 &= \omega_0^2 m_0 \frac{1}{T} \int_0^T \text{Re} \left[u_3 e^{-i\omega t} \right] \text{Re} \left[u_4 e^{-i\omega t} \right] dt \\ &= \frac{1}{2} \omega_0^2 m_0 \{ \text{Re}(u_3) \text{Re}(u_4) + \text{Im}(u_3) \text{Im}(u_4) \} \end{aligned} \tag{8.31}$$

The mean of the lateral force is non-dimensionalized in the form:

$$\bar{C}_F = \frac{1}{\frac{1}{2}\rho g A^2}\bar{f}_y(x) = \sum_{j=1}^{4}\bar{C}_{F_j} \qquad (8.32)$$

The asymptotic behavior of each component for infinite frequency:

$$\left.\begin{array}{l}\bar{C}_{F_1}\\[6pt]\bar{C}_{F_2}\end{array}\right\} = \ \to 0,$$

$$\bar{C}_{F_4} \to 2, \qquad\qquad\qquad (8.33)$$

$$\bar{C}_F \to 1,$$

$$\bar{C}_{F_3} = \ \to -1, \quad \text{for } \omega \to \infty$$

In Eq. (8.33) the first two components approach zero, because the motion terms approach zero at infinite frequency. The 4th term approaches two (2) according to the result of analysis given in the appendix II (Kim and Dalzell, 1980), while the asymptotic expression of the resultant force coefficient is unity according to Eq. (8.6). Thus, the 3rd-coefficient approaches minus one. All the asymptotic expressions of the non-dimensional force coefficients are illustrated in Fig. 8.8(a). We compare the measured (Koterayama, 1976) and computed mean drift force coefficients (Kim and Dalzell, 1980) as shown in Fig. 8.8(b). The resultant mean force coefficient is also shown in Fig. 8.8(b).

The computation in the foregoing employs both near- and far-field methods, which are both in excellent agreement. This is attributed to both solutions being derived from the same source intensities that have been determined by satisfying the same boundary conditions.

As we see Fig. 8.8(b), the theory and experiment are in excellent agreement. Compared to the far-field solution, the near-field solution has an advantage in providing information of the effects of the motions on the mean force.

8.3.1.2 The vertical relative motion

As we have analyzed the vertical relative motion in the foregoing, its contribution to the 2nd-order force is the largest among the four components.

Hence, we have computed the vertical relative motions of a Lewis-form section and the motion of the body in the beam seas employing the close-fit method as function of non-dimensional wave frequency $vB/2$, where v is the deepwater wave number as shown in Fig. 8.9.

Lewis form ship section has half beam draft ratio $H = 1.464$ and fullness coefficient $\beta = 0.942$. We have compared and analyzed the vertical relative motions of a fixed and freely floating ship section as function of wave frequency. The result is summarized:

1) The relative motion for the fixed body at the infinitely high frequency is about 2.0 on the weather side, while zero on the lee side. Same value of 2.0 is found in the reflected wave from a fixed structure at infinitely high frequency (Newman, 1977).

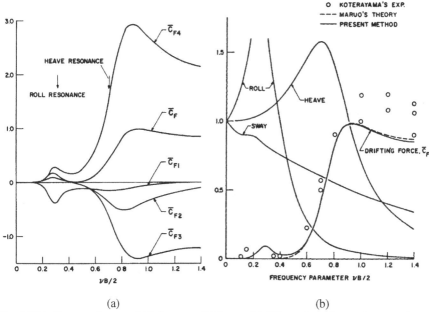

(a)   (b)

Fig. 8.8(a) Computed mean drift force coefficients. (b) Effect of motion and comparison of the far- and near-field method and comparison of analytical mean drift force and experiment by Koterayama (1976).

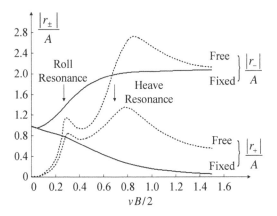

Fig. 8.9. Vertical relative motion RAO of a Lewis-form section (H = 1.464, β = 0.942) in beam seas (close-fit method); which will be used for the mean lateral drift force to compare with experiment by Koterayama(1976) in Fig. 8.8(b).

2) When the body is fixed at zero frequency or in the infinitely long wave, the RAOs for relative motions at the weather and lee side become both unity.

3) When the body is free at zero frequency or infinitely long wave, the relative motions RAOs both at the weather and lee side become zero because we interpret that the wave particle and body move together when the wavelength is infinitely long compared to the ship size. The same is observed in RAOs for heave, roll and sway near zero frequency when the body behaves as the wave particle.

4) The natural heaving and rolling motions cause significant effects on the vertical relative motions, giving much higher relative motion RAO such as 2.7 on the weather side while 1.3 on the lee side.

8.3.1.3 QTFs for lateral drift force

We have computed the complete QTFs of the lateral drift force on a ship at an oblique heading due to bichromatic waves as derived in the added resistance in Sec. 8.2. The QTFs for the swaying drift force at 60° heading are presented in the isometric view of moduli in Fig. 8.9.

We see the difference- and sum-frequency QTFs in the two octant regions. The sum frequency-QTFs appear to be greater than the difference-frequency

QTFs. It may be due to that the body being at zero speed of advance while allowing the motions due to the linear waves. The mean drift force and moment on the series 60 ship at beam seas computed by the far-field approach (Maruo, dotted line) and near-field result (solid line) are illustrated in Fig. 8.11. Comparison of the two force coefficients is in excellent agreement. Here we observe that both near- and far-field computations are nearly identical. If the comparisons turn out to be different, they should be questioned.

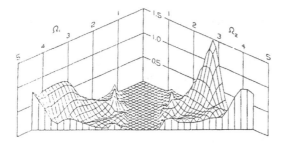

Fig. 8.10. Analytical QTFs for lateral drift force at 60° heading.

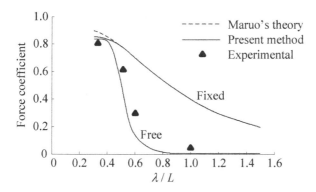

Fig. 8.11. Analytical mean drift force and moment in beam sea. Experimental symbols are for the mean drift force, while the moment is not experimentally measured.

8.4 Simulation of Slowly-Swaying Drift Force

Since the number of oscillations of the slow drift force in the experimental record is much less than those of the linear random wave excitation and response, the 2nd-order statistical convergence requires much longer sample length than the linear one. For instance Dalzell (1974) recommends that one may need to increase 10–12 times the linear sample length. Since such experimental data was not available, one had attempted to resort to simulation of the lateral drift response with long enough sample time (Kim and Boo, 1991; Kim and Kee, 1992).

From the analytical QTFs of the swaying drift force we obtain low pass QTFs in the range of the output frequency 0.0–0.15 rad/s, in the similar way as slow surge shown in Fig. 8.5 in Sec. 8.3.

The basic approach involves simulating first the Gaussian random wave long enough given the wave energy spectrum (Chap. 3) and next applying inverse Fourier transform of the low pass filter frequency response to determine the quadratic impulse response function QIRF for the slowly-varying drift force. Then the drift force in time domain is obtained through double convolution integral as shown in Eq. (7.1).

8.4.1 An example of slowly-varying lateral drift force

We consider 152.4 m long series 60 of C_B 0.6 that was used in the research of the added resistance in Sec. 8.2. First we determine the time-domain system characteristics QIRF $g_2 \Delta t^2$ corresponding to the quadratic frequency response function G_2 in the low output frequency region as shown in Fig. 8.12(a). Now we simulate the slowly-varying lateral drift force of a series in the random beam seas of $H_S = 5.5$ m as given in Fig. 8.12(b) employing the QIRF and the double convolution according to Eq. (7.1). The simulated slowly-varying sway drift force of the ship in the beam sea is obtained as shown in Fig. 8.12(c). It is interesting to note that the slowly-varying sway drift force in the large wave groups become large. The same behaviors were observed in the two examples; in the slowly-varying large drift force of model tanker in the regular wave group as shown in Sec. 7.1.5 and the slowly-varying large surge-drift force in the large random wave groups in Sec. 8.3.

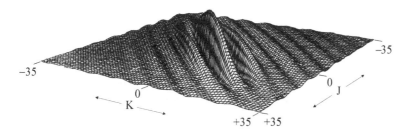

Fig. 8.12(a) Isometric sketch of magnitude of QIRF g_2 ($j\Delta r$, $k\Delta r$) Δt^2, $j=k$ (-35, 35).

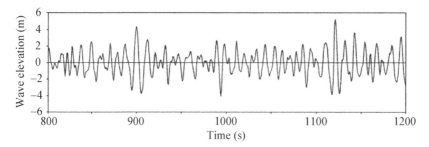

Fig. 8.12(b) Wave elevation.

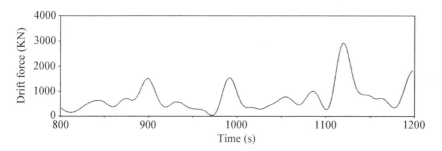

Fig. 8.12(c) Drift force.

We recall that the slowly-varying large surge drift force occurs in the large random wave groups as observed in Sec. 8.3. In the present case the large sway drift force occurs at the large random group waves. These two cases have the same conditions except the direction of the wave to the ship. The reason of large drift force occurring in the large linear wave groups will be examined by observing the large 2nd-order wave group in Chap. 9.

8.4.2 Convergence of the mean drift force

The cumulative mean of the slowly-varying drift force is computed referring to Sec. 7.3.2.3 and the cumulative mean was taken until 40 hrs as illustrated Fig. 8.13. It is interesting to observe that the cumulative mean fluctuates largely in the beginning and moves into the converging region.

In the example, 10 hrs may be a reasonable convergence limit. It is to be noted that using different hrs in simulations may give different results.

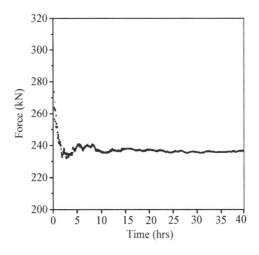

Fig. 8.13. The cumulative mean drift force of series 60 ship.

8.5 Wave Drift Damping

Recall that we have compared the analytical and experimental mean wave added resistance (or mean wave drift force) of a ship running in head seas in Sec. 8.2.1 from which the mean drift force is a function of speed and monochromatic wave frequency. If the ship is moored in a monochromatic wave it will oscillate slowly a long distance in the alternate directions of motion due to the mean wave drift force. Such a wave drift oscillation will cause the ship to experience damping due to its movement over the train of the monochromatic waves and viscosity of water. Figure 8.14 illustrates results of the free oscillations in still water and in waves indicating the decays due to damping. The decay in the wave train is higher because the wave induced damping is added to the still water viscous damping. It is known (Weichers, 1987) that the total damping is linearly proportional to the slow drift speed, thus the contribution of the quadratic viscous damping to the total damping is negligible. Based on the linearity of damping coefficients, the viscous damping can be separated from the total damping coefficient. The remaining damping coefficient is assumed to be caused by the waves. The experiment (Weichers, 1987) confirms that the wave induced damping coefficient is proportional to the wave amplitude square. Thus the wave induced damping has the same property as the 2nd-order mean wave drift force.

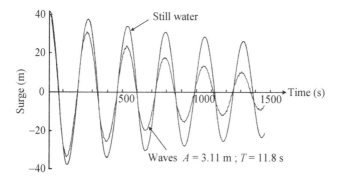

Fig. 8.14. Surge decay curves of a ship model in still water and regular waves (Weichers, 1987).

The mean added wave resistance or mean wave drift force D of a vessel advancing at a fixed speed U may be expressed as a function of monochromatic wave frequency and speed:

$$D = D(\omega, U) \qquad (8.34)$$

Since the monochromatic mean wave drift force at low drift speed may be given in the form of the mean wave drift force at low speed, we may express it approximately in a series expansion retaining the 1st-order term with respect to velocity U:

$$D(\omega,U) = D(\omega,0) + \left(\frac{\partial D(\omega,0)}{\partial U} \right) U \qquad (8.35)$$

The 1st-term is the ordinary mean drift force at zero speed. It is to be noted that the mean wave drift force at zero speed assumes the body undergoes the 1st-order motion. The 2nd term is regarded as the wave drift damping force in which the force derivative represents the linear mean wave drift damping coefficient $B(\omega)$ due to monochromatic wave:

$$B(\omega) = \frac{\partial D(\omega,0)}{\partial U} \qquad (8.36)$$

The coefficients $B(\omega)$ are often determined through experiment. In this regard, studies are conducted by Nakamura (1986), especially Weichers (1987) and Kinoshita and Takaiwa (1990), for example, have been primarily concerned with determination of mean wave drift damping coefficients in monochromatic waves.

Clark et al. (1993) investigated the formulas and computations by others and heuristically derived the theoretical mean wave drift damping, which was later analytically derived (Aranha, 1994; 1996). The mean drift force for two-dimensional body advancing at small U in the following wave of frequency ω by:

$$D(\omega,U) = D(\omega_e,0)\left[1 - 4\left(\frac{U}{c} \right) \right] \qquad (8.37)$$

$$c = \frac{g}{\omega}, \quad \omega_e = \omega\left(1 - \frac{U}{c} \right) \qquad (8.38)$$

Expanding the right hand side with respect to U/c and disregarding the higher-order terms one has:

$$D(\omega,U) = D(\omega,0) + B(\omega) \cdot U \qquad (8.39)$$

with

$$B(\omega) = \frac{\omega^2}{g} \frac{\partial D(\omega,0)}{\partial \omega} + 4\frac{\omega}{g}D(\omega,0) \qquad (8.40)$$

$B(\omega)$ in Eq. (8.40) is the linear mean wave drift damping coefficient derived heuristically by Clark et al. (1993) and the second term in Eq. (8.40) represents the approximate formula used by Wichers and Huijismans (1984). The mean wave drift force of 3-D body in a wave of incidence β, with the wave heading β_1 corrected for the refraction by current U or forward speed of a structure in the positive x-direction (Aranha, 1996):

$$D(\omega,U;\beta) = \left(1 - \frac{4\omega U}{g}\cos\beta\right) \cdot D(\omega_e,0;\beta_1)   \tag{8.41}$$

where

$$\beta_1 = \beta + \frac{2\omega U}{g}\sin\beta, \quad \omega_e = \omega - \left(\frac{\omega^2}{g}\right)U\cos\beta   \tag{8.42}$$

Equation (8.41) can determine the mean wave drift force of a floating body restrained or free to oscillate in deep water monochromatic waves of arbitrary heading, moving forward or backward. The familiar monochromatic mean wave drift force may be hydrodynamically computed, for instance, by HOBEM (Liu et al., 1993) at $U = 0$ in Chap. 9.

For computation of mean wave drift damping as function of monochromatic wave frequency, one needs to determine the mean drift force at $U = 0$ and then the mean drift force at small speed U forward and backward employing Eq. (8.41), respectively. Thus we obtain the mean wave drift damping coefficient in the drift motion in the monochromatic wave in the form:

$$B(\omega) = \frac{D(\omega,U) - D(\omega,-U)}{2U}   \tag{8.43}$$

The expected value of the mean drift damping force in the Gaussian sea can be determined according to the Volterra formula for the expected value of quadratic response in Gaussian sea in Sec. 7.2.4. Since $B(\omega)/A^2$ is regarded the quadratic transfer function of mean drift damping similar to G_2. in Eq. (7.30), and $B(\omega)/A^2\omega$ is the hydrodynamically determined mean $f_2(\omega, -\omega)$ as given in Sec. 9.4.6, the Volterra quadratic mean $G_2(\omega, -\omega)$ is given by $2\times$ $f_2(\omega, -\omega)$. Thus the expected value of potential wave drift damping in the Gaussian sea Eq. (7.30) is given in the form:

$$[B(\omega)] = 2\int_0^\infty \frac{B(\omega)}{A^2}U_{xx}(\omega)d\omega   \tag{8.44}$$

In the preceding we have determined the viscous damping from the decay test of the body in the calm water. See Sec. 6.6.1.2. The total expected value of the mean wave drift damping in the Gaussian sea is the sum of the expected value of potential wave drift damping in the Gaussian sea Eq. (8.44) and the viscous damping. Krafft and Kim (1991) measured the viscous and wave drift damping of a shallow drafted barge slowly-drifting in wave tank and the results were used in the simulation of surge drift motion in random seas. An example of the computation of expected value for TLP will be found in Chap. 9.

8.5.1 Computation of wave drift damping

Zou (1997) computed mean surge wave drift damping of a variety of structures fixed and free in deep water, using Aranha's formula (1996) and computation tool HOBEM (Liu et al., 1995). Figure 8.15 shows a comparison of the wave drift damping of an horizontally oscillating vertical truncated cylinders with draft of three radii (3 × a).the results of 48 and 96 elements by Zou (1997) are in excellent agreement with each other to three decimal digits, justifying that the HOBEM of 48 element is accurate enough. Eatock Taylor et al. (1994), Nossen et al. (1991) and Faltinsen (1990) are compared with the results of HOBEM applying Aranha's theory.

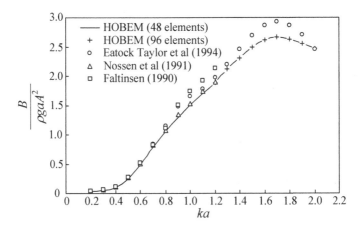

Fig. 8.15 Wave drift damping coefficient of a vertical truncated cylinder k wave number and a is radius of the cylinder. k denotes wavenumber and a is radius of the cylinder.

8.6 Mean Drift Forces on Tandemly Floating Ships

In the design and operations of moored system of multi-floating bodies, it is highly desirable to prevent damages that might occur among the multi-bodies moored in the seas. It is common to the drift forces on the bodies parallel to each other than the longitudinal wave drift force for tandem moored vessels in head seas. Hong et al. (2002) have tested and analyzed the tandem moored ships using the near-field method computed by HOBEM developed by the authors. Figure 8.16 shows a comparison of longitudinal wave drift forces on FPSO (140k tonnes) and shuttle tanker of FPSO. The longitudinal mean drift force on FPSO is much greater than that of the shuttle tanker. This may be attributed to that the beam of FPSO being much greater than that of the shuttle tanker. For details see Hong et al. (2002). Inoue and Ali (2003) presented relative motions between bodies in the vicinity and drift forces of them. It applied far-field method and source distribution method. The computations were carried out for the parallel and tandem positions.

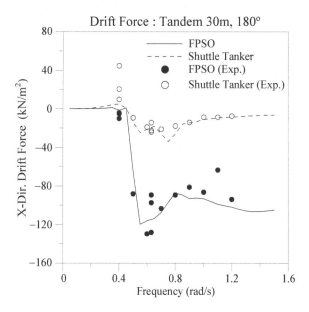

Fig. 8.16. Longitudinal wave drift forces of tandem moored vessels in head sea.

References

Journal references:

Aranha, J.A.P. (1994). A Formula for 'wave damping' in the drift of a floating body, J. Fluid Mech., 275, pp. 147–155.

Aranha, J.A.P. (1996). Second-Order Horizontal Steady Forces and Moment on a Floating Body with Small Forward Speed, J. Fluid Mech., 313, pp. 39–54.

Clark, P.J., Malenica, S. and Molin, B. (1993). An heuristic approach to wave drift damping, App. Ocean. Res., 15, pp. 53–55.

Dalzell, J.F. and Kim, C.H. (1979). Analytical Investigation of the Quadratic Frequency Response for Added Resistance, J. Ship Res., Vol. 23, No. 3, pp. 198–208.

Kim, C.H. and Chou, F. (1973). Prediction of Drifting Force and Moment on an Ocean Platform Floating in Oblique Waves, International Shipbuilding Progress, pp. 78–91.

Kim, C.H. and Dalzell, J.F. (1981). An Analysis of the Quadratic Frequency Response for Lateral Drifting Forces and Moments, J. Ship Res., Vol. 25, No. 2, pp.117–129.

Koterayama, W. (1976). Motions of Moored Floating Body and Tension of Mooring Lines in Waves, J. of Seibu-Zosen-Kai, Japan.

Krafft, M.J. and Kim, C.H. (1991). Experimental Investigation of Drift Damping and Slow Drift Motion in Bi-Frequency Domain, Int. J. Offshore and Polar Eng., Vol. 1, No. 3, pp. 235–238.

Liu, Y.H., Kim, C.H. and Kim, M.H. (1993). The Computation of Mean Drift Forces and Wave Run-Up by Higher-Order Boundary Element Method, Int. J. Offshore and Polar Eng., Vol. 3, No. 2, pp. 101–106.

Maruo, H. (1960). The Drift of a Body Floating on Waves, J. Ship Res., Vol. 4, No. 3, pp. 1–10.

Newman, J.N. (1967). The Drift Force and Moment on Ships in Waves, J. Ship Res., Vol. 11, No. 1, pp. 51–60.

Nossen, J., Grue, J. and Palm, E. (1991). Wave Forces on Three-Dimensional Floating Bodies with Small Forward Speed, J. Fluid Mech., Vol. 227, pp. 135–160.

Ogawa, A. (1967). The drifting force and moment on a ship in oblique regular waves, International Shipbuilding Progress, Vol. 14, No. 149.

Pinkster, J.A. (1980). Low Frequency Second Order Wave Exciting Forces on Floating Structures, Publication No. 650, Netherlands Ship Model Basin, Wageningen-Netherlands.

Soeding, H. (1976). Second Order Forces on Oscillating Cylinders in Waves, Schiffstechnik, Bd. 23.

Weichers, J.E.W. (1987). A Simulation Model for a Single Point Moored Tankers, Marine Research Institute Netherlands, Publication No. 797, Wageningen, The Netherlands.

Book references:

Faltinsen, O.M. (1990) Loads on Ships and Offshore Structures, Cambridge University Press, U.K.

Newman, J.N. (1977) Marine Hydrodynamics, MIT Press, Cambridge, Massachusetts.

Proceedings references:

Dalzell, J.F. and Kim, C.H. (1976). Analytical Investigation of the Quadratic Frequency Response for Added Resistance, SIT-DL-76-1878, AD-A032946/6G1, Stevens Institute of Technology, Hoboken, N.J.

Eatock Taylor, R. and Natvig, B.J. (1994). Effects of Wave Drift Damping on TLP Horizontal Offsets, Proc. 7th BOSS, II, pp.31–41.

Faltinsen, O.M. and Loken, A.E. (1978). Drift Forces and Slowly Varying Forces on Ships and Offshore Structures in Waves, Norwegian Marine Research, No. 1, pp. 2–15.

Hong, S., Kim, J.H., Kim, H.J., and Choi, Y.R. (2002). Experimental Study on Behavior of Tandem and Side-by-side Moored Vessels, Proc 12th Int. Offshore and Polar Eng. Conf., Kitakyushu, Japan, Vol. 1, pp. 841–847.

Inoue, Y. and Ali, M.T. (2003). A Numerical Investigation on the Behaviour of Multiple Floating Bodies of Arbitrary Arrangements in Regular Waves, Proc 13th Int. Offshore and Polar Eng. Conf., Vol. 3, pp. 558–565.

Kim, C.H. and Boo, S.Y. (1990). Statistical Analysis of Slow Drift Forces in Random Seas, Proc. First Pacific/Asia Offshore Mech. Symp., Seoul, Korea, Vol. 2, pp. 169–177.

Kim, C.H. and Breslin, J.P. (1976). Drifting Motion of a Moored Ship in Head Seas, Proc. BOSS'76, Trondheim, Norway, pp. 320–352.

Kim, C.H. and Kee, S.T. (1992). Non-Gaussianity Effect of Input Wave on Cross-Bi-Spectral Estimates, Proc. 2nd Int. Offshore and Polar Eng. Conf., San Francisco, California, Vol. 3, pp. 496–503.

Kim, C.H. (1974). Calculation of Motions and Loads of a Ship Uniformly Advancing in Oblique Waves, Technical Memorandum SIT-DL-74-166, Stevens Institute of Technology, Hoboken, N.J.

Kinoshita, T and Takaiwa, K (1990). A Mathematical Model for Slow Drift Motion of a Vessel Moored in Waves Determined by Oscillation Tests, in Regular Wave Trains, Report of the Institute of Industrial Science, Vol. 35, No. 5, The University of Tokyo.

Lalangas, P. (1963). Lateral and vertical forces and moments on a restrained series 60 ship model in oblique regular waves, Report 920, Davidson Laboratory, Stevens Institute of Technology, Hoboken, N.J.

Newman, J.N. (1974). Second-Order Slowly-varying Forces on Vessels in Irregular Waves, International Symposium on the Dynamics of Marine Vehicles and Structures in Waves, University of College London, pp. 182–186.

Pinkster, J.A. and Van Oortmerssen, G. (1977). Computation of the First and Second Order Wave Forces on Bodies Oscillating in Regular Waves, Proc. 2nd Int. Conf. Numerical Ship Hydrodynamics, University of California, pp. 136–156.

Salvesen, N. (1974). Second-order Steady-state Forces and Moments on Surface Ships in Oblique Regular Waves, Int. Symp. Dynamics of Marine Vehicles and Structures in Waves, University College London, pp. 212–226.

Strom-Tejsen, J., Yeh, H.Y.H., and Moran, D.C. (1973). Added Resistance in Waves, Trans., Soc. of Naval Arch. and Marine Engineers, Vol. 81, pp. 109–143.

Weichers, J.E.W. and Huijismans, R.H.M. (1984). On the low frequency hydrodynamic damping forces acting on offshore moored vessels, OTC Paper No. 4813.

Zou, J. (1997). Investigation of Slowly-Varying Drift Motion and Springing and Ringing of Tension Leg Platform System in Nonlinear Irregular Waves, Ph.D. Dissertation in Ocean Engineering, Texas A & M University.

Chapter 9

2nd-Order Wave and 2nd-Order Force

9.1 Introduction

The 2nd-order nonlinear wave theories were developed by Tick (1961), Longuet-Higgins (1963), Hasselmann (1966), Bowers (1975), Sharma and Dean (1981), and Dalzell (1999). Perturbation expansion methods for treating such problems may be referred to Stoker (1957), Wehausen and Laitone (1960), Newman (1977), Phillips (1977), and Sarpkaya and Isaacson (1981). It is noted that Dalzell (1999) used the computer program in deriving the analytical solution.

We review the 2nd-order wave and 2nd-order force theories in the same fashion. The formulations are basically based on the perturbation expansions. That is, we first solve the linear potential satisfying the linear boundary conditions, with which the 2nd-order potential is determined satisfying the 2nd-order boundary conditions.

The perturbation scheme was originally developed by Stokes (1847). Thus the 2nd-order wave theory is frequently called Stokes 2nd-order wave. The 3rd-order wave developed by Pierson (1993) is similarly called Stokes 3rd-order wave.

There are similarities between the 2nd-order hydrodynamic theories (wave and force) and Volterra quadratic model. This will be demonstrated by comparing the 2nd-order hydrodynamic theories with the Volterra quadratic model.

We will later apply the 2nd-order waves to simulate the field and laboratory waves. The 2nd-order forces and Volterra quadratic model will be applied to estimate the 2nd-order response of typical offshore structures such as TLP to Gaussian seas including the slow drift oscillation and springing response.

However, it is very important to recognize the limitations of the applicability of the 2nd-order theories as well as the Volterra quadratic model to the severer nonlinear seas higher than 2nd-order.

9.2 2nd-Order Wave Theory

9.2.1 Formulation of 2nd-order free surface boundary conditions

Recall that we have written the nonlinear (or exact) boundary conditions on the free surface in Sec. 4.3.1. From those formulas we will derive 1st- and 2nd-order boundary conditions. Let there be a potential flow in the fluid domain that is horizontally infinite and vertically bounded by the constant depth sea bottom and free surface. Then the potential satisfies the continuity condition in the fluids:

$$\frac{\partial^2 \Phi}{\partial x^2} + \frac{\partial^2 \Phi}{\partial z^2} = 0 \tag{9.1}$$

the boundary condition on the uniform flat bottom of depth h:

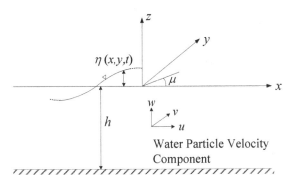

Fig. 9.1. Schematic diagram of coordinate system.

$$\frac{\partial \Phi}{\partial z} = 0, \; z = -h \tag{9.2}$$

and the exact kinematical and dynamical boundary condition on the free surface:

$$\frac{\partial \eta}{\partial t} + \frac{\partial \Phi}{\partial x}\frac{\partial \eta}{\partial x} - \frac{\partial \Phi}{\partial z} = 0 \; \text{ at } z = \eta(x,t) \tag{9.3}$$

$$\frac{\partial \Phi}{\partial t} + \frac{1}{2}\left[\left(\frac{\partial \Phi}{\partial x}\right)^2 + \left(\frac{\partial \Phi}{\partial z}\right)^2\right] + gz = p(x,z,t) \; \text{ at } \; z = \eta \tag{9.4}$$

In Eq. (9.4), it is assumed that density is unit and p represents the uniform atmospheric pressure on the wave surface at any instant t. The linear wave theory provides a first approximation to the exact wave motion. In order to obtain the 2nd-order solutions, we consider a perturbation procedure in which successive approximations are developed. Suppose the disturbance was given to the water, which is initially at rest and the disturbed flow is slightly different from the initial condition. We denote the exact solution the solution for the known state at rest initially, by $\Phi^{(0)}$, $\eta^{(0)}$, $u^{(0)}$ etc. It is assumed that Φ and associated variables such as η, u, w, $\cdots$, may be written in the form:

$$\Phi(x,z,t) = \Phi^{(0)}(x,z,t) + \varepsilon \, \Phi^{(1)}(x,z,t) + \varepsilon^2 \Phi^{(2)}(x,z,t) + O(\varepsilon^3)$$
$$\eta(x,t) \;\; = \eta^{(0)}(x,t) + \varepsilon\eta^{(1)}(x,t) + \varepsilon^2\eta^{(2)}(x,t) + O(\varepsilon^3) \tag{9.5}$$
$$\text{etc.}$$

in which $\Phi^{(0)} = 0$ and $\eta^{(0)} = 0$ and ε is the perturbation parameter such as wave slope. Thus by substituting Eq. (9.5) into the continuity condition Eq. (9.1) and bottom condition Eq. (9.2), and by collecting terms of order ε, ε^2, $\cdots$, we obtain:

$$\frac{\partial^2 \Phi^{(n)}}{\partial x^2} + \frac{\partial^2 \Phi^{(n)}}{\partial z^2} = 0 \; \text{ for } \; n = 1, \, 2, \, \text{in the fluid domain} \tag{9.6}$$

$$\frac{\partial \Phi^{(n)}}{\partial z} = 0 \; \text{ at } \; z = -h \; \text{ for } \; n = 1, \, 2 \tag{9.7}$$

The free surface boundary conditions contain nonlinear terms consisting of quadratic products and are applied to the unknown surface $z = \eta$. We may replace the exact kinematic boundary condition Eq. (9.3) by the statement that the substantial derivative of the pressure is zero over the free surface (Newman, 1977):

$$\frac{Dp}{Dt} = 0 \quad \text{at} \quad z = \eta \tag{9.8}$$

Namely,

$$\left(\frac{\partial}{\partial t} + \nabla\Phi \cdot \nabla\right)\left(\frac{\partial\Phi}{\partial t} + \frac{1}{2}\nabla\Phi \cdot \nabla\Phi + gz\right) = 0 \quad \text{at} \quad z = \eta \tag{9.9}$$

Operating the above derivatives we have:

$$\frac{\partial^2\Phi}{\partial t^2} + g\frac{\partial\Phi}{\partial z} + \left(\frac{\partial}{\partial t} + \frac{1}{2}\nabla\Phi \cdot \nabla\right)(\nabla\Phi \cdot \nabla\Phi) = 0 \quad \text{at} \quad z = \eta \tag{9.10}$$

The exact kinematic boundary condition, Eq. (9.3) has been replaced by Eq. (9.10), which is expressed in terms of one variable Φ instead of two variables η and Φ, which is an advantage in the algebraic works involved. Similar works are given in (Phillips, 1977; Sarpkaya and Isaacson, 1981).

Since pressure $p(x, z, t) = 0$ over the free surface in Eq. (9.4), one has the exact dynamic boundary condition on the free surface in the form:

$$\eta = -\frac{1}{g}\left(\frac{\partial\Phi}{\partial t} + \frac{1}{2}\nabla\Phi \cdot \nabla\Phi\right) \quad \text{at} \quad z = \eta \tag{9.11}$$

Thus, we have two new free surface boundary conditions Eqs. (9.10) and (9.11).

The foregoing free surface boundary conditions have to be satisfied at the unknown wave elevation $z = \eta$. This difficulty is avoided by approximation using Taylor expansions at $z = 0$.

The boundary conditions are presented in the form:

$$F(x, z, t) = z - \eta(x, t) = 0 \text{ at } z = \eta(x, t) \tag{9.12}$$

Then one may expand it in Taylor series about the mean position of oscillation $z = 0$:

$$F(x,z,t) = F(x,0,t) + \eta \frac{\partial}{\partial z} F(x,0,t)$$
$$+ \frac{1}{2!} \eta^2 \frac{\partial^2}{\partial z^2} F(x,0,t) + \cdots = 0 \tag{9.13}$$

where, $F(x, 0, t)$ is the 0th-order of the surface F and the rest represents the 1st- and 2nd-order deviations of the surface F from the equilibrium. By expressing Eqs. (9.10) and (9.11) in the forms of Eqs. (9.12) and (9.13), and by using the series expansions as indicated in Eq. (9.5), we may obtain successive approximations of solutions of multiple order.

Collecting the terms of order ε:

$$\frac{\partial^2 \Phi^{(1)}}{\partial t^2} + g \frac{\partial \Phi^{(1)}}{\partial z} = 0 \text{ at } z = 0 \tag{9.14}$$

$$\eta^{(1)} = \frac{-1}{g} \left(\frac{\partial \Phi^{(1)}}{\partial t} \right) \text{ at } z = 0 \tag{9.15}$$

Collecting the terms of order ε^2 we have:

$$\frac{\partial^2 \Phi^{(2)}}{\partial t^2} + g \frac{\partial \Phi^{(2)}}{\partial z} = \frac{1}{g} \frac{\partial \Phi^{(1)}}{\partial t} \frac{\partial}{\partial z} \left(\frac{\partial^2 \Phi^{(1)}}{\partial t^2} + g \frac{\partial \Phi^{(1)}}{\partial z} \right)$$
$$- \frac{\partial}{\partial t} \left(\nabla \Phi^{(1)} \right)^2 \qquad \text{at } z = 0 \tag{9.16}$$

$$\eta^{(2)} = -\frac{1}{g} \left[\frac{\partial \Phi^{(2)}}{\partial t} - \frac{1}{g} \frac{\partial \Phi^{(1)}}{\partial t} \frac{\partial^2 \Phi^{(1)}}{\partial z \partial t} + \frac{1}{2} (\nabla \Phi^{(1)})^2 \right] \text{ at } z = 0 \tag{9.17}$$

Substituting the first order solution $\Phi^{(1)}$ into Eq. (9.16), we may solve for 2nd-order potential $\Phi^{(2)}$ at $z = 0$ and consequently 2nd-order wave elevation $\eta^{(2)}$ at $z = 0$ using Eq. (9.17).

9.2.2 2nd-order wave potential in bichromatic wave

Following the procedure in the foregoing section we will determine the 1st- and 2nd-order potentials as well as wave elevations. Let the bichromatic wave be given in the form:

$$\eta^{(1)}(x,y,t)=\sum_{i=1}^{2} A_i \cos\psi_i \qquad (9.18)$$

where

$$\begin{aligned}
&\psi_i = (\mathbf{k}_i \cdot \mathbf{x} - \omega_i t + \varepsilon_i), \\
&\mathbf{k}_i = (k_x, k_y) = (k_i \cos\mu_i, k_i \sin\mu_i) \\
&k_i = |\mathbf{k}_i|, \quad \mathbf{x} = (x,y)
\end{aligned} \qquad (9.19)$$

Then the corresponding 1st-order solution that has already been given in Chap. 4 is written for the bichromatic wave:

$$\Phi^{(1)}(x,y,z,t)=\sum_{i=1}^{2} B_i \frac{\cosh k_i(h+z)}{\cosh k_i h}\sin\psi_i \qquad (9.20)$$

with

$$\omega_i^2 = g|\mathbf{k}_i|\tanh|\mathbf{k}_i|h$$

$$B_i = \frac{gA_i}{\omega_i} \qquad (9.21)$$

Substituting Eq. (9.20) into the 2nd-order free surface boundary condition Eq. (9.16), we determine the 2nd-order wave potential due to bichromatic wave in the form (Dean and Sharma, 1981):

$$\begin{aligned}
\Phi^{(2)}(x,y,z,t) = &\sum_{i=1}^{2}\sum_{j=1}^{2} \frac{B_i B_j}{4} \frac{\cosh k_{ij}^-(h+z)}{\cosh k_{ij}^- h} \frac{D_{ij}^-}{(\omega_i - \omega_j)}\sin(\psi_i - \psi_j) \\
&+ \sum_{i=1}^{2}\sum_{j=1}^{2} \frac{B_i B_j}{4} \frac{\cosh k_{ij}^+(h+z)}{\cosh k_{ij}^+ h} \frac{D_{ij}^+}{(\omega_i + \omega_j)}\sin(\psi_i + \psi_j)
\end{aligned} \qquad (9.22)$$

with

$$k_{ij}^+ = |\mathbf{k}_i + \mathbf{k}_j|, \quad k_{ij}^- = |\mathbf{k}_i - \mathbf{k}_j| \qquad (9.23)$$

$$D_{ij}^{-} = \frac{\left(\sqrt{R_i} - \sqrt{R_j}\right)\left\{\sqrt{R_j}(k_i^2 - R_i^2) - \sqrt{R_i}(k_j^2 - R_j^2)\right\}}{\left(\sqrt{R_i} - \sqrt{R_j}\right)^2 - k_{ij}^{-}\tanh k_{ij}^{-}h}$$

$$+ \frac{2\left(\sqrt{R_i} - \sqrt{R_j}\right)^2 (\mathbf{k}_i \cdot \mathbf{k}_j + R_i R_j)}{\left(\sqrt{R_i} - \sqrt{R_j}\right)^2 - k_{ij}^{-}\tanh k_{ij}^{-}h} \qquad (9.24)$$

$$D_{ij}^{+} = \frac{\left(\sqrt{R_i} + \sqrt{R_j}\right)\left\{\sqrt{R_i}(k_j^2 - R_j^2) + \sqrt{R_j}(k_i^2 - R_i^2)\right\}}{\left(\sqrt{R_i} + \sqrt{R_j}\right)^2 - k_{ij}^{+}\tanh k_{ij}^{+}h}$$

$$+ \frac{2\left(\sqrt{R_i} + \sqrt{R_j}\right)^2 (\mathbf{k}_i \cdot \mathbf{k}_j - R_i R_j)}{\left(\sqrt{R_i} + \sqrt{R_j}\right)^2 - k_{ij}^{+}\tanh k_{ij}^{+}h} \qquad (9.25)$$

and

$$R_i \equiv k_i \tanh k_i h \qquad (9.26)$$

It is noted that the difference frequency term in Eq. (9.24) tends to infinity when $i = j$. However, it may easily be avoided by using a control in the computer program.

Substituting the 1st- and 2nd-order potentials Eqs. (9.20) and (9.22) into 2nd-order elevation formula Eq. (9.17), we obtain the 2nd-order wave elevation:

$$\eta_2(x,y,t) = \frac{1}{4}\sum_{i=1}^{2}\sum_{j=1}^{2} A_i A_j \left\{ K_{ij}^{-}\cos(\psi_i - \psi_j) + K_{ij}^{+}\cos(\psi_i + \psi_j) \right\} \qquad (9.27)$$

with

$$K_{ij}^{-} = \frac{D_{ij}^{-} - (\mathbf{k}_i \cdot \mathbf{k}_j + R_i R_j)}{\sqrt{R_i R_j}} + (R_i + R_j) \qquad (9.28)$$

$$K_{ij}^{+} = \frac{D_{ij}^{+} - (\mathbf{k}_i \cdot \mathbf{k}_j - R_i R_j)}{\sqrt{R_i R_j}} + (R_i + R_j) \qquad (9.29)$$

The total wave elevation is the sum of the 1st- and 2nd-order wave:

$$\eta(x,y,t) = \eta_1(x,y,t) + \eta_2(x,y,t) \qquad (9.30)$$

which is called the 2nd-order wave elevation. The foregoing formulas have been coded and extensively applied by the author and their validity has been confirmed privately with Forristall (2000).

In Eq. (9.27) one designates the kernels $K_{ij}^{\pm}$ as the hydrodynamic QTFs of the 2nd-order wave corresponding to QTFs defined in the Volterra quadratic model Eq. (7.11).

9.2.2.1 Similarity between Volterra quadratic model and 2nd-order wave theory

The 2nd-order wave in Eq. (9.27) consists of six components; two mean terms, two double frequency terms, and difference and sum frequency terms. The comparison of Eq. (9.27) and Volterra quadratic model Eq. (7.11) leads to:

$$LTF = G_1 = 1, \; QTF = G_2(\omega_i, \pm\omega_j) = \frac{K_2(\omega_i, \pm\omega_j)}{2} \equiv \frac{K_{ij}^{\pm}}{2} \qquad (9.31)$$

It is noted that one half of the hydrodynamic QTFs are identical to those of Volterra quadratic model. Hence according to Eq. (7.30), the expected value of the 2nd-order random wave in the given sea:

$$E[y(t)] = \frac{1}{2} \int_0^\infty K_2(\omega, -\omega) U_{xx}(\omega) d\omega \qquad (9.32)$$

and the energy spectral density of 2nd-order random wave is given according to Eq. (7.35) in the form:

$$\begin{aligned} U_{\eta\eta}(\omega) &= U_{\eta_1\eta_1}(\omega) \\ &+ \frac{1}{2} \int_0^\infty |K_2(\omega - \xi, \xi)|^2 \, U_{\eta_1\eta_1}(|\omega - \xi|) U_{\eta_1\eta_1}(|\xi|) d\xi \end{aligned} \qquad (9.33)$$

We can also derive the 2nd-order wave elevation time series from the Gaussian sea using Volterra quadratic model Eq. (7.11) by replacing the LTFs and QTFs in the Volterra model by hydrodynamic LTFs and QTFs as given in Eq. (9.31). The foregoing derivation is apparently trivial, but it is important to distinguish the Volterra quadratic model from the 2nd-order hydrodynamic theory.

We will examine the distribution of the hydrodynamic kernels $K_2(\omega_i, \omega_j)$ of a long-crested sea of relatively shallow water defined by the B-spectrum; $E = 10.56 \text{ m}^2$, $H_s = 13$ m and modal frequency $\omega_m = 0.5$ rad/s in the water of 30 m depth. Figure 9.2 shows the distribution of $K_2(\Omega_1, \Omega_2)$ in the difference and sum frequency domain. It appears that the difference frequency terms are all negative while the sum frequency terms are all positive and these become minimum and maximum along the diagonal ω_1-axis in the Ω_1-Ω_2 domain respectively. Thus, according to Eq. (9.32), we will have negative expected value of the 2nd-order wave elevation.

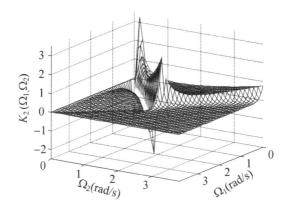

Fig. 9.2. $K_2(\Omega_1, \Omega_2)$ of a long-crested sea wave derived from a B-spectrum.

9.2.2.2 Energy spectrum of 2nd-order random wave

We consider energy density spectrum of the above 2nd-order random wave containing the 1st- and 2nd-order terms. Obviously there are two ways of determining the energy density spectrum. First one determines the energy density spectrum from the simulated 2nd-order wave elevation time series Eq. (9.30) as shown in Fig. 9.3(a). Second one may employ the energy density spectral formula given in Eq. (9.33) as illustrated in Fig. 9.3(b). Since the energy spectrum of the 2nd-order terms is very small, the semi-log scale is used. In the low and high and wave frequency regions, the effect of the 2nd-order wave appears to be significant, though their magnitudes are small compared to the linear wave energy spectrum as shown in Fig. 9.3(a).

The results shown in Figs. 9.3(a) and (b) are in excellent agreement. Thus, the validity of the formula Eq. (9.33) has been reassured.

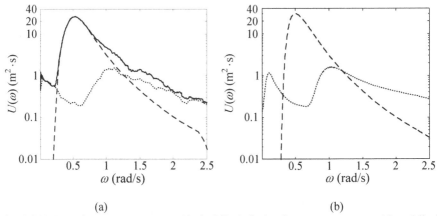

(a)   (b)

Fig. 9.3(a) 1st-order energy spectrum (dashed line), 2nd-order energy spectrum (dotted line) and 1st + 2nd -order energy spectrum (solid line). (b) presents Eq. (9.33).

9.2.3 Self-interacting Stokes 2nd-order wave

From Eqs. (9.22) and (9.27), we may extract the self-interacting Stokes 2nd-order wave , by setting $i = j$ and dropping the suffix i:

$$\Phi = \frac{gA}{\omega} \frac{\cosh k(z+h)}{\cosh kh} \sin(kx - \omega t + \varepsilon)$$
$$+ A^2 \frac{3\omega}{8} \frac{\cosh 2k(z+h)}{\sinh^4 kh} \sin 2(kx - \omega t + \varepsilon)$$

$$(9.34)$$

$$\eta(x,t) = -\frac{1}{2} A \frac{k}{\sinh 2kh} + A\cos(kx - \omega t + \varepsilon)$$
$$+ A^2 \frac{k}{4} \frac{\cosh kh}{\sinh^3 kh} (2 + \cosh 2kh)\cos 2(kx - \omega t + \varepsilon)$$

$$(9.35)$$

where the mean elevation term in Eq. (9.35) is evidently negative for finite water while zero in the deep water. The term self-interacting wave is named in view of bichromatic interaction. The self-interacting Stokes 2nd-order wave is also frequently called Stokes 2nd-order regular wave.

Dean and Dalrymple (1991) derived the same mean term in the following way; it first formulated the 2nd-order wave elevation in a general form, where the mean is determined by specifying Bernoulli's constant to be zero corresponding to no set-down in deep water. It is to be noted that in the derivation of 2nd-order wave Eq. (9.27), we have used Bernoulli's equation of zero constant $C(t) = 0$ as shown in Eq. (9.11). The negative mean elevation is also called mean sea-level set-down by Mei (1989).

9.2.3.1 Phase-locked 2nd-order wave

In referring to Eq. (9.35), the 1st-order wave satisfies the homogeneous linear boundary condition and the dispersion relation, thus it is called free wave (Chap. 4). On the other hand, the 2nd-order wave is derived from the inhomogeneous free surface boundary condition in Eq. (9.16), which does not satisfy the dispersion. The linear wave phase speed is proportional to square root of wavelength λ, while the 2nd-order wave phase velocity is not proportional to the square root of $\lambda / 2$. In fact we find from the phase function of the 2nd-order wave that the phase speed of 2nd-order wave is identical to 1st-order wave. In other words the 2nd-order wave travels with the 1st-order wave. Because of this, the Stokes 2nd-order wave is phase-locked or bounded.

If we take zero-mean in the Stokes 2nd-order wave Eq. (9.35), the 1st- and 2nd-order waves are superimposed with zero-phase difference or phase-locked. The superimposed wave in such manner is slightly elevated from the position of the 1st-order wave. In other words, the resulting 2nd-order wave becomes asymmetric with respect to MWL, in contrast to the 1st-order wave. The slight elevation of the 2nd-order wave from the 1st-order wave is commonly described as the high crest and flatter trough that demonstrates the nonlinearity feature of the Stokes 2nd-order regular wave.

9.2.3.2 Stokes 2nd-order progressing wave in deep water

The bichromatic progressing wave in deep water generates Stokes 2nd-order wave in the forms:

$$\Phi^{(1)} + \Phi^{(2)} = \sum_{j=1}^{2} \left(\frac{gA_j}{\omega_j} e^{k_j z} \sin\theta_j \right) \tag{9.36}$$

$$
\begin{aligned}
\eta^{(1)}&(x,t) + \eta^{(2)}(x,t) \\
&= A_1 \cos\theta_1 + A_2 \cos\theta_2 + \frac{A_1^2 k_1}{2} \cos 2\theta_1 + \frac{A_2^2 k_2}{2} \cos 2\theta_2 \\
&\quad - A_1 A_2 \frac{|k_1 - k_2|}{2} \cos(\theta_1 - \theta_2) + A_1 A_2 \frac{k_1 + k_2}{2} \cos(\theta_1 + \theta_2)
\end{aligned}
\tag{9.37}
$$

$$\theta_j = k_j x - \omega_j t + \varepsilon_j, \quad k_j = \frac{\omega_j^2}{g}, \quad j = 1,2$$

$$\eta^{(1)}_{\text{mean}}(x,t) = 0, \quad \eta^{(2)}_{\text{mean}}(x,t) = 0 \tag{9.38}$$

Here, we note that the 2nd-order potential does not exist. However, the 2nd-order wave elevation is derived from the quadratic interaction of the linear bichromatic waves according to Eq. (9.17). In the deep water the mean elevation is zero as seen in Eq. (9.35).

9.2.3.3 2nd-order wave group

We consider the linear amplitude modulated wave from interference of the two nearly similar sinusoids of $A_1 = 3$ m and $A_2 = 3.5$ m amplitudes, with frequencies $\omega_1 = 0.6$ rad/s and $\omega_2 = 0.9$ rad/s, both moving in x-direction in the water of constant depth 30 m.

The wave modulated by these sinusoids appears to be much elongated as shown in Fig. 9.4(b). We recall the formula of amplitude modulation in Chap. 5

$$\eta(t) = 2A \cos\left(\frac{\Delta\omega}{2} \right) \cos \omega t \tag{9.39}$$

where the modulated and carrier wave period are estimated as 41.9 s and 8.4 s, respectively, because $\Delta\omega = 0.3$ rad/s, and $\omega \approx (\omega_1 + \omega_2)/2 = 0.75$ rad/s. The foregoing estimations are in good agreement with the figure computed.

Employing Eq. (9.27), one computes the sum and difference frequency waves, respectively as shown in Figs. 9.4(b) and 9.4(c).

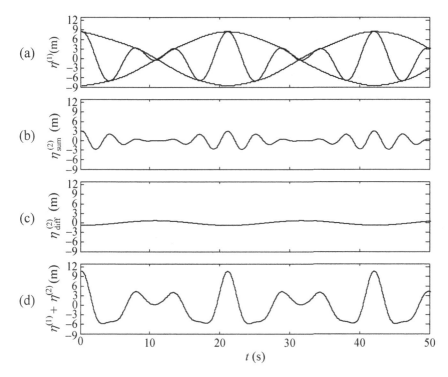

Fig. 9.4(a) The 1st-order amplitude modulated waves. (b) The 2nd-order sum frequency or 2nd-order amplitude modulated waves. (c) The 2nd-order difference frequency wave. (d) The sum of linear and 2nd-order wave group.

The 2nd-order sum frequency wave behaves similarly to the linear amplitude modulated wave. The only difference is in the carrier wave frequency, which doubles the 1st-order carrier wave.

The 2nd-order difference frequency wave elevation in Eq. (9.27) omitting the mean terms may be expressed:

$$\eta(t) = \frac{1}{2} A_1 A_2 K_{12}^- \cos \Delta \omega t \qquad (9.40)$$

Thus the period of the 2nd-order difference frequency wave is equal to $2\pi / \Delta \omega = 20.94$ s. This is in good agreement with the computation as shown in Fig. 9.4(c). It is noted that the negative mean is negligibly small.

Addition of these sum and difference 2nd-order waves to the linear amplitude modulated wave results in the 2nd-order amplitude modulated wave as illustrated in Fig. 9.4(d), which is slightly elevated from the position of the linear amplitude modulated wave. The same 2nd-order characteristic is observed as in the self-interacting Stokes 2nd-order wave in Sec. 9.2.3.

The large amplitude of 2nd-order amplitude modulated wave looks precisely similar to the 1st-order amplitude modulated wave. Thus, we may safely conclude that the large amplitude of the 2nd-order modulated wave should occur at the large amplitude of the linear modulated wave. The large slowly-varying drift forces occur in the linear large wave groups (see Secs. 7.1.5 and 8.2.1.3 and 8.4.1). Thus we may safely conclude that the large slowly-varying drift forces occur at the large amplitude of the 2nd-order amplitude modulated wave as well. If the 2nd-order wave amplitude is large, the 2nd-order drift force amplitude is also large.

9.2.4 Simulation of 2nd-order random waves

9.2.4.1 Long-crested sea

To avoid ambiguity, we redefine a new terminology. The total sum frequency term is the sum of the double frequency and sum frequency components, whereas the total difference frequency term is the sum of the mean and difference frequency components. The total 2nd-order wave is the sum of total sum and total difference frequency terms. The sum of total 1st-order and total 2nd-order wave is the result of the 2nd-order interaction of the bichromatic waves.

A long-crested relatively shallow water 2nd-order wave is simulated based on the B-spectrum with total energy $E = 10.56$ m^2, significant wave height $H_s = 13$ m, and the modal frequency $\omega_m = 0.5$ rad/s. Here the random phase simulation method is employed. The total record length is taken 1800 s with the sample interval 0.5 s and frequency increment 0.00349 rad/s, respectively. Note that the 2nd-order kernels of the long-crested sea are shown in Fig. 9.2.

Figure 9.5(a) presents the total sum frequency wave, and Fig. 9.5(b) represents the total difference frequency wave, which is depressed significantly. It is noted that the large depression occurs in the large group of linear random waves.

The total 2nd-order wave is shown in Fig. 9.4(c). Addition of the total 2nd-order wave to the total 1st-order wave is compared with the 1st-order random wave as shown in Fig. 9.5(d). The resultant 2nd-order random wave appears to be slightly elevated from the position of the 1st-order random wave. Although negative mean is present, it is negligibly small compared to the other wave elevations. Such addition of the phase-locked waves to the free waves creates the 2nd-order wave of high crests and flatter troughs as discussed in the foregoing sections. It is noted that the property of the regular wave similarly holds in the random waves.

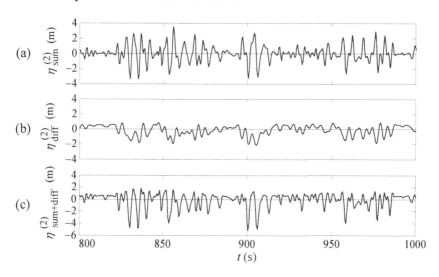

Fig. 9.5(a) Total sum frequency 2nd-order wave. (b) Total difference frequency 2nd-order wave (depression). (c) Total 2nd-order wave with mean = −0.067 m.

9.2.4.2 2nd-order short-crested sea waves

The effect of directional waves on the 2nd-order wave elevation was investigated by simulations (Sharma and Dean, 1981). The wave is simulated by using the following equation:

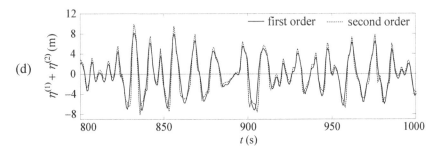

Fig. 9.5(d) The sum of 1st- and 2nd-order waves.

$$\eta(x,y,t) = \sum_{i=1}^{\infty}\sum_{l=1}^{\infty} A_{il}\cos\psi_{il}$$
$$+\frac{1}{4}\sum_{i=1}^{\infty}\sum_{j=1}^{\infty}\sum_{l=1}^{\infty} A_{il}A_{jl}\left\{K_{ijl}^{-}\cos(\psi_{il}-\psi_{jl})+K_{ijl}^{+}\cos(\psi_{il}+\psi_{jl})\right\} \tag{9.41}$$

where

$$A_{il}=\sqrt{2U(\omega_i,\mu_l)\Delta\omega_i\Delta\mu_l}, \tag{9.42}$$

$$\psi_{il}=(k_i x\cos\mu_l+k_i y\sin\mu_l-\omega_i t+\varepsilon_i) \tag{9.43}$$

$K_{ijl}^{\pm}$ denote hydrodynamic QTFs including wave heading.

The directional wave spectrum $U(\omega,\mu)$ is assumed in the form of separation:

$$U(\omega,\mu)=U(\omega)D_\nu(\mu) \tag{9.44}$$

with cosine power spreading function:

$$D_\nu(\mu)=\begin{cases}C_\nu\cos^{2\nu}(\mu-\mu_0) & \text{for } \dfrac{-\pi}{2}\leq\mu\leq\dfrac{\pi}{2}\\ 0 & \text{for otherwise}\end{cases} \tag{9.45}$$

The foregoing authors employed the directional wave of B-spectrum of total energy of 2.05 m^2 with $H_s = 5.72$ m, for the water depth of 30.5 m and investigated the effects of the directionality on the behavior of the waves and on the forces acting on a vertical bottom-fixed column. Two cases were studied in directional seas ± 90° with $\nu = 1$ and the other is $\nu = 5000$, which gives approximately unidirectional sea. The directional sea waves and loads are less than those of the unidirectional ones. Since the field waves are generally directional rather than unidirectional, the conventional unidirectional experiment in the wave tank and simulation in the unidirectional random waves would result in overestimation.

Dalzell (1999) simulated directional wave in the range of ± 22° instead of ± 90° and unidirectional sea, with the water depth of 15 m. The total difference-frequency term (depression) corresponding to the large wave groups is larger in the unidirectional sea than in the directional sea.

Forristall (2000) also investigated the effects of directionality on the depression in shallow water.

9.2.4.3 Comparison of 2nd-order wave to field wave

Forristall (2000) analyzed applicability of the 2nd-order wave theory for simulation of random seas measured in the specified region of the North Sea near the coastal zone of the Netherlands and suggested its applicability for the estimation of wave-over-topping on the fixed structures.

A variety of wave gauges were employed to measure the elevations, and formidable discrepancies were found between the data measured with different wave gauges.

The study found that the 2nd-order wave theory was good for the crest heights distribution of the measured waves.

The probability of peaks exceeding the crest heights of the narrow-band Gaussian seas is given in Eq. (1.95):

$$\Pr\{\text{peaks} \geq H_c\} = \exp\left(-\frac{H_c^2}{H_s^2/8}\right), \qquad 0 \leq H_c \leq \infty \qquad (9.46)$$

where H_c is the crest height and H_s is the significant wave height that can be given in terms of energy of the spectrum $H_s = 4(m_0)^{0.5}$ as derived in

Eq. (1.92). The solid line in Fig. 9.6 represents the Rayleigh probability of exeedence Eq. (9.46).

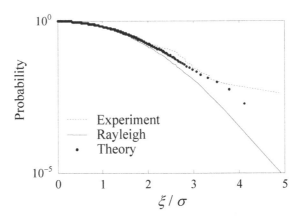

Fig. 9.6. Probability of exceedence of the crest height H_c of experimental and 2nd-order wave theory.

The research employs the following idea: it compares the ratio of the measured peak crest height to Rayleigh Eq. (9.46) at a given probability as illustrated in Fig. 9.6, to the ratio of the theoretical crest height of 2nd-order wave obtained from the same spectrum of the measured data to Rayleigh at the same probability. If these ratios are in good agreement the 2nd-order theory is regarded a good model for the simulation of the measured random wave. As an example, Hurricane Camille waves were investigated with two scanning rates 1.0 Hz and 4.0 Hz as shown in Fig. 9.7.

The measured Camille wave and simulated 2nd-order Camille wave are definitely non-Gaussian. Use of higher rate gives higher nonlinearity than the lower scan rate. It may be due to picking up truer nonlinear wave elevation from the closer discretization of the analog data. The 2nd-order simulated waves approximately model the measured data. Many detailed investigations are found in Forristall (2000).

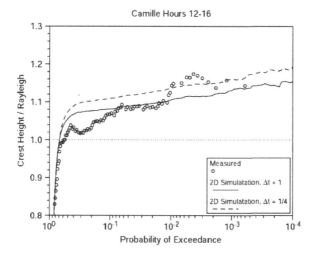

Fig. 9.7. Hurricane Camille (1200 hrs – 1600 hrs) G. Forristall (2000).

9.3 2nd-Order Wave Force on Bottom-Mounted Column

We will derive the 2nd-order wave force in a similar manner to the 2nd-order wave in Sec. 9.2 on a bottom-mounted column in the finite water. One first solves the linear diffraction satisfying the 1st-order boundary conditions. Next one determines the 2nd-order diffraction satisfying the 2nd-order boundary conditions. With the 2nd-order diffraction potential one determines the 2nd-order diffraction wave-exciting force.

9.3.1 1st-order diffraction

The solution of 1st-order diffraction has been widely employed (MacCamy and Fuchs, 1954) in coastal engineering. The body and space coordinate systems are denoted by $o\text{-}xyz$ and $O\text{-}XYZ$ respectively, and Z and z are vertically upward. These are in coincidence for the bottom-mounted column. The 1st-order wave-exciting force for the bottom mounted column is determined with the 1st-order incident wave and diffraction wave potentials:

$$\Phi_I^{(1)}(x,y,z,t)=\text{Re}\left\{-i\frac{gA}{\omega}\frac{\cosh k(h+z)}{\cosh kh}e^{ikR\cos\theta}e^{-i\omega t}\right\}$$

$$=\text{Re}\left\{-i\frac{gA}{\omega}\frac{\cosh k(h+z)}{\cosh kh}\left[\sum_{m=0}^{\infty}\beta_m J_m(kR)\cos(m\theta)\right]e^{-i\omega t}\right\}$$

(9.47)

where

$$\beta_m=\begin{cases}1 & m=0\\2i^m & m\geq 0\end{cases},\text{ referring to Abramowitz and Stegun (1987) and}$$

$$\Phi_D^{(1)}(x,y,z,t)=\text{Re}\left\{\begin{array}{l}\displaystyle\sum_{m=0}^{\infty}A_m\cos m\theta\left[J_m(kR)+iY_m(kR)\right]\times\\[2mm]\dfrac{\cosh k(h+z)}{\cosh kh}e^{-i\omega t}\end{array}\right\}$$

(9.48)

In Eq. (9.48) we have considered the diffraction potential satisfying radiation condition. Fulfilling the linear diffraction boundary condition on the body surface one determines the unknowns A_m of the diffraction potential. Then the sum of the incident and diffraction potential is given:

$$\Phi_I^{(1)}+\Phi_D^{(1)}=\text{Re}\left[(\varphi_I^{(1)}+\varphi_D^{(1)})e^{-i\omega t}\right]$$

(9.49)

with

$$\varphi_I^{(1)}+\varphi_D^{(1)}$$

$$=\left\{\begin{array}{l}-i\dfrac{gA}{\omega}\dfrac{\cosh k(h+z)}{\cosh kh}\\[4mm]\left\{\left[J_0(kR)-\dfrac{J_1'(ka)}{J_1'(ka)-iY_1'(ka)}\left(J_0(kR)+iY_0(kR)\right)\right]\right.\\[4mm]\left.+2\sum_{m=1}^{\infty}i^m\left[J_m(kR)-\dfrac{J_m'(ka)}{J_m'(ka)-iY_m'(ka)}\left(J_m(kR)+iY_m(kR)\right)\right]\right\}\cos m\theta\end{array}\right\}$$

(9.50)

Integration of the 1st-order pressure over the body surface gives the 1st-order wave-exciting force on the bottom-mounted column.

The asymptotic expression of the diffraction potential Eq. (9.48) is obtained by making use of the formulas of circular wave theory in Sec. 4.3.6.2 in the form:

$$\varphi_D^{(1)}(x,y,z) \approx \mathrm{Re}\left\{ \sum_{m=0}^{\infty} A_m \frac{\cos(m\theta)}{\sqrt{\pi k R}} e^{i\left(kR-\frac{\pi}{4}-\frac{m\pi}{2}\right)} \frac{\cosh k(h+z)}{\cosh kh} \right\}, \quad R \to \infty \quad (9.51)$$

Or it may be written in an alternative form:

$$\varphi_D^{(1)}(x,y,z) = \frac{f^{(1)}(\theta)}{\sqrt{R}} \cosh k(h+z) e^{ikR} + O\left(R^{-5/2}\right) \quad (9.52)$$

The foregoing diffraction potential presents the outward progressing wave, if we consider the potential multiplied with the time factor $e^{-i\omega t}$.

9.3.2 2nd-order diffraction

We consider the Stokes 2nd-order incident wave Eq. (9.34):

$$\Phi_I(x,y,z,t) = \mathrm{Re}\left\{ -i\frac{gA}{\omega}\frac{\cosh k(h+z)}{\cosh kh} e^{ikx} e^{-i\omega t} \right\}$$

$$+ \mathrm{Re}\left\{ -i\frac{3A^2\omega}{8}\frac{\cosh 2k(h+z)}{\sinh^4 kh} e^{i2kx} e^{-i2\omega t} \right\} \quad (9.53)$$

and the 2nd-order diffraction potential as given below:
Expanding the resultant velocity potential to the 2nd-order:

$$\Phi = \varepsilon\Phi^{(1)} + \varepsilon^2\Phi^{(2)} \quad (9.54)$$

$$\Phi^{(k)} = \mathrm{Re}\left[\left(\varphi_I^{(k)} + \varphi_D^{(k)}\right) e^{-i2\omega t} \right], \quad k=1,2 \quad (9.55)$$

The 2nd-order boundary conditions for the velocity potential are:

$$\frac{\partial \Phi_D^{(2)}}{\partial n} = -\frac{\partial \Phi_I^{(2)}}{\partial n} \quad \text{on } S \quad (9.56)$$

$$\frac{\partial \Phi_D^{(2)}}{\partial n} = 0 \quad \text{on } z = -h \quad (9.57)$$

$$\frac{\partial^2 \Phi^{(2)}}{\partial t^2} + g\frac{\partial \Phi^{(2)}}{\partial z} =$$

$$\frac{1}{g}\frac{\partial \Phi^{(1)}}{\partial t}\frac{\partial}{\partial z}\left(\frac{\partial^2 \Phi^{(1)}}{\partial t^2} + g\frac{\partial \Phi^{(1)}}{\partial z}\right) - \frac{\partial}{\partial t}(\nabla \Phi^{(1)})^2, \quad z = 0 \tag{9.58}$$

$$\text{Radiation condition}\left\{\Phi^{(2)} - \Phi_I^{(2)}\right\}, \quad \sqrt{x^2 + y^2} \to \infty \tag{9.59}$$

For the 1st-order diffraction potential the radiation condition is given:

$$\lim_{R\to\infty}\sqrt{R}\left(\frac{\partial \phi_{cD}^{(1)}}{\partial R} - \nu\phi_{sD}^{(1)}\right) = 0, \quad R = \sqrt{x^2 + y^2} \to \infty \tag{9.60}$$

but the 1st-order radiation cannot be applied to the 2nd-order radiation because it is phase-locked. Thus, we consider the 2nd-order free-surface condition Eq. (9.58) in the form:

$$\frac{\partial^2 \Phi_D^{(2)}}{\partial t^2} + g\frac{\partial \Phi_D^{(2)}}{\partial z} = -\frac{\partial^2 \Phi_I^{(2)}}{\partial t^2} - g\frac{\partial \Phi_I^{(2)}}{\partial z}$$

$$+\frac{1}{g}\frac{\partial \Phi^{(1)}}{\partial t}\frac{\partial}{\partial z}\left(\frac{\partial^2 \Phi^{(1)}}{\partial t^2} + g\frac{\partial \Phi^{(1)}}{\partial z}\right) - \frac{\partial}{\partial t}(\nabla \Phi^{(1)})^2 = q \tag{9.61}$$

where q consists of the periodic terms in the double frequency 2ω and the time invariant term q_0:

$$q = \text{Re}\left[\alpha^{(2)}e^{-i2\omega t}\right] + q_0 \tag{9.62}$$

Consequently, we may assume $\Phi_D^{(2)}$ in two components:

$$\Phi_D^{(2)} = \text{Re}\left[\varphi_{D2}^{(2)}e^{-i2\omega t}\right] + \Phi_{D0}^{(2)} \tag{9.63}$$

$\Phi_{D0}^{(2)}$ represents a stationary flow since it contributes to forces of only order ε^3 on the body, and thus we shall discard it in the following. Because the general solution of the diffraction potential consists of the solutions of the homogeneous and inhomogeneous equations on the free surface, and these solutions give the free and phase-locked diffraction potentials respectively, we write:

$$\varphi_{D2}^{(2)} = \varphi_{D2F}^{(2)} + \varphi_{D2L}^{(2)} \tag{9.64}$$

where in the subscript, F stand for free and L for phase-locked, respectively. Thus the homogeneous and inhomogeneous boundary conditions are written in the forms:

$$-4\omega^2\varphi_{D2F}^{(2)} + g\frac{\partial\varphi_{D2F}^{(2)}}{\partial z} = 0$$

$$-4\omega^2\varphi_{D2L}^{(2)} + g\frac{\partial\varphi_{D2L}^{(2)}}{\partial z} = \alpha^{(2)}$$

(9.65)

Both potentials $\varphi_{D2F}^{(2)}$ and $\varphi_{D2L}^{(2)}$ satisfy jointly the body boundary condition Eq. (9.56). It is of interest to examine the asymptotic expression of the foregoing potentials in the far-field. We may write the asymptotic expression of the 2nd-order diffraction potential similar to the 1st-order potential Eq. (9.52) in the form:

$$\varphi_{D2F}^{(2)} \simeq \frac{f_F^{(2)}(\theta)}{\sqrt{R}}\cosh k_2(z+h)e^{ik_2R} + O(R^{-5/2})$$

(9.66)

where k_2 satisfies the dispersion for the linear wave of frequency 2ω:

$$4\omega^2 = gk_2\tanh k_2 h$$

(9.67)

Equation (9.66) being multiplied with the time factor $e^{-i2\omega t}$ represents the 2nd-order free diffraction wave progressing outward from the body.

Meanwhile, the asymptotic expression of the phase-locked diffraction potential is determined from the inhomogeneous boundary condition Eq. (9.65) with the time factor omitted:

$$\alpha^{(2)} = 4\omega^2\varphi_I^{(2)} - g\frac{\partial\varphi_I^{(2)}}{\partial z}$$

$$-\frac{i\omega}{2g}(\varphi_I^{(1)} + \varphi_D^{(1)})\frac{\partial}{\partial z}\left[-4\omega^2(\varphi_I^{(1)} + \varphi_D^{(1)}) + g\frac{\partial}{\partial z}(\varphi_I^{(1)} + \varphi_D^{(1)})\right]$$

(9.68)

$$+ i\omega(\nabla\varphi_I^{(1)} + \nabla\varphi_D^{(1)})^2, \quad z = 0$$

Making use of the following relation:

$$\text{Re}\left[ae^{-i\omega t}\right]\text{Re}\left[be^{-i\omega t}\right] = \text{Re}\left[\frac{ab^*}{2}\right] + \text{Re}\left[\frac{ab}{2}e^{-i2\omega t}\right]$$

(9.69)

$\alpha^{(2)}$ is written in the form:

$$\alpha^{(2)} = \alpha_{ID} + \alpha_{DD}$$

$$\alpha_{ID} = \frac{-i\omega}{2g}\left[\begin{array}{l} \varphi_I^{(1)}\dfrac{\partial}{\partial z}\left(g\dfrac{\partial \varphi_D^{(1)}}{\partial z} - 4\omega^2\varphi_D^{(1)}\right) \\[4mm] +\varphi_D^{(1)}\dfrac{\partial}{\partial z}\left(g\dfrac{\partial \varphi_I^{(1)}}{\partial z} - 4\omega^2\varphi_I^{(1)}\right)\end{array}\right] + 2i\omega\nabla\varphi_I^{(1)}\nabla\varphi_D^{(1)} \quad \text{at} \quad z = 0 \qquad (9.70)$$

$$\alpha_{DD} = \frac{-i\omega}{2g}\varphi_D^{(1)}\frac{\partial}{\partial z}\left(g\frac{\partial \varphi_D^{(1)}}{\partial z} - 4\omega^2\varphi_D^{(1)}\right)$$

$$\frac{-i\omega}{2g}\varphi_I^{(1)}\frac{\partial}{\partial z}\left(g\frac{\partial \varphi_I^{(1)}}{\partial z} - 4\omega^2\varphi_I^{(1)}\right) + i\omega\left(\nabla\varphi_D^{(1)2} + \nabla\varphi_I^{(1)2}\right) \quad \text{at} \quad z = 0 \qquad (9.71)$$

We first evaluate the foregoing equations by employing the 1st-order incident wave potential Eq. (9.47) and asymptotic expression of the 2nd-order diffraction potential Eq. (9.52), in the forms:

$$\alpha_{ID} = \frac{g^{(2)}(\theta)}{\sqrt{R}}e^{ikR(1+\cos\theta)} + O(R^{-3/2}) \qquad (9.72)$$

and

$$\alpha_{DD} = O(R^{-1}) \qquad (9.73)$$

Then, we look for the asymptotic expression of phase-locked 2nd-order diffraction potential assuming the solution in the form:

$$\varphi_{D2L}^{(2)} = \frac{h^{(2)}(\theta,z)}{\sqrt{R}}e^{ikR(1+\cos\theta)} + O(R^{-1}) \qquad (9.74)$$

satisfying the continuity and bottom boundary condition:

$$\nabla^2\varphi_{D2L}^{(2)} = 0 \text{ for } z \leq 0 \text{ and } \frac{\partial \varphi_{D2L}^{(2)}}{\partial z} = 0 \text{ for } z = -h \qquad (9.75)$$

we obtain:

$$\varphi_{D2L}^{(2)} = \frac{f_L^{(2)}(\theta)}{\sqrt{R}}\cosh\left[k\sqrt{2+2\cos\theta}(z+h)\right]e^{ikR(1+\cos\theta)} + O(R^{-1}) \qquad (9.76)$$

Multiplying the time factor $e^{-i2\omega t}$ to the above potential, we obtain the progressing wave outward from the body. Thus the above phase-locked 2nd-order diffraction potential satisfies the radiation condition.

9.3.3 2nd-order wave force on bottom-mounted column

We consider the wave-exciting force in x- or surge-direction:

$$F_x(t) = -\int_{S(t)} p(t) n_1 dS \tag{9.77}$$

The unit normal is outward from the body surface into the water. $S(t)$ is the instantaneous wetted surface. The Bernoulli's equation:

$$p = -\rho g z - \rho \frac{\partial \Phi}{\partial t} - \rho \frac{(\nabla \Phi)^2}{2} \tag{9.78}$$

Expanding F_x up to the 2nd-order:

$$F_x = F_x^{(0)} + \varepsilon F_x^{(1)} + \varepsilon^2 F_x^{(2)} \tag{9.79}$$

$$F_x^{(0)} = \int_S -\rho g z n_1 dS \tag{9.80}$$

$$F_x^{(1)} = \int_S -\rho \frac{\partial \Phi^{(1)}}{\partial t} n_1 dS \tag{9.81}$$

$$F_x^{(2)} = \int_S \left(-\rho \frac{\partial \Phi^{(2)}}{\partial t} - \rho \frac{(\nabla \Phi^{(1)})^2}{2} \right) n_1 dS$$

$$+ \int_{\Delta S(t)} \left(-\rho g z - \rho \frac{\partial \Phi^{(1)}}{\partial t} \right) n_1 dS \tag{9.82}$$

where S denotes the mean wetted body surface and $\Delta S(t)$ is the alternating wetted part of $S(t)$, extending from $z = 0$ to $z = -1 / g \, (\partial \Phi^{(1)}/\partial t)$. Assume the body and free surface intersect at a right angle. Then, since the last term in the parenthesis represents the vertical displacement of the water surface at a point of the intersection, we may introduce a variable $z_w = -\rho g [(\partial \Phi^{(1)}/\partial t)/g]$ and integrate it in the vertical direction and simultaneously along the circular arc, i.e., $\int dS = \int dz_w \times d\Gamma$. Then the last integral with $z = 0$ results in the form:

$$F_x^{(2)} = \int_S \left(-\rho \frac{\partial \Phi^{(2)}}{\partial t} - \rho \frac{(\nabla \Phi^{(1)})^2}{2} \right) n_1 dS + \int_{\Gamma_C} \frac{\rho}{2g} \left(\frac{\partial \Phi^{(1)}}{\partial t} \right)^2 n_1 d\Gamma \qquad (9.83)$$

where Γ_C is the intersection of S with $z = 0$.

One may write the foregoing force in the form:

$$F_x^{(2)} = F_{x1}^{(2)} + F_{x2}^{(2)} \qquad (9.84)$$

The 2nd -order force is due to the products of the 1st-order potential:

$$F_{x1}^{(2)} = \int_S -\rho \frac{(\nabla \Phi^{(1)})^2}{2} n_1 dS + \int_{\Gamma_C} \frac{\rho}{2g} \left(\frac{\partial \Phi^{(1)}}{\partial t} \right)^2 n_1 d\Gamma \qquad (9.85)$$

The 2nd-order force in Eq. (9.85) consists of the Bernoulli's quadratic term of the 1st-order potential and the 2nd-order term due to the square of vertical relative displacement of the water surface when the body is fixed. See the similar result in Sec. 8.3.1.1. The mean 2nd-order force may be obtained from the quadratic products of 1st-order potential in Eq. (9.85).

The 2nd-order force due to the 2nd-order incident and diffraction potential is:

$$F_{x2}^{(2)} = \int_S -\rho \frac{\partial \Phi^{(2)}}{\partial t} n_1 dS \qquad (9.86)$$

which is written in the form:

$$F_{x2}^{(2)} = \int_S -\rho \frac{\partial \Phi^{(2)}}{\partial t} n_1 dS = \mathrm{Re} \left\{ \int_S \rho 2 i\omega (\varphi_I^{(2)} + \varphi_{D2}^{(2)}) n_1 dS e^{-2i\omega t} \right\}$$

$$F_{xI}^{(2)} = \mathrm{Re} \left\{ \int_S \rho 2 i\omega \varphi_I^{(2)} n_1 dS e^{-2i\omega t} \right\} \qquad (9.87)$$

$$F_{xD}^{(2)} = \mathrm{Re} \left\{ \int_S \rho 2 i\omega \varphi_{D2}^{(2)} n_1 dS e^{-2i\omega t} \right\}$$

To obtain the 2nd-order diffraction force we use the same method as proposed by Molin (1979) and Lighthill (1979). The basic idea is to imitate Haskind (Chap. 6) by making use of a fictitious linear radiation potential at double frequency 2ω for instance in surge mode:

$$\Psi = \mathrm{Re}[\psi e^{-2i\omega t}] \qquad (9.88)$$

where ψ satisfies:

$$\nabla^2 \psi = 0 \quad \text{in fluid domain } D$$

$$\frac{\partial \psi}{\partial n} = n_1 \quad \text{on } S \tag{9.89}$$

$$g \frac{\partial \psi}{\partial z} - 4\omega^2 \psi = 0 \quad \text{on } z = 0$$

Then, the 2nd-order diffraction force Eq. (9.87) may be written:

$$F_{xD}^{(2)} = \text{Re}\left[\int_S 2i\omega\rho\, \varphi_{D2}^{(2)} \frac{\partial \psi}{\partial n} dS\, e^{-2i\omega t} \right] \tag{9.90}$$

Consider the Green's 2nd-identity for the diffraction and radiation potentials:

$$\int_S \varphi_{D2}^{(2)} \frac{\partial \psi}{\partial n} dS = \int_S \psi \frac{\partial \varphi_{D2}^{(2)}}{\partial n} dS + \int_{S_F + S_B + S_\infty} \left(\psi \frac{\partial \varphi_{D2}^{(2)}}{\partial n} - \varphi_{D2}^{(2)} \frac{\partial \psi}{\partial n} \right) dS \tag{9.91}$$

where S is the mean wetted body surface, S_∞ represents a circular cylinder of radius R enclosing the body. S_F and S_B are the free surface and bottom portions limited by S_∞.

Making use of Eqs. (9.56) through (9.62) and Eq. (9.89), we can reduce the integral over the free surface to:

$$\int_{S_F} \left(\psi \frac{\partial \varphi_{D2}^{(2)}}{\partial n} - \varphi_{D2}^{(2)} \frac{\partial \psi}{\partial n} \right) dS = \int_{S_F} \frac{\alpha^{(2)}}{g} \psi\, dS \tag{9.92}$$

Evidently the integral over the bottom surface is zero:

$$\int_{S_B} \left(\psi \frac{\partial \varphi_{D1}^{(2)}}{\partial n} - \varphi_{D1}^{(2)} \frac{\partial \psi}{\partial n} \right) dS = 0 \tag{9.93}$$

Using the asymptotic expressions for the free and phase-locked waves of $\varphi_{D2}^{(2)}$ Eqs. (9.66) and (9.76) and the theorem of stationary phase, one can show that the integral over S_∞ oscillates and reduces to zero when R goes to ∞:

$$\int_{S_\infty} \left(\psi \frac{\partial \varphi_{D2}^{(2)}}{\partial n} - \varphi_{D2}^{(2)} \frac{\partial \psi}{\partial n} \right) dS = \text{oscillation} \to \text{zero} \quad \text{as } R \to \infty \tag{9.94}$$

Accordingly one can determine the 2nd-order diffraction force by making use of Eqs. (9.90) through (9.94) in the form:

$$F_{xD}^{(2)} = \mathrm{Re}\left\{ \left[\int_S -2i\rho\omega \frac{\partial \varphi_I^{(2)}}{\partial n}\psi dS + \int_{S_F} 2i\rho\omega \frac{\alpha^{(2)}}{g}\psi dS \right]e^{-2i\omega t} \right\} \qquad (9.95)$$

The 2nd-order diffraction force consists of the integrals over the body and free surface. Obviously, the Froude-Krylov force has to be added to obtain the resultant wave-exciting force.

The oscillation of surface integral in the far-field Eq. (9.94) was discussed in Appendix 2 of Molin (1979):

It considers an integral on the surface S_∞:

$$I = \int_{S_\infty}\left(\psi \frac{\partial \phi_{D1}^{(2)}}{\partial n} - \varphi_{D1}^{(2)}\frac{\partial \psi}{\partial n} \right)dS = I_F + I_L \qquad (9.96)$$

$$I_F = \int_{-h}^0 dz \int_0^{2\pi} Rd\theta \left(\psi \frac{\partial \varphi_{D1F}^{(2)}}{\partial n} - \varphi_{D1F}^{(2)}\frac{\partial \psi}{\partial n} \right)$$

$$I_L = \int_{-h}^0 dz \int_0^{2\pi} Rd\theta \left(\psi \frac{\partial \varphi_{D1L}^{(2)}}{\partial n} - \varphi_{D1L}^{(2)}\frac{\partial \psi}{\partial n} \right) \qquad (9.97)$$

When R goes to infinity, $\psi, \varphi_{D2F}^{(2)}, \varphi_{D2L}^{(2)}$ admit the following asymptotic expressions:

$$\psi = \frac{\chi(\theta)}{\sqrt{R}}\cosh k_2(z+h)e^{ik_2 R} + O(R^{-5/2}) \qquad (9.98)$$

$$\varphi_{D2F}^{(2)} = \frac{f_F^{(2)}(\theta)}{\sqrt{R}}\cosh k_2(z+h)e^{ik_2 R} + O(R^{-5/2}) \qquad (9.99)$$

$$\varphi_{D2L}^{(2)} = \frac{f_L^{(2)}(\theta)}{\sqrt{R}}\cosh\left[k\sqrt{2+2\cos\theta}(z+h) \right]e^{ikR(1+\cos\theta)} + O(R^{-1}) \qquad (9.100)$$

Both $\varphi_{D2F}^{(2)}$ and $\varphi_{D2L}^{(2)}$ were derived before as shown in Eqs. (9.66) and (9.74).

It can be shown that I_F behaves like R^{-2} when R goes to infinity. The integral I_L requires the use of the theorem of the stationary phase. Then, I_L appears to behave like $R^{-1/2}e^{i\lambda R}$.

Therefore, while computing the free surface integral in Eq. (9.95) up to some finite distance R, one must expect such oscillations. They become quite sensitive in very shallow water and damp out slowly. In such cases the mean value of the 2nd-order diffraction force on the free surface over one period of oscillation may be taken so that at some far enough distance, this mean value may be kept constant with R.

Molin and Marion (1986) further determined the 2nd-order diffraction force of a bottom-mounted column and 2nd-order radiation force by solving the 2nd-order radiation potential.

Kim and Yue (1989) solved the same problem as above in a different way giving a semi-analytical formula for the wave-exciting force of a bottom-mounted column in monochromatic wave. The semi-analytical formulas for the surge-exciting force were compared for validation of computation by other numerical methods. Kim and Yue (1990) extended the previous work to the response of axisymmetric floating bodies in the bichromatic waves. Ring-source integral equation method in conjunction with a new analytic free surface integration in the entire-local-wave-free domain was developed. The semi-analytical solution was also obtained for the bottom-mounted column in the bichromatic waves.

9.4 2nd-Order Force on Main-Platform of ISSC TLP by HOBEM

The computation of the 2nd-order wave-exciting force and radiation force of the main-platform of ISSC TLP (Eatock Taylor and Jefferys, 1986) by HOBEM (Liu et al., 1991) will be reviewed (Liu et al., 1995). These authors developed formulas for 2nd-order force on the oscillatory body referring to Molin (1979), Molin and Marion (1986), and Kim and Yue (1989, 1990) and others.

9.4.1 General formulation of 2nd-order force

Consider the 2nd-order interaction of monochromatic wave with the flow due to 3-D floating body assuming potential flow and weak wave nonlinearities. Then the velocity potential Φ is given in a perturbation series up to 2nd-order:

$$\Phi = \varepsilon \, \Phi^{(1)} + \varepsilon^2 \Phi^{(2)} \tag{9.101}$$

At each order, the velocity potentials can be decomposed into incident, diffraction and radiation potentials:

$$\Phi^{(1)} = \Phi_I^{(1)} + \Phi_D^{(1)} + \Phi_R^{(1)}, \; \Phi^{(2)} = \Phi_I^{(2)} + \Phi_D^{(2)} + \Phi_R^{(2)} \tag{9.102}$$

where

$$\Phi^{(1)} = \mathrm{Re}\left[\varphi^{(1)} e^{-i\omega t} \right], \Phi^{(2)} = \mathrm{Re}\left[\varphi^{(2)} e^{-i2\omega t} \right] \tag{9.103}$$

In the presence of a monochromatic wave with amplitude A and frequency ω we have, at 1st-order, $\varphi^{(1)}$ that is frequency dependent and linearly proportional to wave amplitude. While at the 2nd-order, we have the mean and double frequency potentials whose amplitudes are proportional to the square of the wave amplitude A^2. Similar expressions can be used for the forces:

$$
\begin{pmatrix} \Phi(\mathbf{x},t) \\ \mathbf{F}(t) \end{pmatrix} = \mathrm{Re}\left[\varepsilon \begin{pmatrix} \varphi^{(1)}(\mathbf{x}) \\ \mathbf{f}^{(1)}(t) \end{pmatrix} e^{-i\omega t} \right.
$$
$$
\left. + \varepsilon^2 \left\{ \begin{pmatrix} \overline{\varphi}^{(2)}(\mathbf{x}) \\ \overline{\mathbf{f}}^{(2)}(t) \end{pmatrix} + \begin{pmatrix} \varphi^{(2)}(\mathbf{x}) \\ \mathbf{f}^{(2)}(t) \end{pmatrix} e^{-i2\omega t} \right\} \right] \tag{9.104}
$$

The 2nd-order forces can be obtained directly from the integration of the pressure over the instantaneous wetted body surface. To illustrate this process, we define the space- and body-fixed coordinate systems; O-XYZ and o-xyz. We then consider the transformation of the coordinates $\mathbf{X}$ and $\mathbf{x}$ and unit normal vectors $\mathbf{N}$ and $\mathbf{n}$ between the two systems in the forms:

$$\mathbf{X} = \mathbf{x} + \varepsilon\left(\Xi^{(1)} + \Omega^{(1)} \times \mathbf{x} \right) + \varepsilon^2 \left(\Xi^{(2)} + \Omega^{(2)} \times \mathbf{x} + H \, \mathbf{x} \right) \tag{9.105}$$

$$\mathbf{N} = \mathbf{n} + \varepsilon\left(\Omega^{(1)} \times \mathbf{n} \right) + \varepsilon^2 \left(\Omega^{(2)} \times \mathbf{n} + H\mathbf{n} \right) \tag{9.106}$$

where $\Xi^{(i)} = \left(\Xi_1^{(i)}, \Xi_2^{(i)}, \Xi_3^{(i)} \right), \Omega^{(i)} = \left(\Omega_1^{(i)}, \Omega_2^{(i)}, \Omega_3^{(i)} \right), i = 1,2$ denote ith-order translational and rotational motions, respectively. And the matrix H of

2nd-order elements is defined by:

$$H = -\frac{1}{2}\begin{pmatrix} \Omega_2^{(2)} + \Omega_3^{(2)} & 0 & 0 \\ -2\Omega_1^{(1)}\Omega_2^{(1)} & \Omega_1^{(2)} + \Omega_3^{(2)} & 0 \\ -2\Omega_1^{(1)}\Omega_3^{(1)} & -2\Omega_2^{(1)}\Omega_3^{(1)} & \Omega_1^{(2)} + \Omega_2^{(2)} \end{pmatrix} \quad (9.107)$$

which depends on the sequence of the rotation of axis, that is, 1st-order roll-pitch-yaw.

We represent the pressure on the instantaneous body surface in terms of Taylor expansion on the mean-wetted body surface:

$$P(t) = -\rho\left\{ gz + \left[\frac{\partial\Phi(\mathbf{x},t)}{\partial t} + g\left(\Xi_3 + \Omega_1 y - \Omega_2 x\right)\right] \right.$$

$$+ \left[\frac{1}{2}\nabla\Phi(\mathbf{x},t)\cdot\nabla\Phi(\mathbf{x},t)\right. \quad (9.108)$$

$$\left.\left. + \left(\Xi + \Omega\times\mathbf{x}\right)\cdot\nabla\Phi_t(\mathbf{x},t) + gH\ \mathbf{x}\cdot\nabla z \right]\right\}$$

The integration of the pressure $P(t)$ over the instantaneous body surface $S(t)$ can be expressed as a sum of the mean wetted surface integral and fluctuating surface integral:

$$\mathbf{F} = \int_{S(t)} P(t)\ \mathbf{n}dS = \int_S P(t)\ \mathbf{n}dS + \oint_l dl \int_0^{\varsigma_r + \cdots} P(t)\frac{\mathbf{n}}{\sqrt{1-n_3^2}}dz \quad (9.109)$$

After collecting terms of equal order in Eq. (9.109), we obtain the 1st- and 2nd-order forces:

$$\mathbf{F}^{(1)} = -\rho\int_S \frac{\partial\left(\Phi_I^{(1)} + \Phi_D^{(1)} + \Phi_R^{(1)}\right)}{\partial t}\mathbf{n}dS$$

$$- \rho g A_{WP}\left(\Xi_3^{(1)} + y_f\Omega_1^{(1)} - x_f\Omega_2^{(1)}\right)\mathbf{k} \quad (9.110)$$

$$
\mathbf{F}^{(2)} = -\rho \int_S \frac{\partial \left(\Phi_I^{(2)} + \Phi_D^{(2)} + \Phi_R^{(2)} \right)}{\partial t} \mathbf{n} dS
$$

$$
- \rho g A_{WP} \left(\Xi_3^{(2)} + y_f \Omega_1^{(2)} - x_f \Omega_2^{(2)} \right) \mathbf{k}
$$

$$
+ \frac{\rho g}{2} \oint_{WL} \varsigma_r^{(1)^2} \mathbf{N} dl
$$

$$
- \rho \int_S \left(\frac{1}{2} \nabla \Phi^{(1)} \cdot \nabla \Phi^{(1)} \right) \mathbf{n} \, dS \tag{9.111}
$$

$$
- \rho \int_S \left(\Xi^{(1)} + \Omega^{(1)} \times \mathbf{x} \right) \cdot \frac{\partial}{\partial t} \nabla \Phi^{(1)} \mathbf{n} dS
$$

$$
+ \Omega^{(1)} \times \mathbf{F}^{(1)}
$$

$$
- \rho g A_{WP} \left(\Omega_1^{(1)} \Omega_3^{(1)} x_f + \Omega_2^{(1)} \Omega_3^{(1)} y_f \right) \mathbf{k}
$$

where $\varsigma_r^{(1)}$ is the 1st-order vertical relative elevation:

$$
\varsigma_r^{(1)} = \varsigma^{(1)} - \Xi_3^{(1)} - y \Omega_1^{(1)} + x \Omega_2^{(1)} \text{ and } \mathbf{N} = \frac{\mathbf{n}}{\sqrt{1 - n_3^2}} .
$$

and where ρ is the fluid density, g the gravitational acceleration, A_{WP} the waterplane area, (x_f, y_f) the center of floatation of the waterplane, and $\mathbf{k}$ the unit vector in z-direction. The expression for the moments can be obtained by replacing $\mathbf{n}$ with $\mathbf{x} \times \mathbf{n}$ in Eq. (9.105). Details are referred to Ogilvie (1983).

9.4.2 Formulation of 2nd-order diffraction potential force

We consider the 2nd-order diffraction boundary conditions on the free surface and body surface in the forms:

$$
\frac{\partial^2 \Phi_D^{(2)}}{\partial t^2} + g \frac{\partial \Phi_D^{(2)}}{\partial z} = -\frac{\partial^2 \Phi_I^{(2)}}{\partial t^2} - g \frac{\partial \Phi_I^{(2)}}{\partial z}
$$

$$
+ \frac{1}{g} \frac{\partial \Phi^{(1)}}{\partial t} \frac{\partial}{\partial z} \left(\frac{\partial^2 \Phi^{(1)}}{\partial t^2} + g \frac{\partial \Phi^{(1)}}{\partial z} \right) - \frac{\partial}{\partial t} \left(\nabla \Phi^{(1)} \right)^2 = Q, \, (Z = 0) \tag{9.112}
$$

$$\frac{\partial \Phi_D^{(2)}}{\partial n} = -\frac{\partial \Phi_I^{(2)}}{\partial n} + B^{(2)} \quad \text{on } S \tag{9.113}$$

where $B^{(2)}$ is the effect of the quadratic product of linear rotations on the diffraction (Ogilvie, 1983):

$$
\begin{aligned}
B^{(2)} = \mathbf{n} \cdot \left(\frac{\partial H}{\partial t} \mathbf{x} \right) &- \mathbf{n} \cdot \left[(\mathbf{\Xi}^{(1)} + \mathbf{\Omega}^{(1)} \times \mathbf{x}) \cdot \nabla \right] \nabla \Phi^{(1)} \\
&+ (\mathbf{\Omega}^{(1)} \times \mathbf{n}) \cdot \left[\mathbf{V}^{(1)} - \nabla \Phi^{(1)} \right]
\end{aligned}
\tag{9.114}
$$

We define:

$$\mathbf{\Xi}^{(1)} = \text{Re}\left[\mathbf{\xi}^{(1)} e^{-i\omega t} \right], \mathbf{\Omega}^{(1)} = \text{Re}\left[\mathbf{\alpha}^{(1)} e^{-i\omega t} \right]$$

$$\mathbf{\xi}^{(1)} = \left\{ \xi_j^{(1)} \right\} = \left\{ \xi_1^{(1)}, \xi_2^{(1)}, \xi_3^{(1)} \right\} \text{ (surge, sway, heave) and}$$

$$\mathbf{\alpha}^{(1)} = \left\{ \alpha_j^{(1)} \right\} = \left\{ \alpha_1^{(1)}, \alpha_2^{(1)}, \alpha_3^{(1)} \right\} \text{ (roll, pitch and yaw)}$$

$$\Xi_j^{(1)} = \text{Re}\left\{ \xi_j^{(1)} e^{-i\omega t} \right\}, \ \Omega_j^{(1)} = \text{Re}\left\{ \alpha_j^{(1)} e^{-i\omega t} \right\}$$

$$\mathbf{V}^{(1)} = \frac{\partial}{\partial t} (\mathbf{\Xi}^{(1)} + \mathbf{\Omega}^{(1)} \times \mathbf{x}) \tag{9.115}$$

Similarly define B^+ and H^+ as $B^{(2)} = \text{Re}[B^+ e^{-i2\omega t}]$ and $H = \text{Re}[H^+ e^{-i2\omega t}]$

$$
H^+ = -\frac{1}{2} \begin{bmatrix} (\alpha_2^{(2)} + \alpha_3^{(2)}) & 0 & 0 \\ -2\alpha_1^{(1)}\alpha_2^{(1)} & (\alpha_1^{(2)} + \alpha_3^{(2)}) & 0 \\ -2\alpha_1^{(1)}\alpha_3^{(1)} & -2\alpha_2^{(1)}\alpha_3^{(1)} & (\alpha_1^{(2)} + \alpha_2^{(2)}) \end{bmatrix} \tag{9.116}
$$

Liu et al. (1995) derived the 2nd-order diffraction force in a similar manner to Eqs. (9.95) and (9.68) for the bottom-mounted column and modified the diffraction boundary condition affected by the body motion in the form:

$$F_{Dk} = \mathrm{Re}\left(-\rho \int_S \frac{\partial \Phi_D^{(2)}}{\partial t} n_k dS \right)$$

$$= \mathrm{Re}\left(-\rho \int_S \frac{\partial \varphi_D^{(2)}}{\partial t} n_k dS e^{-i2\omega t} \right) \tag{9.117}$$

$$= \mathrm{Re}\left\{ \rho i2\omega \left[\int_S \left(B^+ - \frac{\partial \varphi_I^{(2)}}{\partial n} \right) \varphi_k^+ dS + \frac{1}{g} \int_{S_F} \Omega^+ \varphi_k^+ dS \right] e^{-i2\omega t} \right\}$$

where F_{Dk} denotes 2nd-order diffraction force in k-direction. φ_k^+ represents a fictitious linear radiation potential at double frequency 2ω, for instance, in surge mode $k = 1$:

And Q^+ and B^+ are as given below:

$$Q^+ = -\frac{i\omega}{2g}\left[\varphi^{(1)}\left(\frac{\partial^2 \varphi^{(1)}}{\partial z^2} - v\frac{\partial \varphi^{(1)}}{\partial z} \right) - \varphi_I^{(1)}\left(\frac{\partial^2 \varphi_I^{(1)}}{\partial z^2} - v\frac{\partial \varphi_I^{(1)}}{\partial z} \right) \right]$$

$$+ i\omega\left[\nabla \varphi^{(1)} \cdot \nabla \varphi^{(1)} - \nabla \varphi_I^{(1)} \cdot \nabla \varphi_I^{(1)} \right], z = 0 \tag{9.118}$$

and

$$B^+ = \frac{1}{2}\left\{ \begin{array}{l} -i\omega \mathbf{n} \cdot \left(H^+ \mathbf{x} \right) - \mathbf{n} \cdot \left[\mathbf{X}^{(1)} \cdot \nabla \right] \nabla \varphi^{(1)} \\ -\left(i\omega \mathbf{X}^{(1)} + \nabla \varphi^{(1)} \right) \cdot \left(\boldsymbol{\alpha}^{(1)} \times \mathbf{n} \right) \end{array} \right\} \tag{9.119}$$

In the above

$$\mathbf{X}^{(1)} = \boldsymbol{\xi}^{(1)} + \mathbf{x} \times \boldsymbol{\alpha}^{(1)}$$

It is to be noted that the 2nd-order diffraction potential force Eq. (9.117) is given in terms of 1st-order responses in the formulation except the 2nd-order incident wave potential. Thus the 2nd-order motion in the diagonal terms in H^+ has to be zero.

It is more numerically involved than the simple bottom-mounted column in determining the integral over the free surface because of the multiple columns intersecting the waterline in the case of TLP.

9.4.3 The 1st- and 2nd-order force

From Secs. 9.4.1 and 9.4.2, we obtain the 1st- and 2nd-order order wave exciting force and 2nd-order radiation force:

$$\mathbf{F}_{w1} = -\rho \int_S \frac{\partial(\Phi_I^{(1)} + \Phi_D^{(1)} + \Phi_R^{(1)})}{\partial t} \mathbf{n} dS$$
$$- \rho g A_{WP} \left(\Xi_3^{(1)} + y_f \Omega_1^{(1)} - x_f \Omega_2^{(1)} \right) \mathbf{k}$$

(9.120)

$$\mathbf{F}_{w2} = -\rho \int_S \mathbf{n} \frac{\partial(\Phi_I^{(2)} + \Phi_D^{(2)})}{\partial t} dS$$
$$+ \frac{1}{2} \rho g \int_{WL} \frac{\mathbf{n}}{\sqrt{1-n_3^2}} (\zeta_r^{(1)})^2 \, dl$$
$$- \rho \int_S \mathbf{n} \left[\frac{1}{2} [\nabla \Phi^{(1)}]^2 \right] dS$$
$$- \rho \int_S \mathbf{n} \left[(\Xi^{(1)} + \Omega^{(1)} \times \mathbf{x}) \cdot \nabla \Phi_t^{(1)} \right] dS$$
$$+ \Omega^{(1)} \times \mathbf{F}_1$$
$$- \rho g A_{WP} [\Omega_1^{(1)} \Omega_3^{(1)} x_f + \Omega_2^{(1)} \Omega_3^{(1)} y_f] \mathbf{k}$$

(9.121)

$$\mathbf{F}_{R2} = -\rho \int_S \frac{\partial \Phi_R^{(2)}}{\partial t} \mathbf{n} \, dS$$
$$- \rho g A_{wp} \left(\Xi_3^{(2)} + y_f \Omega_1^{(2)} - x_f \Omega_2^{(2)} \right) \mathbf{k}$$

(9.122)

The unit normal is directed into the fluid; and $\zeta_r^{(1)}$ denotes the 1st-order relative wave elevation:

$$\zeta_r^{(1)} = \zeta_w - (\Xi_3^{(1)} + \Omega_1^{(1)} y_f - \Omega_2^{(1)} x_f)$$

(9.123)

ζ_w is the first-order wave elevation, A_{WP} is the waterplane area, and x_f and y_f are the coordinates of the center of floatation respectively.

The 2nd-order wave-exciting force on the body due to the monochromatic wave is given in terms of the 1st-order motions, excluding the 2nd-order incident wave potential. 2nd-order radiation force is given in terms of 2nd-order motion.

Equation (9.121) consists of the pure 2nd-order potential due to 2nd-order incident and diffraction potential and those of the quadratic products of 1st-order potential:

$$\mathbf{F}_{2p} = -\rho \int_S \mathbf{n} \frac{\partial(\Phi_I^{(2)} + \Phi_D^{(2)})}{\partial t} \, dS \qquad (9.124)$$

and

$$\mathbf{F}_{2q} = \frac{1}{2}\rho g \int_{WL} \frac{\mathbf{n}}{\sqrt{1-n_3^2}} (\varsigma_r^{(1)})^2 \, dl$$

$$-\rho \int_S \mathbf{n} \left[\frac{1}{2}[\nabla\Phi^{(1)}]^2 \right] dS$$

$$+\Omega^{(1)} \times \mathbf{F}^{(1)} \qquad (9.125)$$

$$-\rho \int_S \mathbf{n} \left[(\Xi^{(1)} + \Omega^{(1)} \times \mathbf{x}) \cdot \nabla\Phi_t^{(1)} \right] dS$$

$$-\rho g A_{WP} [\Omega_1^{(1)}\Omega_3^{(1)} x_f + \Omega_2^{(1)}\Omega_3^{(1)} y_f]\mathbf{k}$$

The 2nd-order force components in Eq. (9.125) consist of the four meaningful components as found in Sec. 8.4.1.1 and in Eqs. (9.85) and (9.86). These components are the quadratic products of linear potential that can give the mean 2nd-order force in the monochromatic wave, while the pure 2nd-order potential does not.

9.4.4 2nd-order radiation potential

We have solved the 1st-order radiation and diffraction potentials and 2nd-order diffraction potential in the monochromatic wave allowing the 1st-order motions.

The 2nd-order motions $u_k^{(2)} = \left\{ \xi_1^{(2)}, \xi_2^{(2)}, \xi_3^{(2)}, \alpha_1^{(2)}, \alpha_2^{(2)}, \alpha_3^{(2)} \right\}$ are determined by solving matrix equation written in the form omitting the time factor $e^{-i2\omega t}$:

2nd-Order Wave and 2nd-Order Force

$$\sum_{k=1}^{6} \Big[- 4\omega^2 \left(m_{srk} \left(2\omega \right) + m_{frk} \left(2\omega \right) \right)$$

$$- i2\omega \left(C_{srk} \left(2\omega \right) + C_{frk} \left(2\omega \right) \right) + \left(K_{srk} + K_{frk} \right) u_k^{(2)} \left(\omega \right) \Big] \qquad (9.126)$$

$$= F_r^{(2)} \left(\omega \right), \, r = 1, \, 2, \cdots, 6$$

Since the virtual mass m_{srk}, damping and stiffness and wave-exciting force at double frequency are known, we can determine the 2nd-order motions. The radiation potential at the double frequency is known, thus one has the 2nd-order radiation potential in the form:

$$\Phi_R^{(2)} = \sum_{k=1}^{6} \text{Re} \Big[-i2\omega \, u_k^{(2)} \varphi_R^{(2)} e^{-i2\omega t} \Big] \qquad (9.127)$$

where $\varphi_R^{(2)}$ represents the double frequency radiation potential per unit velocity of the double frequency amplitude. Thus we have determined the 2nd-order radiation potential in the monochromatic wave.

9.4.5 2nd-order diffraction and radiation in bichromatic waves

The 1st- and 2nd-order wave-exciting forces due to the bichromatic wave, in Sec. 9.4.3, can be expressed in the form:

$$y(t) = F_1(t) + F_2(t)$$

$$= \text{Re} \sum_{j=1}^{2} A_j f_1(\omega_j) e^{i(\omega_j t - \varepsilon_j)} \qquad (9.128)$$

$$+ \text{Re} \sum_{j=1}^{2} \sum_{k=1}^{2} A_j A_k \begin{Bmatrix} f_2(\omega_j, \omega_k) e^{i[(\omega_j + \omega_k)t - (\varepsilon_j + \varepsilon_k)]} \\ + f_2(\omega_j, -\omega_k) e^{i[(\omega_j - \omega_k)t - (\varepsilon_j - \varepsilon_k)]} \end{Bmatrix}$$

where A_i indicates incident wave amplitude and ε_i random phase angle. The 2nd-order coefficients consist of the sum and difference frequency components $f_2(\omega_j, \pm\omega_k)$, where $f_2(\omega_j, -\omega_j)$ is the mean term due to monochromatic wave.

The foregoing force kernels may be designated the hydrodynamic QTFs for the 2nd-order force equivalent to $G_2(\omega_i, \pm\omega_j)$ of the Volterra quadratic model. It is to be noted that the 2nd-order wave forces in the foregoing are determined under the condition that only the 1st-order response motions are allowed. The 2nd-order response motions are not allowed. This seems to contradict to the definition of the 2nd-order radiation force that requires the 2nd-order response motions. However, 2nd-order motion in monochromatic wave can be separately determined given the 2nd-order diffraction and radiation potential. In the bichromatic wave, radiation forces are actually a linear sum of two double frequency components.

9.4.6 Similarity between Volterra quadratic model and 2nd-order force

Comparing the LTFs and QTFs of Volterra quadratic model Eq. (7.11) with those of the hydrodynamic force, we obtain the following relations:

$$LTF = G_1 = f_1, \quad QTF = G_2\left(\omega_i, \ \pm\omega_j\right) = 2f_2\left(\omega_i, \ \pm\omega_j\right) \qquad (9.129)$$

Equation (9.129) states that the LTFs are identical, while QTFs of Volterra quadratic model equals twice the hydrodynamic QTFs.

Thus the expected value of the output in the Gaussian sea is obtained by employing the hydrodynamic QTFs in Eq. (7.32):

$$E[y] = 2\int_0^\infty f_2(\omega, \ -\omega)U_{xx}(\omega)d\omega \qquad (9.130)$$

Similarly the energy density spectrum of the Volterra quadratic response Eq. (7.35) is given in terms of the hydrodynamic QTFs of the 2nd-order force:

$$U_{yy}(\omega) = U_{xx}(\omega) + 8\int_0^\infty | f_2(\omega - \xi, \xi)|^2 \, U_{xx}\left(|\omega - \xi|\right)U_{xx}\left(|\xi|\right)d\xi \qquad (9.131)$$

It is common practice to simulate the wave-exciting force on a structure in the given Gaussian sea employing Volterra quadratic model Eq. (7.11) in the form:

$$y(t) = y_1(t) + y_2(t)$$

$$= \mathrm{Re}\sum_{j=1}^2 A_j G_1(\omega_j)e^{i(\omega_j t - \varepsilon_j)} \qquad (9.132)$$

$$+ \mathrm{Re}\frac{1}{2}\sum_{j=1}^2\sum_{k=1}^2 A_j A_k \left\{ \begin{array}{l} G_2(\omega_j, \omega_k)e^{i[(\omega_j + \omega_k)t - (\varepsilon_j + \varepsilon_k)]} \\ + G_2(\omega_j, -\omega_k)e^{i[(\omega_j - \omega_k)t - (\varepsilon_j - \varepsilon_k)]} \end{array} \right\}$$

where G_1 and G_2 represent the LTFs and QTFs and A_j and ε_j denote the amplitude spectrum and random phase of the given wave energy spectrum. However, these frequency response functions can only be determined either by experiment or hydrodynamic calculation. If one substitutes the hydrodynamic LTFs and QTFs given in Eq. (9.129) into Eq. (9.132), one obtains the hydrodynamic wave-exciting force in the random sea in the form:

$$y(t) = y_1(t) + y_2(t)$$

$$= \operatorname{Re} \sum_{j=1}^{2} A_j f_1(\omega_j) e^{i(\omega_j t - \varepsilon_j)} \tag{9.133}$$

$$+ \operatorname{Re} \sum_{j=1}^{2} \sum_{k=1}^{2} A_j A_k \left\{ \begin{array}{l} f_2(\omega_j, \omega_k) e^{i[(\omega_j + \omega_k)t - (\varepsilon_j + \varepsilon_k)]} \\ + f_2(\omega_j, -\omega_k) e^{i[(\omega_j - \omega_k)t - (\varepsilon_j - \varepsilon_k)]} \end{array} \right\}$$

which is in fact identical to the original hydrodynamic response Eq. (9.128). Because of the foregoing result, we often say that we have used Volterra quadratic model for simulation of the 2nd-order force in the given Gaussian random sea.

The hydrodynamic LTFs and QTFs of the main-part of ISSC TLP may be computed for instance by HOBEM (Liu et al., 1995). These hydrodynamic data will be used in the simulation of the wave exciting-force time series using Volterra quadratic model.

9.5 HOBEM for Main-Platform of ISSC TLP

HOBEM (Liu et al., 1995) was employed for computation of the hydrodynamic LTFs and QTFs on the main-platform of ISSC TLP in monochromatic waves. We will introduce significant features of HOBEM compared with other analyses such as CPM and experiments.

9.5.1 The features of HOBEM

9.5.1.1 HOBEM discretization

Let us consider computation of QTFs of bottom-mounted column using Eq. (9.128). Discretization by HOBEM is illustrated in Fig. 9.8. Note that the integral is over the body and free surface. Applying the symmetry property of geometry, we take one quarter of the entire body, which may also be referred to Sec. 6.2.4.

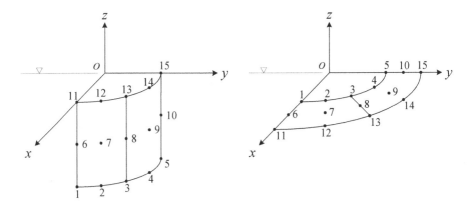

Fig. 9.8. Discretization of body surface (left) and free surface (right).

Figure 9.8 illustrates the nodes on the body and free surface of a bottom mounted cylinder, where we realize the nodes that are common for the body and free surface. The right-handed rule is used in taking the order of nodes. For instance, it is applied on the first quadrilateral element of the body surface with global node numbers of 11, 12, 13, 8, 3, 2, 1, 6, and 7. Each element consists of, a total of nine nodes including the center one.

9.5.1.2 Comparison of HOBEM with other analyses

The HOBEM calculation of the bottom-mounted column shown in Fig. 9.8 is compared with the semi-analytical formula in Fig. 9.9, showing excellent

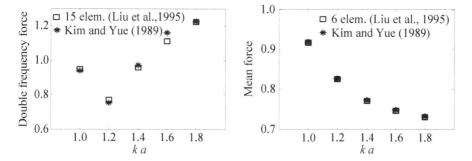

Fig. 9.9. Comparison of HOBEM with semi analytical formula (Kim and Yue, 1989). k and a are wavenumber and radius of the column, respectively.

agreement of the force by HOBEM and semi-analytical solution (Kim and Yue, 1989), which gives support for HOBEM as a useful tool for arbitrary shape body.

Another example of the HOBEM is illustrated below by comparing with the result of CPM used by Zhao and Faltinsen (1989).

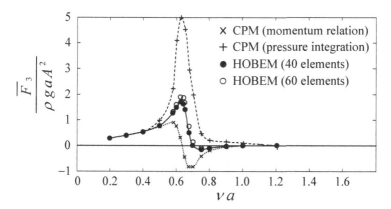

Fig. 9.10. Comparison of far- and near-field solution (Zhao and Faltinsen,1989) and HOBEM (Liu et al., 1993).

The vertical mean drifting force on a floating vertical truncated cylinder that is free in the swaying and heaving motion was computed using far- and near-field method with CPM. A large discrepancy was found between these methods, which Zhao and Faltinsen (1989) attributed to the effect of CPM applied to the edge between the bottom and cylinder surface. Liu et al. (1993) computed the same problem employing HOBEM based on the near-field approach which resulted in the uniform convergence giving a unique result. HOBEM in fact can avoid the singularity due to the edge along the bottom of the flush-cut cylinder as it was analytically explained in Sec. 6.2.4.1.

9.5.1.3 Double derivatives of the linear potential

Equations (9.118) and (9.119) contain double derivatives of the linear velocity potential that cannot accurately be computed by CPM. Liu et al. (1995) presented an accurate method using HOBEM.

The tangential derivatives of the linear velocity potential along ξ and η; $\partial\varphi / \partial\xi$, $\partial\varphi / \partial\eta$, and normal derivative $\partial\varphi / \partial n$ on the body surface are expressed in the matrix form:

$$
\begin{pmatrix} \dfrac{\partial\varphi}{\partial\xi} \\[2mm] \dfrac{\partial\varphi}{\partial\eta} \\[2mm] \dfrac{\partial\varphi}{\partial n} \end{pmatrix} = \begin{pmatrix} x_\xi \, y_\xi \, z_\xi \\ x_\eta \, y_\eta \, z_\eta \\ n_x \, n_y \, n_z \end{pmatrix} \begin{pmatrix} \dfrac{\partial\varphi}{\partial x} \\[2mm] \dfrac{\partial\varphi}{\partial y} \\[2mm] \dfrac{\partial\varphi}{\partial z} \end{pmatrix}
\tag{9.134}
$$

Thus, the gradient of the velocity potential $\nabla\varphi$ on the body surface can simply be obtained using the above equation and the body boundary condition. After evaluating the velocity vector:

$$
\mathbf{V} = \left(V_x, V_y, V_z \right) = \nabla\varphi = \left(\frac{\partial\varphi}{\partial x}, \frac{\partial\varphi}{\partial y}, \frac{\partial\varphi}{\partial z} \right)
\tag{9.135}
$$

one uses shape functions to represent its distribution on each element:

$$
\mathbf{V} = \sum_{j=1}^{S_1} N_j \left(\xi, \eta \right) \mathbf{V}_j
\tag{9.136}
$$

Here the velocity vector is interpreted to a higher degree than the geometric coordinates i.e., $S_1 \geq S$, where S is the number of nodes used for coordinates (x, y, z) interpolation. Then, the tangential derivatives of the velocity vector, V along ξ and η coordinates on the body surface can be calculated by differentiating the shape functions in the form:

$$\frac{\partial \mathbf{V}}{\partial \xi} = \nabla(\nabla\varphi)\cdot\mathbf{r}_\xi, \frac{\partial \mathbf{V}}{\partial \eta} = \nabla(\nabla\varphi)\cdot\mathbf{r}_\eta \qquad (9.137)$$

where $\mathbf{r}_\xi$ and $\mathbf{r}_\eta$ represent tangential vectors of $\mathbf{r}$ along ξ and η coordinates, respectively.

Considering the Laplace's equation for the velocity potential φ, one has the following matrix equation for the double derivatives of the potential:

$$\begin{pmatrix} \dfrac{\partial V_x}{\partial \xi} \\[2mm] \dfrac{\partial V_x}{\partial \eta} \\[2mm] \dfrac{\partial V_y}{\partial \xi} \\[2mm] \dfrac{\partial V_y}{\partial \eta} \\[2mm] \dfrac{\partial V_z}{\partial \xi} \\[2mm] \dfrac{\partial V_z}{\partial \eta} \end{pmatrix} = \begin{bmatrix} x_\xi & y_\xi & z_\xi & 0 & 0 & 0 \\ x_\eta & y_\eta & z_\eta & 0 & 0 & 0 \\ 0 & x_\xi & 0 & y_\xi & z_\xi & 0 \\ 0 & x_\eta & 0 & y_\eta & z_\eta & 0 \\ 0 & 0 & x_\xi & 0 & y_\xi & z_\xi \\ 0 & 0 & x_\eta & 0 & y_\eta & z_\eta \end{bmatrix} \begin{pmatrix} \varphi_{xx} \\ \varphi_{xy} \\ \varphi_{xz} \\ \varphi_{yy} \\ \varphi_{yz} \\ \varphi_{zz} \end{pmatrix} = [B] \begin{pmatrix} \varphi_{xx} \\ \varphi_{xy} \\ \varphi_{xz} \\ \varphi_{yy} \\ \varphi_{yz} \\ \varphi_{zz} \end{pmatrix} \qquad (9.138)$$

$$
\begin{pmatrix} \varphi_{xx} \\ \varphi_{xy} \\ \varphi_{xz} \\ \varphi_{yy} \\ \varphi_{yz} \\ \varphi_{zz} \end{pmatrix} = [A] \begin{pmatrix} \dfrac{\partial V_x}{\partial \xi} \\[6pt] \dfrac{\partial V_x}{\partial \eta} \\[6pt] \dfrac{\partial V_y}{\partial \xi} \\[6pt] \dfrac{\partial V_y}{\partial \eta} \\[6pt] \dfrac{\partial V_z}{\partial \xi} \\[6pt] \dfrac{\partial V_z}{\partial \eta} \end{pmatrix} \qquad (9.139)
$$

where $[A] = \left\{ [B]^T [B] \right\}^{-1} [B]^T$

An example of the double derivatives computed for the velocity potential on a sphere surface was in excellent agreement with the analytical solution (Liu et al., 1995).

9.5.1.4 Computation of the 2nd-order force of main platform of ISSC TLP

Liu et al. (1995) computed the 2nd-order wave-exciting forces of the main platform of ISSC TLP for three cases: fixed in waves, free to move in waves and free to move with tendons attached (but not structurally jointed). In the 3rd case above, an approximate method is applied. It assumes the tendons to function as a mass less spring. Therefore the wave actions on the tendons are neglected, which is a common assumption in uncoupled analyses. This means that the main-platform and tendons are not jointed as an integrated system. Restricting the vertical plane motions and providing some station-keeping forces in the horizontal plane are two primary roles of the tendons. This assumption is not certainly justifiable for deep water TLP, for which we expect appreciable inertia effects of the tendons. In that case, the dynamic coupled analysis program is necessary as discussed in Sec. 9.6.

9.5.1.5 Comparison of 2nd-order force of main-platform of ISSC TLP with experiment

The foregoing computations of the ISSC TLP were compared with the results of Matsui et al. (1992) for the fixed case. The experiment was conducted in a small wave tank of 2 m wide and 15 m long, and 1 m deep with a length scale of 1/168. The correlation between theory and experiment appears to be fair considering the fact that the experiment was carried out with a very small scale model and that the magnitude of the measured 2nd-order force was basically small compared to the 1st-order wave load. A larger scale experiment is desirable to see more reliable correlation between theory and experiment.

9.6 The Impulse Response Function Method for Simulation

We have reviewed the six-degree freedom linear equation of motion of a floating body in the frequency domain in Chap. 6. The six-degree freedom motion equation of a floating body in time domain is derived from the linear impulse response function method (Cummins, 1962):

$$\sum_{k=1}^{6}\left\{ \begin{array}{l} (M_{S_{rk}} + M_{f_{rk}}(\infty))\ddot{u}_k + (C_{v_{rk}} + C_{S_{rk}})\dot{u}_k + b_{w_{rk}}\dot{u}_k \\ + \int_0^t L_{rk}(t-\tau)\dot{u}_k(\tau)d\tau + (K_{S_{rk}} + K_{f_{rk}})u_k \end{array} \right\} \qquad (9.140)$$

$$= F_{1r}(t) + F_{2r}(t), \quad r = 1,\ 2,\cdots,6$$

where M_{srk}, C_{srk}, and K_{srk} indicate the mass, damping and stiffness matrices of structure: $M_{frk}(\infty)$ is the asymptotic values of added mass when $\omega \rightarrow \infty$. K_{frk} represents the hydrostatic restoring force matrix. C_{vrk} is the viscous damping of fluid, b_{wrk} is wave drift damping, $L_{rk}(t)$ denotes a retardation function and $u_k(t)$ is the generalized displacement in the time domain. The retardation function $L_{rk}(t)$ is evaluated using linear radiation damping coefficient C_{frk} in the form:

$$L_{rk}(t) = \frac{2}{\pi}\int_0^\infty C_{frk}(\omega)\cos \omega t d\omega \qquad (9.141)$$

The 1st-and 2nd-order wave loads $F_{1r}(t) + F_{2r}(t)$ are the result of the simulation of the wave exciting-force time series using Volterra quadratic model as discussed in Sec. 9.4.3. Since Eq. (9.140) is the linear equation of motion being excited by the 2nd-order force, the solution of the equation gives the 2nd-order response.

The foregoing equation is called Cummins' time domain equation. It can be applied for simulation of motion of floating body affected by the 2nd-order random wave. This model was applied for instance to the slowly-varying surge drift motion of a moored barge in head seas experimentally studied by Krafft and Kim (1991). The simulation was in excellent agreement with the model test including the offset. It is to be noted that the 2nd-order wave force was determined under the condition that 1st-order motions are only allowed. The above model cannot be applied to the dynamic coupled TLP system that will be discussed below.

9.7 Response of Coupled TLP System

The previous study (Liu et al., 1995) is an uncoupled analysis of TLP. The studies are on the 2nd-order wave force and motion of the main-platform of ISSC in the monochromatic waves including the effect of the mass less tendons attached to the main-platform of the TLP as an uncoupled model. However, the real TLP consists of the main-platform and tendons being structurally jointed in a certain fashion both to the bottom of the main-platform and to the sea bed. Hence we have completely departed from the studies of uncoupled analysis method given in Secs. 9.4 and 9.5.

The most important responses in the design of TLP are the tendon's structural responses, especially springing and ringing of tendons in random seas. The springing and ringing cause tendons to fail that is very costly. In order to analyze the structural response, the coupled dynamic equation is with finite element method (FEM) technique which was first developed by Kim et al. (1994) and further employed to simulate the wave drift damping, and springing and ringing response as described below.

9.7.1 Introduction to coupled TLP system

We review the basic features of the TLP design including the elements of environmental loading and the range of the natural frequencies due to the design concept.

Figure 9.11 illustrates the TLP configuration, environmental loads and the primary TLP responses such as steady offset, maximum tether tension, and tether angle and deck clearance. The environmental parameters affecting TLP response are wind, wave, current, tide, and marine growth. For design, the effects of these parameters must be known. However, we are here mainly interested in the effects of the nonlinear waves. If we assume there is only the wave, it is presumed that the response motion will be reached the steady state oscillating around the position of the offset and set-down.

The resonance period of floating structure is the most important design parameter to be considered in the design and operation. The resonance periods of offshore floating structures are determined to avoid the large wave energy. As shown in Fig. 9.12, TLP has the short resonance period, while

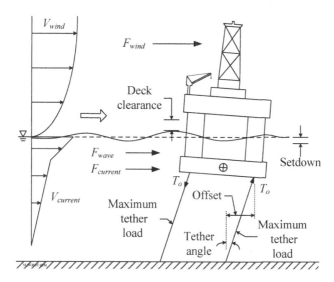

Fig. 9.11. Sketch of a TLP in natural environment (Demiabilek, 1989).

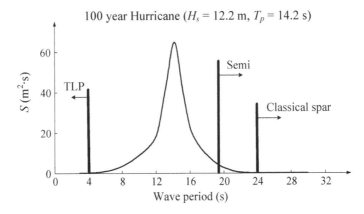

Fig. 9.12. Sketch of one-sided wave spectrum against wave period and regions of heave natural periods of TLP, Spar, and Semi-submersible.

semi-submersible and spar have longer periods. The high pretension of tendons only change surge/sway natural periods. High tendon cross sectional area (= high tendon vertical stiffness) makes natural frequency very high compared to spar and semisubmersible. There are differences in the structural stiffness: semisubmersible and spar have mooring lines of very soft stiffness compared to the TLP that has large cross section tendons creating high structural stiffness in the vertical direction.

As such, TLP resonance frequency is high compared to the other structures. Although not shown in the foregoing figure, the slow drift resonance motion of soft moored long tendons may present. Thus, we have two important high and low resonance frequency regions that are affected by the 2nd-order wave forces on TLP in the nonlinear random seas. Between these low and high frequency resonance motions we have the major wave frequency responses due to linear waves. These are usually designated low, high and wave frequency responses.

The basic rules for preliminary designs of TLP given the configuration and structural stiffness is given in Demiabilek (1987).

9.7.2 The basic strategy to deal with environmental loads

Kim et al. (1994) developed the coupled analysis of ISSC TLP designed by ISSC (Eatock Taylor and Jefferyes, 1986). The particulars of the environment and the tendons of ISSC TLP are shown in Tables 9.1 and 9.2. The environmental fluid load on the coupled TLP system consists of the long-crested random wave force and the concurrent action of the steady wind and current. The effects of the unsteady wind and current are not considered in the study (Kim et al., 1994). The wave impacts and vortex shedding are also excluded. For the mean wind load and the mean current distribution, we employ the recommended formulas by API (1987).

The wave force is assumed to be unaffected by the current and is calculated using the HOBEM code. The viscosity effect is estimated by the Morison's equation considering the relative particle velocity of the current and wave to the moving body. However, the viscous effect is found to be negligibly small compared to the hydrodynamic wave-exciting forces.

All the environmental loads are computed at the upright position of the TLP. The LTFs and QTFs for the forces are computed in the bi-frequency domain by HOBEM and converted into the time series as given in Eq. (9.128).

The resultant environmental load is then used as the external fluid loading in the incremental equation of the coupled TLP system as will be discussed in the next section.

Table 9.1. Particulars of the environment

Water depth	450 m
Wind velocity	100 km/hr
Shape coefficient for wind	1
Platform projected area	2000 m^2
Wind loading point	C.G. of the platform
Current velocity at SWL (u_0)	1 m/s
JONSWAP spectrum (100 year return sea in Gulf of Mexico)	
Significant wave height	12.5 m
Modal period	12 s
Peakedness parameter	2
Minimum and maximum frequency	0.2 – 1.9 rad/s
Wave heading	0
C_M in Morison's equation	2.0
C_D in Morison's equation	1.0
$u = u_0(z/h)^{1/7}$ current velocity profile	

Table 9.2. Particulars of the ISSC TLP tendons

Number of tendons at each leg	1
Space between the tendon (column) centers	86.25 m
Outer radius	0.5 m
Inner radius	0.3434 m
Tendon cross sectional area	0.415 m^2
Tendon length	415 m
Pretension per tendon	3.43×10^{7}N
Tendon mass density	2.992×10^3 kg/m
Young's modulus	2.1×10^{11} N/m^2
Axial stiffness per tendon	0.1×10^8 Ns/m
Tendon material damping	2.1×10^{-6} Ns/m
Top joint	fixed
Bottom joint	hinged

9.7.3 Dynamic equation of the coupled TLP system

The response of TLP platform and tendons to the wave force is determined in terms of the displacement of structure in time. The method of analysis of such integrated structure is known as coupled analysis. When the motion of TLP and tendons are treated separately, it is called uncoupled analysis. Here one uses coupled analysis.

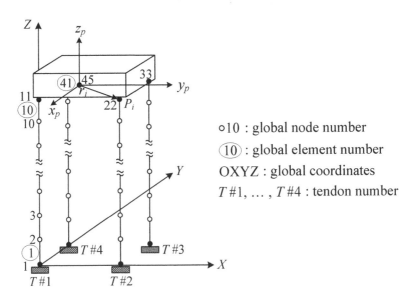

∘10 : global node number

⑩ : global element number

OXYZ : global coordinates

T #1, ... , T #4 : tendon number

Fig. 9.13. Schematic diagram of finite element system.

The dynamic response of the coupled TLP system can be simulated by integrating the incremental dynamic equation of the system which will be represented shortly. Analysis is based on the three-dimensional hybrid element method and updated Lagrangian formulation, for instance by Bathe and Bolourchi (1975) and Bathe et al. (1979), that allows us to address the tendons undergoing large displacement with high pretensions.

One assumes the platform is connected to the top ends of the tendons whose bottom ends are connected to the seabed. The connecting conditions can be arbitrary, i.e., hinged or elastic joint. See Table 9.2.

Before dealing with the dynamic equation one has to define three constraint conditions at the connections referring to Fig. 9.13.

The kinematic constraint condition at the top interface node is given by:

$$\Delta U_I = T_I \Delta U_P \qquad (9.142)$$

ΔU_I and ΔU_P are the incremental displacement vectors of the interface node and the C.G. of the platform (node number 45), respectively, where T_I represents the relation matrix.

The kinematic constraint condition at the bottom node is given by:

$$U_R = \bar{U}_R \qquad (9.143)$$

where $\bar{U}_B$ is the prescribed translational or rotational displacement vectors at the bottom node.

The dynamic constraint condition at the interface node is given by:

$$F_{PT} = T_I^T F_I \qquad (9.144)$$

where F_{PT} denotes the force exerted by the tendon on the platform, at C.G., while F_I denotes the force exerted by the tendon at the interface node, whereas T_I^T is the transposed matrix of T_I defined in Eq. (9.142).

With the foregoing setup one writes the dynamic incremental equation of motion of the platform:

$$M_R \ddot{U}_P + C_P \dot{U}_P + K_f \Delta U_P = F_f - K_f U_P + F_{PT} \qquad (9.145)$$

where M_R denotes the inertia matrix of the platform in the air plus the added inertia of the platform, C_R represents the damping matrix due to fluid and inertial gyroscopic effect, K_f, represents the stiffness due to the fluid, F_f represents the fluid external load on the platform, U_P represents the displacement vector at time t, and ΔU_P is the incremental displacement during the time increment Δt.

One considers next the incremental dynamic equation of the tendons and connecting elements at the interfaces with nodes. Here the tendons are modeled by 3-D beam element and the connecting elements by spring-dampers. The large platform displacement will induce both the finite rigid-body motion displacement and the elastic deformation. The latter can be assumed to be very small. Therefore one can treat the two motions separately by applying the updated Lagrangian formulation. On the other hand the effects of the rotational motion of the tendons and the connecting elements on the TLP system dynamics are much less than those of the translational motions. Thus one can use the lumped mass method, which gives the incremental equation of the tendons and connecting elements in the form:

$$
\begin{pmatrix} M_{JJ} & 0 \\ 0 & M_{II} \end{pmatrix} \begin{pmatrix} \ddot{U}_J \\ \ddot{U}_I \end{pmatrix} + \begin{pmatrix} C_{JJ} & C_{JI} \\ C_{IJ} & C_{II} \end{pmatrix} \begin{pmatrix} \dot{U}_J \\ \dot{U}_I \end{pmatrix}
$$

$$
+ \begin{pmatrix} K_{JJ} & K_{JI} \\ K_{IJ} & K_{II} \end{pmatrix} \begin{pmatrix} \Delta U_J \\ \Delta U_I \end{pmatrix} = \begin{pmatrix} F_J \\ F_I \end{pmatrix} + \begin{pmatrix} 0 \\ F_{TP} \end{pmatrix}
$$

(9.146)

where the subscripts I and J denote the quantities associated with the connecting elements of tendons, respectively. M, C and K are the inertia, damping, and stiffness matrix, respectively. F_{TP}, different from F_{PT}, denotes the forces exerted by the platform on the connecting elements at the interface nodes.

By combining Eqs. (9.145) and (9.146) one has the dynamic coupled equation of the platform and tendons:

$$
\begin{bmatrix} M_S & 0 & 0 \\ 0 & M_{JJ} & 0 \\ 0 & 0 & M_{II} \end{bmatrix} \begin{pmatrix} \ddot{U}_P \\ \ddot{U}_J \\ \ddot{U}_I \end{pmatrix} + \begin{bmatrix} C_S & 0 & 0 \\ 0 & C_{JJ} & C_{JI} \\ 0 & C_{IJ} & C_{JJ} \end{bmatrix} \begin{pmatrix} \dot{U}_P \\ \dot{U}_J \\ \dot{U}_I \end{pmatrix}
$$

$$
+ \begin{bmatrix} K_f & 0 & 0 \\ 0 & K_{JJ} & K_{JI} \\ 0 & K_{IJ} & K_{JJ} \end{bmatrix} \begin{pmatrix} \Delta U_P \\ \Delta U_J \\ \Delta U_I \end{pmatrix} = \begin{pmatrix} F_f + F_{PT} \\ F_J \\ F_I + F_{TP} \end{pmatrix}
$$

(9.147)

The above equation has redundant degree of freedoms, which can be eliminated by employing the constraint conditions. It results in the following coupled dynamic equation:

$$M\ddot{U} + C\dot{U} + K\Delta U = F \tag{9.148}$$

The dynamic response of ISSC TLP system is determined by solving Eq. (9.148).

The ISSC TLP is discretized by 44 elements as shown in Fig. 9.13. The equation of the system elements are given in terms of Lagrangian large displacement (Houbolt, 1950). Kim et al. (1994) demonstrated that the finite element solution was in excellent agreement with the analytical solution for a long column. The effectiveness of the Houbolt integration scheme (HIS) on the accuracy of computation was investigated by varying the time increment.

We confirmed through a few exercises that the HIS was stable without conditions. But its integration accuracy depends on the magnitude of the time increment. If the increment chosen is too large, the error increases. If too small, an increment results in unnecessary waste of computer resources. Thus, it is important to choose an optimum increment. A widely used general rule is that the increment should not be greater than $T_{min}/20 - T_{min}/30$, where T_{min} denotes the minimum period of excitation or the minimum natural period of the system within a meaningful frequency band. To test the foregoing rule a study was conducted by computing a simple harmonic motion of a mass supported by a cantilever beam in the vacuum and comparing with the exact solution, which gives following result (Kim et al., 1994).

$\Delta t(s)$	$\Delta t / T_{min}$	accuracy
0.02	1/100	very good
0.06	1/30	good
0.10	1/20	fair

In dealing with the foregoing mass-cantilever, we confirmed that the response to a dual frequency excitation is identical to the linear superposition of the two independent harmonic responses when the displacement is small. We validated also for a free rigid-body motion affected by an initial external force it can generate a very long time series without any numerical problem.

9.7.4 Simulation of coupled ISSC TLP response in random sea

Employing the above dynamic coupled equation, developed by Kim et al. (1994) computed the responses of ISSC TLP in irregular waves of the 100 year return sea spectrum in the Gulf of Mexico. The environmental and the structural particulars are given in Tables 9.1 and 9.2.

The 2nd-order wave-exciting-and radiation-force in the random seas is theoretically computed according to the Volterra quadratic model as described in Sec. 9.5 using the particulars of the environment and structures. The computation gives the time histories of surge, heave, and pitch motion of the TLP and tension force in an element for instance, element 10 of tendon #1 on the sea side. The three modes of motion have the natural frequencies that are found in the simulated time series. The natural frequencies are in good agreement with the conventional estimates as shown in Table 9.3. It was found from the time series of the tension that the natural high frequency pitching motion could be the main contributor to the high frequency tension creating high frequency springing.

Table 9.3. Estimate of natural frequencies of ISSC TLP

	Virtual mass [kg]	Damping Coefficient [Ns]	Virtual stiffness[N/m]	Natural frequency [rad/s]
Surge	0.866×10^8	0.949×10^3	0.279×10^6	0.057
Heave	0.658×10^8	0.349×10^1	0.822×10^9	3.53
Pitch	0.133×10^{12} *	0.189×10^6	0.151×10^{13}	3.37
Surge-pitch			-0.106×10^8 **	

*[kg] , **[N]

9.7.5 The wave drift damping of the coupled ISSC TLP

The monochromatic mean surge wave drift damping of a variety of structures, fixed and free in deep water, were computed by Zou (1997) using Aranha (1996) and HOBEM (Liu et al., 1995). The structures include a vertical truncated cylinder, Snorre-TLP and others. The damping of the Snorre TLP was also calculated by Eatock Taylor and Natvig (1994) using a slightly different HOBEM from (Liu et al., 1995). These two computations are found to be in an excellent agreement as shown in Fig. 9.14.

Convergence tests were conducted by varying number of elements followed by a decision to take 38 HOBEM elements in Zou (1997).

The study also estimated the expected value of surge wave drift damping of ISSC TLP due to the 100 year return sea in the Gulf of Mexico. The estimation gives 3.62×10^5 Nsm^{-1}, which is interestingly in the same order of magnitude as the tested and calculated data of Hutton-TLP, i.e., 3.34×10^5 Nsm^{-1} to 1.38×10^5 Nsm^{-1} as reported by Le Boulluec et al. (1994).

9.7.6 Springing and fatigue analysis

Ringing is a bursting behavior occurring in the phenomenal sea (Stansberg et al., 1995) typified by a ringing (extreme) factor (peak tension/std) greater than 7 and kurtosis greater than 5. Springing is a steady phenomenon occurring in the weakly nonlinear random sea, with the ringing (extreme) factor of less than 5 (Jefferys and Rainey, 1994). In reality these springing and ringing occur simultaneously (Natvig, 1994) when the sea is very high.

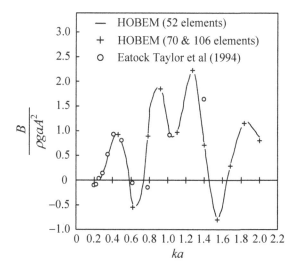

Fig. 9.14. The surge wave drift damping coefficient of floating Snorre TLP in waves.

However, springing alone can be treated in the weakly nonlinear fatigue sea being below $H_s = 9$ m (high sea, Table 2.1). Specifically the springing analysis was made for a North Sea TLP at the typical fatigue sea state with JONSWAP spectrum, $H_s = 8$ m, $T_m = 12$ s, and $\gamma = 1.7$, the result of which was in excellent agreement with model test data (Zou et al., 1999).

For the fatigue analysis, firstly one computes hydrodynamic LTFs and QTFs for the wave-exciting forces on the TLP applying HOBEM. Then these hydrodynamic responses are Fourier transformed inversely to determine the hydrodynamic LIRFs and QIRFs (Zhao, 1996; Zhao and Kim, 1998), with which the response force is simulated long time according to the Volterra quadratic model. Inputting the foregoing simulated 2nd-order force time series one simulates the coupled dynamic equation, which gives the time series of tension in the tendons. The high frequency tension time series at a typical element is non-dimensionalized with the std of the high frequency part of the time series. The result agrees with the statistical criterion for springing (Jefferys and Rainey, 1994). The experimental energy density spectrum of the tension is in excellent comparison with the result of computation as shown in Fig. 9.15.

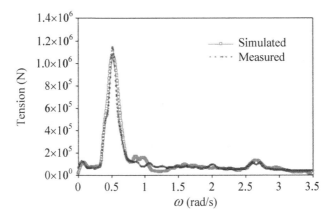

Fig. 9.15. Comparison of measured and simulated tendon tension energy density spectrum.

It is absolutely necessary to note that the simulation here is a purely theoretical application of the Volterra quadratic model for the given particulars of TLP and wave energy spectrum. The springing and ringing together may be searched using experimental data as will be discussed in Sec. 9.9. The analysis of the fatigue due to the foregoing tension was made in Zou et al. (1999).

The fatigue analysis is based on the weakly nonlinear seas as mentioned earlier. We have to consider sea severities lower than $H_s = 9$ m for the estimation of fatigue analysis. Zou et al. (1999) dealt with the fatigues of the tendons employing the simulated tension (API, 1987).

9.8 Limitedness of 2nd-Order Theories

9.8.1 Limitedness of 2nd-order wave theory

Kumar et al. (2002C) were interested in finding the approximate criterion of limitedness of the 2nd-order random wave theory for modeling the waves produced in the 2-D wave tanks. The basic approach was similar to Forristall (2000) in Sec. 9.2.4.3. The data from the world renounced wave basin were employed. Twelve laboratory wave data out of twenty were selected based on the comparison of the variances of the measured wave to the target energy spectrum. When the difference of the above was more than 10% of the target spectrum, the data was discarded.

In the simulation the angular frequency was chosen to be twice the Nyquist frequency, $\omega_{max} = 2\pi / \Delta t$, where Δt was the prototype sample time step. The size of $\Delta \omega$ was given by $\Delta \omega = 2\pi / T$ where T was the total sample length which was chosen as one hour (3600 s). The study simulated forty independent realizations, each one hour long for each target spectrum until the statistical convergence of the ratio of the crest-height to the Rayleigh was guaranteed. This was done by averaging the ratio of crest height to Rayleigh by consecutively adding the newly computed value from the consecutive realizations distributed over a very small band of probability of exceedence.

For the waves of significant wave height $H_s = 3$ m, 4 m, 6 m and 9 m, the comparison of the ratios indicates that the measured waves are approximately similar to the 2nd-order waves. This kind of conclusion is obtained by observing Figs. 9.16(a), (b) etc. Beyond $H_s = 9$ m, the 2nd-order theory highly underestimates the laboratory data.

The likelihood of the threshold for the limitation of the 1st-order theory seems to be 4.0 m.

Similar comparison study was made for the waves of $H_s = 7$ m, 8 m, 9 m, 11 m and 15.4 m together. The comparisons of the data above 9 m of H_s show that the 2nd-order theory generally underestimates. Based on the above conclusion, the study attempted to draw a conclusion that the 2nd-order theory could be valid approximately up to the waves of $H_s = 9$ m.

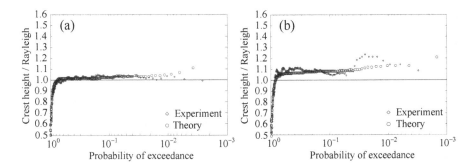

Fig. 9.16(a) Comparisons of ratios of crest height to Rayleigh for H_s = 3 m. (Left fig.). (b) Comparisons of ratios of crest height to Rayleigh for H_s = 11 m (Right fig.).

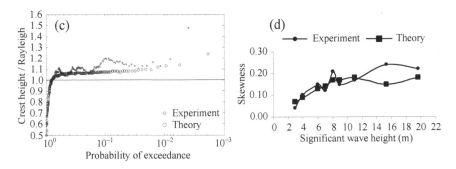

Fig. 9.16(c) Comparisons of ratios of crest height to Rayleigh for ringing wave of Heidrun TLP, for H_s = 15.4 m (Left fig.). (d) skewness versus significant wave height (Right fig.).

The conclusions drawn from the behavior of the crest height/Rayleigh such as given in Figs. 9.16(a)-(c) could be subjective. Thus the skewness was compared as shown in Fig. 9.16(d). The skewness generally tends to increase with the significant wave height, though some fluctuations are present in the test data. Discrepancy of skewness between the theory and experiment occurs around H_s = 9 m and 11 m. These comparisons partially support the foregoing assessment that is based on the observation of the ratios of crest height to the Rayleigh. The sudden fluctuation of the average deviation occurring in the range of rough to high sea (H_s = 4.0 m to 9.0 m) in Fig. 9.16(d) may be attributed to the limited sample length of experimental data. If many longer test data were available and the same analyses could be conducted, there would be a better converging statistical trend of many data within the same fluctuation range.

The variances of the target spectrum and the measured wave of the above 12 waves are compared in Fig. 9.17, where the narrow-band Gaussian wave satisfies $m_0 = H_s^2/16$. It demonstrates that the 12 waves approximately satisfy the equal variances of the target and measured wave spectrum.

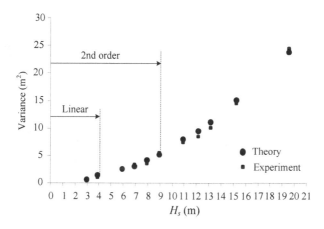

Fig. 9.17. Comparison of variance against significant wave height and assumed range of validity of theories.

Thus, one might tentatively conclude from the specific study that the 1st-order theory would be valid up to rough sea ($H_s = 2.5$ m–4 m) and 2nd-order wave theory valid up to high sea ($H_s = 6$ m–9 m) in the long-crested sea waves. These conclusions cannot yet be generalized. It should be noted that the foregoing conclusion is soly on the very limited data of long-crested seas generated in wave tanks. Also the effect of directionality is not taken into account. Thus it is difficult to have a firm conclusion of the criteria for practical application.

9.8.2 Limitedness of 2nd-order wave force

We have confirmed that the QTFs of Volterra quadratic model is exactly identical to the hydrodynamic QTFs of the added resistance in the Gaussian waves. There is also similarity between the QTFs of the Volterra quadratic model and the hydrodynamic QTFs for the 2nd-order wave and force theory respectively, with different multiplication factor of ½ and 2, as shown in Secs. 9.2.2.1 and 9.4.6. Thus the application of 2nd-order wave force theory for the 2nd-order response is bounded to be similarly limited as the 2nd-order wave theory. In other words, 2nd-order force theory may be applied up to the high seas ($H_s = 6$ m–9 m). This criterion is also not a general one because the limitation of the 2nd-order wave theory was based on the limited data and analyses.

9.8.3 Limitedness of Volterra quadratic model

We have the same degree of limitedness in the applicability of the Volterra quadratic model to the hydrodynamic theories because the hydrodynamic theories and quadratic Volterra model are in the precise similarity relations as proved in Secs. 9.2.2.1 and 9.4.6.

It is important to recognize that Volterra quadratic model is only applicable to the 2nd-order level of waves as defined in Sec. 9.8.2.

Beyond the 2nd-order nonlinearity, one might mistakenly exercise to choose arbitrarily large energy density spectra as an input. The simulation would then result in unrealistically large responses leading to false judgment, for instance, ringing. However, it is to be noted that the ringing of the specific Heidrun TLP occurred in the specific Heidrun ringing wave such as the phenomenal sea that is far beyond the 2nd-order random seas (Stansberg et al., 1995; Statoil, 1996).

9.9 Brief Review of Springing and Ringing

Springing and ringing have long been research targets in ocean engineering. Its importance lies on the effects of sea severity on the tendon response, the failure of which is certainly costly. No mathematical models are generally available for the nonlinear waves and loads in the field and wave tank. Therefore, it is essential to generate and store the measured nonlinear waves to be used in the later analyses for application of models. Zou et al. (1998) generated the laboratory storm seas and measured the springing and ringing on ISSC TLP model, which could not be achieved in the Heidrun ringing wave. Instead of carrying out model test in the wave tank, it is desirable to employ a model that bridges between the measured wave and unknown force. ULSM (universal linear system model) was developed by Kim and Zou (1995). Two types of ULSM were developed. One relates the given wave and wave kinematics (ULSM-kinematics) and consequently determines the wave force by Morison's equation. The other model relates the given wave to the unknown force employing diffraction theory (ULSM-diffraction).

9.9.1 Pure theoretical approach

We usually employ a pure theoretical method called Volterra quadratic model for simulation of Gaussian sea loads for the weakly nonlinear waves. Zhao and Kim (1998) developed a technique to simulate the 2nd-order force on ISSC TLP in Gaussian seas for very long time (Chap. 8), which was used for practical application for fatigue analysis by Zou et al. (1999).

9.9.2 Experimental approach with model

Since the springing and ringing occur simultaneously (Natvig, 1994 and others) in high seas, and ringing was overwhelmingly believed to be due to highly nonlinear horizontal wave kinematics, research was focused in creating strongly nonlinear waves such as large 2-D transient in the beginning of research (Kim et al., 1992). Later the technique was further advanced to include the strongly nonlinear transient wave to be a part of the random sea in natural mode (Zou and Kim, 1999). Thus many researches were interested in finding the effects of the nonlinear kinematics on the springing and ringing. These experimental approaches have to employ the foregoing models ULSM-kinematics and ULSM-diffraction to bridge the measured wave to predict the unknown force.

Another experimental approach is to use experimentally measured wave and force directly without using the model for assessment of springing and ringing.

9.9.2.1 Extreme kinematics and impact

Zou and Kim (1995) measured extreme kinematics and impact load on a vertical truncated column in the strongly asymmetric transient wave group and equivalent-height large regular wave. This study attempted to relate the extreme kinematics (Kim et al., 1992) to the impact. Also Zou and Kim (1996) have conducted careful laboratory tests employing LDA (Laser Doppler Anemometer), dynamic pressure gauges and electronic sensing tape pasted on the cylinder surface to measure wave elevation simultaneously with the others.

An experimental study (Zou and Kim, 1995) was conducted to investigate the process of impact on a vertical truncated cylinder due to a non-breaking steep wave in deep water. The simultaneous measurements of wave elevation, particle kinematics, wave elevation around the cylinder, dynamic pressure distribution and impact force, made it possible to analyze the impact process due to the steep-front asymmetric wave. The strong nonlinear particle kinematics in front of the structure might have produced air pocket, impacting pressure, and impacting force. Maximum run-up reached about 1.34 times the wave height. The rise time was much longer than drop time.

9.9.2.2 ULSM-kinematics

Kim and Zou (1995) proposed the 1st-version of ULSM-kinematics for simulation of a weak impacting force due to slightly stronger nonlinear wave. When it was applied to a strongly nonlinear wave (Kim et al., 1996), the model's ability to predict the response was unsatisfactory. In the following year Zou et al. (1997) proposed the 2nd-version of ULSM-kinematics in addressing the simulation of the wave force on the ISSC TLP in the highly nonlinear wave kinematics. The foregoing work revealed that the springing was due to the weakly nonlinear wave, while the ringing due to the strongly nonlinear wave.

The 2nd version ULSM-kinematics was employed in the investigation of extremely nonlinear kinematics, and impact and non-impact load on the series of cylinders with various radii (Kim et al., 1997). Employing further the 2nd-version of ULSM-kinematics, Kim et al. (1997) simulated the springing and ringing of ISSC TLP due to isolated transient waves. It was again found that the weakly nonlinear wave produced springing and strongly nonlinear wave ringing. It is to be noted that the assessment made in the foregoing on the springing or ringing was not based on the statistical criteria because the sample lengths were short.

9.9.2.3 ULSM-diffraction

The ULSM-kinematics could produce impact or non-impact depending on the degree of nonlinearity of the incident wave on the ISSC TLP.

Another approach for simulation of the diffraction force was developed by Kim and Zou (1998), which is called ULSM-diffraction. Validity of this approach was demonstrated by comparing with the experiment of the single large column affected by a replica of Heidrun TLP ringing wave (Stansberg, 1995; Statoil, 1996).

Wang and Kim (1999A) applied the foregoing ULSM-diffraction ULSM-(L), ULSM-(L + Q), ULSM-(L + Q + C) to simulate the wave load on a large column in the replica Heidrun TLP ringing wave. It was found that the ULSM-(L + Q) was the best predictor among them. In the above L, Q and C represent LTF, QTF and CTF, respectively.

An analysis of the horizontal force on the large column in the replica of the Heidrun ringing wave (Stansberg et al., 1995; Statoil, 1996) was made by Kim and Wang (1999B). The steep regular wave has 30 m wave height and 15.7 s period. The length scale 1 / 100 was used in order to transform into the Stokes 5th-order wave (Alex and Kim, 1999). It was necessary to define a quasi LTF, or LTF^*. LTF^* is the ratio of output force amplitude to the input wave and is computed by ULSM-(L + Q). The LTF and LTF^* were compared as a function of wave frequency. The discrepancy between them is large at the low frequency and small in the high frequency component. The nonlinearity of the component wave forces were investigated varying wave steepness.

9.9.2.4 Progress report of springing and ringing of ISSC TLP

The progress report of springing and ringing (Kim, 1998) was necessary to find where the research stood on the ringing wave load model. To date, the 3rd-order mathematical theories were not readily applicable. The ULSM-kinematics and ULSM-diffraction developed so far have been applied for simulation of the nonlinear wave forces.

9.9.2.5 Ringing of ISSC TLP in a storm sea

Zou et al. (1998) investigated ringing of ISSC TLP in a storm sea generated in wave tank at Texas A & M University (TAMU). The observed ringing was due to the directly measured force data on the ISSC TLP model in the storm sea. The foregoing ringing wave is similar to the transient wave that can be produced through distortion in the irregular wave train (Funke and Mansard, 1982), by taking the appropriate vertical asymmetry parameter (Myrhaug and Kjeldsen, 1984), and high peakedness parameter γ (Isherwood, 1986). MARINTEK wave has 3 hr long record of JONSWAP $H_s = 15.4$ m, $T_m = 17.8$ s, and $\gamma = 1.7$ and vertical asymmetry of $\lambda = 1.50$. The storm sea generated at TAMU tank had the JONSWAP of $H_s = 9.88$ m, $T_m = 11.02$, $\gamma = 6.5$ and the vertical asymmetry of $\lambda = 2.15$. The test measured 20 s for each of a total of 8 runs to make 27 min long ensemble. At each run the random seed number was changed. The foregoing TAMU storm wave had stronger nonlinearity than the Statoil wave or the replica of Heidrun ringing wave. Thus, the ringing found in ISSC TLP, was caused by the stronger nonlinearity of the TAMU laboratory wave. The ringing of ISSC TLP was not found in the replica of Heidrun ringing wave.

9.9.2.6 Simulation of ringing of ISSC TLP in Heidrun ringing wave

Wang and Kim (2001) applied ULSM-diffraction to simulate wave load on ISSC TLP in the replica of Heidrun ringing wave hoping that it would produce ringing. However, ringing was not found in the simulation. This may be attributed to the nonlinearity of the replica of Heidrun ringing wave being weaker than the TAMU ringing wave in Sec. 9.9.2.5.

9.9.2.7 Ringing of Heidrun TLP due to Heidrun ringing wave

Kumar and Kim (2002A; 2002B) have observed ringing in the Heidrun TLP caused by the replica of Heidrun ringing wave. ULSM-diffraction was applied to simulate the wave load. In the paper, ULSM-diffraction has been renamed UNIOM-diffraction because it is more appropriate than ULSM-diffraction.

References

Journal references:
Aranha, J.A.P. (1996). Second-Order Horizontal Steady Forces and Moment on a Floating Body with Small Forward Speed, J. Fluid Mech., 313, pp. 39–54.
Cummins, W.E. (1962). The Impulse Response Function and Ship Motions, David Taylor Model Basin Report 1661
Dalzell, J.F. (1999). A note on finite depth second-order wave-wave interactions, Appl. Ocean Res., 21. pp. 105–111.
Eatock Taylor, R. and Hung, S.M. (1987). Second order diffraction forces on a vertical cylinder in regular waves, Appl. Ocean Res., Vol. 9, pp. 19–30.
Eatock Taylor, R. and Jefferys, E.R. (1986). Validity of Hydrodynamic Load Predictions for a Tension Leg Platform, Ocean Eng., Vol. 13, No. 5, pp. 449–490.
Forristall, G.Z. (2000). Wave Crest Distributions: Observations and Second-Order Theory, J. Phys. Oceanogr., American Meteorological Society, Vol. 30, pp. 1931–1943.
Hasselmann, K. (1966). On Nonlinear Ship Motions in Irregular Waves, J. Ship Res., Vol. 10, No. 1, pp. 64–68.

Isherwood, R.M. (1986). A Revised Parameterization of the JONSWAP Spectrum, Appl. Ocean Res., Vol. 9, No. 1, pp. 47–50.

Kim, C.H., Randall, R.E., Boo, S.Y. and Krafft, M.J. (1992). Kinematics of 2-D Transient Water Waves Using Laser Doppler Anemometry, J. Waterways, Port, Coastal Ocean Eng., ASCE, Vol. 118, No. 2, pp. 147–165.

Kim, C.H. and Zou, J. (1995). A Universal Linear System Model for Kinematics and Forces Affected by Nonlinear Irregular Waves, Int. J. Offshore Polar Eng., Vol. 5, No. 3, pp. 166–170.

Kim, C.H., Xu, Y. and Zou, J. (1997). Impact and Non-Impact on Vertical Truncated Cylinder Due to Strong and Weak Asymmetric Wave, Int. J. Offshore Polar Eng., ISOPE, Vol. 7, No. 3, pp. 161–167.

Kim, C.H. and Wang, Z.M. (1999A). Universal Linear System Model For Storm Sea Loads on Column, Int. J. Offshore Polar Eng., Vol. 9, No. 1, pp. 48–54.

Kim, C.H. and Wang, Z.M. (1999B). Horizontal Force of Vertical Truncated Column in Stokes 5th-Order Waves, Int. J. Offshore Polar Eng., Vol. 9, No. 3, pp. 167–174.

Kim, M.H. and Yue, K.P. (1989). The second-order diffraction solution for an axisymmetric body — Part 1. Monochromatic waves, J. Fluid Mech., Vol. 200, pp. 235–264.

Kim, M.H. and Yue, K.P. (1990). The complete second-order diffraction solution for an axisymmetric body — Part 2. Bichromatic waves and body motions, J. Fluid Mech., Vol. 211, pp. 547–593.

Kim, C.H., Kim, M.H., Liu, Y.H. and Zhao, C.T. (1994). Time Domain Simulation of Nonlinear Response of a Coupled TLP System, Int. J. Offshore Polar Eng., Vol. 4, No. 4, pp. 281–291.

Kumar, A. and Kim, C.H. (2002B). Ringing of Heidrun TLP in High and Steep Random Waves, Int. J. Offshore Polar Eng., Vol. 12, No. 3, pp. 189–195.

Kumar, A., Kim, C.H. and Zou, J. (2002C). Limitedness of 2nd-order Theories for Laboratory High Sea Waves and Forces on Structures, Int. J. Offshore Polar Eng., Vol. 12, No. 4, pp. 243–248.

Liu, Y.H., Kim, C.H. and Lu, S.X. (1991). Comparison of Higher-Order Boundary Element and Constant Panel Method for Hydrodynamic Loadings, Int. J. Offshore and Polar Eng., Vol. 1, No. 1, pp. 8–17.

Liu, Y.H., Kim, C.H. and Kim, M.H. (1993). The Computation of Mean Drift Forces and Wave Run-Up by Higher-Order Boundary Element Method, Int. J. Offshore and Polar Eng., Vol. 3, No. 2, pp. 101–106.

Liu, Y.H., Kim, M.H. and Kim, C.H. (1995). The Computation of Second-Order Mean and Double Frequency Wave Loads on a Compliant TLP by HOBEM, Int. J. Offshore and Polar Eng., Vol. 5, No. 2, pp. 111–119.

Longuet-Higgins, M.S. (1963). The Effect of Nonlinearities on Statistical Distribution in the Theory of Sea Waves, J. Fluid Mech., Vol. 17, pp. 459–480.

Matsui, T., Suzuki, T. and Sakoh, Y. (1992). Second-Order Diffraction Forces on Floating Three Dimensional Bodies in Regular Waves, Int. J. Offshore Polar Eng., ISOPE, Vol. 2, No. 3, pp. 175–185.

Molin, B. (1979). Second-order diffraction loads upon three-dimensional bodies, Appl. Ocean Res., Vol. 1, pp. 197–202.

Myrhaug, D. and Kjeldsen, S.P. (1984). Parametric Modeling of Joint Probability Density Distribution for Steepness and Asymmetry in Deep Water Waves, Appl. Ocean Res., Vol. 6, No. 4, pp. 207–220.

Pierson, W. J. (1993). Oscillatory 3rd-Order Perturbation Solutions for Sums of Interacting Long-Crested Stokes Waves on Deep Water, J. Ship Res., Vol. 37, No. 4, pp. 354–383.

Sharma, J.N. and Dean, R.G. (1981). Second-Order Directional Seas and Associated Wave Forces, J. Soc. Petroleum Eng., pp. 129–140.

Stokes, G.G. (1847). On the theory of oscillatory waves, Cambridge Trans. VIII (Reprinted in Mathematical and Physical Paper, I, 1880), Cambridge University Press, London

Telste, J.G. and Noblessse, F. (1986). Numerical Evaluation of the Green Function of Water Wave Radiation and Diffraction, J. Ship Res., Vol. 30, No. 2, pp. 69–84.

Tick, L.J. (1961). The Estimation of Transfer Functions of Quadratic System, Technometrics., Vol. 3, pp. 563–567.

Zou, J. and Kim, C.H. (1996). Experimental Study of Impacting Wave Force on Vertical Truncated Cylinder, Int. J. Offshore Polar Eng., ISOPE, Vol. 6, No. 4, pp. 291–293.

Zou, J., Xu, Y., Kim, C.H. and Zhao, C.T. (1998). Ringing of ISSC TLP Due to Laboratory Storm Seas, Int. J. Offshore Polar Eng., Vol. 8, No. 2, pp. 81–89.

Book references:

Abramowitz, M. and Stegun, A. (1972) Handbook of Mathematical Functions, Dover Publications, Inc., New York.

Chakrabarti, S.K. (1987) Hydrodynamics of Offshore Structures Computational Mechanics, Publications, Southhamption Boston.

Chakrabarti, K.S. (1990) Nonlinear Methods in Offshore Engineering, Elsevier.

Dean, Robert G. and Dalrymple, Robert A. (1991) Water wave Mechanics For Engineers and Scientists, Advanced Series on Ocean Engineering — Volume 2.

Demiabilek, Z. (1989) TENSION LEG PLATFORM — A State of the Art Review, American Society of Civil Engineers, 345 East 47th Street, New York, New York. 10017–2398.

Mei, C.C. (1989) The applied dynamics of ocean surface waves, Vol.1, World Scientific Publishing Co., Singapore.

Newman, J.N. (1977) Marine Hydrodynamics, MIT Press, Cambridge, Massachusetts.

Phillips, O.M. (1977) The Dynamics of the Upper Ocean, 2nd edition, Cambridge Univ. Press.

Sarpkaya T. and Isaacson, M. (1981) Mechanics of Wave Forces on Offshore Structures, Van Nostrand Reinhold Co, New York.

Stoker, J.J. (1957) Water Waves, The Mathematical Theory with Applications, Interscience Publishers, Inc., New York.

Wehausen, J.V. and Laitone, E.V. (1960) Surface Waves, in Encyclopedia of Physics, Vol. IX, Springer-Verlag, Berlin, Goettingen, Heidelberg.

Volterra, V. (1959) Theory of Functionals and Integral and Integro-Differential Equations, Dover Publications Inc., New York.

Proceedings references:

API RP 2T (1987). Recommended Practice for Planning, Designing and Constructing Tension Leg Platforms, The API Recommended Practice 2T (RP 2T), American Petroleum Institute, Washington, D.C.

Bathe, K.J. and Bolourchi, S. (1979). Large Displacement Analysis of Three-Dimensional Beam Structures, Int. J. Num. Method in Eng., Vol. 14, pp. 961–986.

Bathe, K.J., Ramm, E. and Wilson, E.L. (1975). Finite Element Formulations for Large Deformation Dynamic Analysis, Int. J. Num. Method in Eng., Vol. 9, pp. 353–386.

Cummins, W.E. (1962). The Impulse Response Function and Ship Motions, David Taylor Model Basin, Hydrodynamics Lab., Report 1661.

Dean, R.G. and Sharma, J.N. (1981). Simulation of Wave Systems Due to Nonlinear Directional Spectra, Int. Symp. Hydrodynamics Ocean Eng, Norwegian Inst. of Tech., Vol. 2, pp. 1211–1222.

Eatock Taylor, R. and Natvig, B.J. (1994). Effects of Wave Drift Damping on TLP Horizontal Offsets, Proc. 7th BOSS, Vol. 2, pp. 31–41.

Houbolt, J.C. (1950). A Recuurence Matrix Solution for the Dynamic Response of Elastic Aircraft, J. Aero. Scie., pp.540–550.

Jefferys, E.R. and Rainey, R.C.T. (1994). Slender Body Models of TLP and GBS 'Ringing', Proc. BOSS, Vol. 2, pp. 587–605.

Kim, C.H., Xu, Y., Zou, J. and Won, Y.S. (1996). A Model for Weak Impacting Force on Vertical Truncated Cylinder Due to Steep Asymmetric Wave, Proc. 6th Int. Offshore Polar Eng. Conf., Los Angeles, ISOPE, Vol. 3, pp. 215–220.

Kumar, A. and Kim, C.H. (2002A). Time Domain Simulation of Ringing of Heidrun TLP, Proc. 12th Int. Offshore Polar Eng. Conf., Kitakyushu, Japan, Vol. 1, pp. 176–183.

Le Boulluec, M., Le Buhan, P., Chen, X.B., Deleuil, G., Foulhoux, L., Molin, B., and Villeger, F. (1994). Recent Advances on the Slow-Drift Damping of Offshore Structures, Proc. 7th BOSS, Vol. 2, pp. 9–30.

Lighthill, M.J. (1979). Waves and hydrodynamic loading, Proc. 2nd Int. Conf. Behavior Offshore Structures, BOSS, London, pp. 1–40.

MacCamy, R.C. and Fuchs, R.A. (1954). Wave forces on piles: A diffraction theory, Technical Memorandum No. 69, Beach Erosion Board, Corps of Engineers.

Molin, B. and Marion, A. (1986). Second-order loads and motions for floating bodies in regular waves, Proc. Offshore Mechanics Arctic Eng., Vol. 1 , pp. 253–360.

Funke, E.R. and Mansard, E.P.D. (1982). The Control of Wave Asymmetries in Random Waves, Proc. 18th Int. Conf. Coastal Eng., Cape Town, South Africa, pp. 725–744.

Natvig, B.J. (1994). A Proposed Ringing Analysis Model for Higher Order Tether Response, Proc. 4th Int. Offshore Polar Eng. Conf., Osaka, ISOPE, Vol. 1, pp. 40–51.

Ogilvie, T.F. (1983). Second order hydrodynamic effects on ocean platforms, Int. Workshop Ship and Platform Motion, Berkeley, pp. 205–265.

Stansberg, C.T., Huse, E, Krogstad, J.R. and Lehn, E. (1995). Experimental Study of Non-Linear Loads on Vertical Cylinders in Steep Random Waves, Proc. 5th Int. Offshore Polar Eng. Conf., ISOPE, Vol. 1, pp. 75–82.

Statoil (1996). Single Column Test Data Produced at MARINTEK, Stavanger, Norway.

Wang, Z.M. and Kim, C.H. (2001). Nonlinear Response of ISSC TLP in High and Steep Random Waves, Proc. 11th Int. Offshore Polar Eng. Conf., Vol. 3, pp. 440–446.

Zhao, C.T. (1996). Theoretical Investigation of Springing-Ringing Problems in Tension Leg Platforms, Ph.D. Dissertation in Ocean Engineering, Texas A & M University.

Zhao, R. and Faltinsen, O.M. (1989). Interaction between current, waves and marine structures, Proc. of Num. Ship Hydrodynamics, Hiroshima, Japan.

Zhao, C.T. and Kim, C.H. (1998). Second-Order Nonlinear Wave Force in Random Seas, Proc. 8th Int. Offshore Polar Eng. Conf., Montreal, Canada, Vol. 3, pp. 535–542.

Zou, J. and Kim, C.H. (1995). Extreme Wave Kinematics and Impact Loads on a Fixed Truncated Circular Cylinder, Proc. 5th Int. Offshore Polar Eng. Conf., The Hague, Netherlands, Vol. 3, pp. 216–225.

Zou, J., Xu, Y. and Kim, C.H. (1997). Time Domain Simulation of a TLP's Response Due to Laboratory-Generated Asymmetric Irregular Waves, Proc. 7th Int. Offshore Polar Eng. Conf., Honolulu, ISOPE, Vol. 1, pp. 193–200.

Zou, J. (1997). Investigation of Slowly-Varying Drift Motion and Springing and Ringing of Tension Leg Platform System in Nonlinear Irregular Waves, Ph.D. Dissertation in Ocean Engineering, Texas A & M University.

Zou, J. and Kim, C.H. (2000). Generation of Strongly Asymmetric Wave in Random Seaway, Proc. 11th Int. Offshore Polar Eng. Conf., Vol. 3, pp. 95–102.

Division IV

Universal Nonlinear Input-Output Model

Chapter 10

Volterra Cubic Model
and
3rd-Order Wave and Force

10.1 Introduction

In Chaps. 7-9, we have reviewed the Volterra quadratic model, 2nd-order wave theory, and 2nd-order diffraction force theory. Here we will review the 3rd-order Volterra model (Dalzell, 1982), 3rd-order wave theory, and the 3rd-order wave force theory.

10.2 The Volterra Cubic Model

Dalzell (1982) extended the Volterra quadratic model to the cubic input-output model. The objective was to explore the applicability of the third degree functional polynomial model to nonlinear seakeeping problems, and to attempt the development of an approach by which 3rd-degree non-linearity in observed responses of ships to waves might be interpreted. Basic approaches to the deterministic characterization of such systems were discussed, and basic system identification theory for the random sea case was developed in the work. The latter involves the estimation of a new entity called the cross-tri-spectrum. Simulations of the response of a system with quadratic and cubic non-linearity were carried out for random Gaussian excitation. A first attempt was made at cross-tri-spectral identification of cubic response from samples of random excitation and response.

There are similarity relations between 2nd-order hydrodynamic theories and Volterra quadratic model. The 3rd-order theories of the wave and force have not completely been developed for comparison with the Volterra cubic model to find the similarities as the quadratic model and theories. However, there are some correlations among them as will be shown below that help understand the 3rd-order Volterra model and hydrodynamic theories.

10.2.1 Volterra cubic model in time domain

The 3rd-order response to the zero-mean Gaussian input is given by:

$$
\begin{aligned}
y(t) = & \int g_1(\tau_1)x(t-\tau_1)d\tau_1 \\
& + \iint g_2(\tau_1,\tau_2)x(t-\tau_1)x(t-\tau_2)d\tau_1 d\tau_2 \\
& + \iiint g_3(\tau_1,\tau_2,\tau_3)x(t-\tau_1)x(t-\tau_2)x(t-\tau_3)d\tau_1 d\tau_2 d\tau_3
\end{aligned}
\tag{10.1}
$$

The cubic impulse response function (CIRF) and cubic transfer function (CTF) are defined in the similar form as the quadratic system property as treated in Chap. 7:

$$
g_3(\tau_1,\tau_2,\tau_3) = \frac{1}{(2\pi)^3} \iiint G_3(\omega_1,\omega_2,\omega_3)e^{i(\omega_1\tau_1+\omega_2\tau_2+\omega_3\tau_3)}d\omega_1 d\omega_2 d\omega_3
$$

$$
\tag{10.2}
$$

$$
G_3(\omega_1,\omega_2,\omega_3) = \iiint g_3(\tau_1,\tau_2,\tau_3)e^{-i(\omega_1\tau_1+\omega_2\tau_2+\omega_3\tau_3)}d\tau_1 d\tau_2 d\tau_3
\tag{10.3}
$$

The impulse response function is assumed symmetric:

$$
g_3(\tau_1,\tau_2,\tau_3) = g_3(\tau_2,\tau_3,\tau_1) = g_3(\tau_3,\tau_1,\tau_2)
\tag{10.4}
$$

Thus the CTF is also symmetric because of the Fourier transform of Eq. (10.4):

$$
G_3(\omega_1,\omega_2,\omega_3) = G_3(\omega_2,\omega_3,\omega_1) = G_3(\omega_3,\omega_2,\omega_1)
\tag{10.5}
$$

and

$$
G_3(-\omega_1,\ -\omega_2,\ -\omega_3) = G_3^*(\omega_1,\omega_2,\omega_3)
\tag{10.6}
$$

CTF $G_3(\omega_1, \pm\omega_2, \pm\omega_3)$ expresses the normalized steady state response at frequencies $(\omega_1, \pm\omega_2, \pm\omega_3)$ due to excitation of three sinusoids of frequencies $\omega_1, \omega_2, \omega_3$.

10.2.2 Volterra cubic model in frequency domain

The response of the cubic-order nonlinear system to the excitations of single wave can be obtained using Eqs. (10.2) through (10.6). The same can be applied to the responses to bichromatic and trichromatic excitations. The results are usually expressed in terms of electronic engineering LTFs (RAOs), QTFs and CTFs.

We will present the results of theoretical experiments of the cubic system in the monochromatic, bichromatic, and trichromatic waves in the following sections.

10.2.2.1 Result of the experiment in monochromatic wave

Let us consider the experiment of a cubic system defined by Eqs. (10.1) to (10.6) excited by the monochromatic wave $x(t)$ below:

$$x(t) = A_1 \cos(\omega_1 t - \varepsilon_1) = \frac{A_1}{2}\left(e^{i(\omega_1 t - \varepsilon_1)} + e^{-i(\omega_1 t - \varepsilon_1)}\right) \tag{10.7}$$

Then the response $y(t)$ is obtained in the form:

$$y(t) = \mathrm{Re}\left\{
\begin{array}{l}
\left\{A_1 G_1(\omega_1) + \dfrac{3}{4}A_1^3 G_3(\omega_1,\omega_1,-\omega_1)\right\}e^{i(\omega_1 t - \varepsilon_1)} \\[2mm]
+\dfrac{1}{2}A_1^2[G_2(\omega_1,-\omega_1)] \\[2mm]
+\dfrac{1}{2}A_1^2\left[G_2(\omega_1,\omega_1)e^{i(2\omega_1 t - 2\varepsilon_1)}\right] \\[2mm]
+\dfrac{1}{4}A_1^3\left[G_3(\omega_1,\omega_1,\omega_1)e^{i(3\omega_1 t - 3\varepsilon_1)}\right]
\end{array}
\right\} \tag{10.8}$$

In the above experiment we find additional 3rd-order response amplitude at the excitation frequency: The 2nd-order response consists of the familiar mean and double frequency terms. The output frequency of 3rd-order interaction of monochromatic wave (or self-excitation) is found as the last term.

10.2.2.2 Result of the experiment in bichromatic wave

We consider 3rd-order excitation by bichromatic wave $x(t)$.

$$x(t) = \sum_{j=1}^{2} A_j \cos(\omega_j t - \varepsilon_j) \tag{10.9}$$

Then the 3rd-order response $y(t)$ is given in the form:

$$y(t) = \sum_{j=1}^{2} \mathrm{Re}\left[\left\{ A_j G_1(\omega_j) + \frac{3}{4} A_j^3 G_3(\omega_j, \omega_j, -\omega_j) \right\} e^{i(\omega_j t - \varepsilon_j)} \right]$$

$$+ \frac{3}{2} A_1^2 A_2 \mathrm{Re}\left[G_3(\omega_1, -\omega_1, \omega_2) e^{i(\omega_2 t - \varepsilon_2)} \right] + \frac{3}{2} A_2^2 A_1 \mathrm{Re}\left[G_3(\omega_2, -\omega_2, \omega_1) e^{i(\omega_1 t - \varepsilon_1)} \right]$$

$$+ \frac{1}{2} A_1^2 G_2(\omega_1, -\omega_1) + \frac{1}{2} A_2^2 G_2(\omega_2, -\omega_2) + \sum_{j=1}^{2} \frac{1}{2} A_j^2 \mathrm{Re}\left[G_2(\omega_j, \omega_j) e^{i(2\omega_j t - 2\varepsilon_j)} \right]$$

$$+ A_1 A_2 \mathrm{Re}\left[G_2(\omega_1, \omega_2) e^{i\{(\omega_1 + \omega_2)t - (\varepsilon_1 + \varepsilon_2)\}} \right] + A_1 A_2 \mathrm{Re}\left[G_2(\omega_2, -\omega_1) e^{i\{(\omega_2 - \omega_1)t - (\varepsilon_2 - \varepsilon_1)\}} \right]$$

$$+ \sum_{j=1}^{2} \frac{1}{4} A_j^3 \mathrm{Re}\left[G_3(\omega_j, \omega_j, \omega_j) e^{i(3\omega_j t - 3\varepsilon_j)} \right]$$

$$+ \frac{3}{4} A_1^2 A_2 \mathrm{Re}\left[G_3(\omega_1, \omega_1, \omega_2) e^{i\{(2\omega_1 + \omega_2)t - (2\varepsilon_1 + \varepsilon_2)\}} \right]$$

$$+ \frac{3}{4} A_1^2 A_2 \mathrm{Re}\left[G_3(\omega_1, \omega_1, -\omega_2) e^{i\{(2\omega_1 - \omega_2) - (2\varepsilon_1 - \varepsilon_2)\}} \right]$$

$$+ \frac{3}{4} A_1 A_2^2 \mathrm{Re}\left[G_3(\omega_2, \omega_2, \omega_1) e^{i\{(2\omega_2 + \omega_1)t - (2\varepsilon_2 + \varepsilon_1)\}} \right]$$

$$+ \frac{3}{4} A_1 A_2^2 \mathrm{Re}\left[G_3(\omega_2, \omega_2, -\omega_1) e^{i\{(2\omega_2 - \omega_1)t - (2\varepsilon_2 - \varepsilon_1)\}} \right] \tag{10.10}$$

In the above experiment we find additional 3rd-order response amplitudes at each linear excitation frequency. The familiar quadratic terms are found again as in the above section. And the 3rd-order self-interaction of monochromatic wave and 3rd-order interactions due to bichromatic wave excitation are found.

10.2.2.3 Result of the experiment in trichromatic wave

Similarly given the trichromatic wave excitation:

$$x(t) = \sum_{j=1}^{3} A_j \cos(\omega_j t - \varepsilon_j) \qquad (10.11)$$

then we obtain the response in the form:

$$y(t) = \sum_{j=1}^{3} \mathrm{Re}\left[\left\{ A_j G_1(\omega_j) + \frac{3}{4} A_j^3 G_3(\omega_j, \omega_j, -\omega_j) \right\} e^{i(\omega_j t - \varepsilon_j)} \right]$$

$$+ \frac{3}{2} \sum_{k=1}^{3} \sum_{j \neq k}^{3} A_j^2 A_k \, \mathrm{Re}\left[G_3(\omega_j, -\omega_j, \omega_k) e^{i(\omega_k t - \varepsilon_k)} \right] + \sum_{j=1}^{3} \frac{1}{2} A_j^2 G_2(\omega_j, -\omega_j)$$

$$+ \sum_{j=1}^{3} \frac{1}{2} A_j^2 \, \mathrm{Re}\left[G_2(\omega_j, \omega_j) e^{i(2\omega_j t - 2\varepsilon_j)} \right]$$

$$+ \sum_{j=1}^{2} \sum_{k=j+1}^{3} A_j A_k \, \mathrm{Re}\left[G_2(\omega_k, \omega_j) e^{i\{(\omega_k + \omega_j)t - (\varepsilon_k + \varepsilon_j)\}} \right]$$

$$+ \sum_{j=1}^{2} \sum_{k=j+1}^{3} A_j A_k \, \mathrm{Re}\left[G_2(\omega_k, -\omega_j) e^{i\{(\omega_k - \omega_j)t - (\varepsilon_k - \varepsilon_j)\}} \right]$$

$$+ \sum_{j=1}^{3} \frac{1}{4} A_j^3 \, \mathrm{Re}\left[G_3(\omega_j, \omega_j, \omega_j) e^{i(3\omega_j t - 3\varepsilon_j)} \right]$$

$$+ \sum_{k=1}^{3} \sum_{j \neq k}^{3} \frac{3}{4} A_j^2 A_k \, \mathrm{Re}\left[G_3(\omega_j, \omega_j, \omega_k) e^{i\{(2\omega_j + \omega_k)t - (2\varepsilon_j + \varepsilon_k)\}} \right]$$

$$+ \sum_{k=1}^{3} \sum_{j \neq k}^{3} \frac{3}{4} A_j^2 A_k \, \mathrm{Re}\left[G_3(\omega_j, \omega_j, -\omega_k) e^{i\{(2\omega_j - \omega_k)t - (2\varepsilon_j - \varepsilon_k)\}} \right]$$

$$+ \frac{3}{2} A_1 A_2 A_3 \, \mathrm{Re}\left[G_3(\omega_1, \omega_2, \omega_3) e^{i\{(\omega_1 + \omega_2 + \omega_3)t - (\varepsilon_1 + \varepsilon_2 + \varepsilon_3)\}} \right]$$

$$+ \frac{3}{2} A_1 A_2 A_3 \, \mathrm{Re}\left[G_3(\omega_1, \omega_2 - \omega_3) e^{i\{(\omega_1 + \omega_2 - \omega_3) - (\varepsilon_1 + \varepsilon_2 - \varepsilon_3)\}} \right]$$

$$+ \frac{3}{2} A_1 A_2 A_3 \, \mathrm{Re}\left[G_3(\omega_1, -\omega_2, \omega_3) e^{i\{(\omega_1 - \omega_2 + \omega_3)t - (\varepsilon_1 - \varepsilon_2 + \varepsilon_3)\}} \right]$$

$$+ \frac{3}{2} A_1 A_2 A_3 \, \mathrm{Re}\left[G_3(-\omega_1, \omega_2, \omega_3) e^{i\{(\omega_3 + \omega_2 - \omega_1)t - (\varepsilon_3 + \varepsilon_2 - \varepsilon_1)\}} \right] \qquad (10.12)$$

In addition to the familiar linear and quadratic responses the output frequencies of 3rd-order interactions due to mono-, bi- and tri-chromatic wave excitations are found. These are shown in Table 10.1, where p denotes the number of frequency or wave of each excitation, while the 2nd-column presents the system characteristics.

10.3 Deterministic Identification of Frequency Response Functions

In Eqs. (10.8), (10.10), and (10.12), there are three frequency response functions of interest in the present problem, i.e.,

Linear	$\rightarrow$	$G_1(\omega_1)$
Quadratic	$\rightarrow$	$G_2(\omega_1, \omega_2)$
Cubic	$\rightarrow$	$G_3(\omega_1, \omega_2, \omega_3)$

The interpretation of $G_1(\omega_1)$ is identical to that of pure linear theory as given in Chap. 3. It is simply the normalized amplitude and phase of linear steady system response. The quadratic response $G_2(\omega_1, \omega_2)$, has been interpreted in Chap. 7, and what nominally remains is to interpret the cubic frequency response function. In words, $G_2(\omega_1, \pm\omega_2)$ expresses the normalized steady state responses at frequencies $(\omega_1 \pm \omega_2)$ to two sinusoids of frequencies of ω_1 and ω_2, due to interaction of two frequency components. The units of response $G_2(\omega_1, \omega_2)$ are (response unit)/(excitation unit)^2. Similarly, $G_3(\omega_1, \pm\omega_2, \pm\omega_3)$ expresses the normalized steady state response at frequencies $(\omega_1, \pm\omega_2, \pm\omega_3)$ to three sinusoids of frequencies ω_1, ω_2, ω_3 due to interaction of the three frequency components. The units of $G_3(\omega_1, \omega_2, \omega_3)$ are (response unit)/(excitation unit)^3.

Referring to Eq. (10.8), $y(t)$ consists of 1st-order and 2nd-order mean and double frequency and 3rd-order triple frequency term. This is the familiar expression in view of quadratic system analysis. However, there is an exception; that is, the 1st-order term is the sum of pure 1st-order as usual and 3rd-order terms. This complication arises from the cubic system. If it were quadratic, the response would be the sum of pure linear, mean and double frequency terms (Chap. 7).

Table 10.1. Output frequencies and order of interactions, due to 3rd-order interaction of monochromatic, bichromatic and trichromatic waves

$p = 1$ Frequency	Interaction	$p = 2$ Frequency	Interaction	$p = 3$ Frequency	Interaction
ω_1	Linear & Cubic	ω_1	Linear & Cubic	ω_1	Linear & Cubic
0	Quadratic	ω_2	"	ω_2	"
$2\omega_1$	"	0	Quadratic	ω_3	"
$3\omega_1$	Cubic	$2\omega_1$	"	0	Quadratic
		$2\omega_2$	"	$2\omega_1$	"
		$\omega_1 + \omega_2$	"	$2\omega_2$	"
		$\omega_1 - \omega_2$	"	$2\omega_3$	"
		$3\omega_1$	Cubic	$\omega_1 + \omega_2$	"
		$3\omega_2$	"	$\omega_1 + \omega_3$	"
		$2\omega_1 + \omega_2$	"	$\omega_2 + \omega_3$	"
		$2\omega_1 - \omega_2$	"	$\omega_2 - \omega_1$	"
		$2\omega_2 + \omega_1$	"	$\omega_3 - \omega_1$	"
		$2\omega_2 - \omega_1$	"	$\omega_3 - \omega_2$	"
				$3\omega_1$	Cubic
				$3\omega_2$	"
				$3\omega_3$	"
				$2\omega_1 + \omega_1$	"
				$2\omega_1 + \omega_1$	"
				$2\omega_1 + \omega_2$	"
				$2\omega_1 + \omega_1$	"
				$2\omega_3 + \omega_2$	"
				$2\omega_1 + \omega_3$	"
				$2\omega_2 + \omega_3$	"
				$2\omega_2 - \omega_1$	"
				$2\omega_3 - \omega_{-1}$	"
				$2\omega_1 - \omega_2$	"
				$2\omega_3 - \omega_2$	"
				$2\omega_1 - \omega_3$	"
				$2\omega_2 - \omega_3$	"
				$\omega_1 + \omega_2 + \omega_3$	"
				$\omega_1 + \omega_2 - \omega_3$	"
				$\omega_1 - \omega_2 + \omega_3$	"
				$\omega_3 + \omega_2 - \omega_1$	"

The cubic nonlinearity may produce response at the excitation frequency as well as at the 3rd harmonic of the excitation. While the values of the functions $G_2(\omega, \omega)$ and $G_2(\omega, -\omega)$ may in principle be inferred by a frequency analysis of a single experiment, more than one experiment is required to separate $G_1(\omega)$ and $G_3(\omega, \omega, -\omega)$. The complication, however, will not arise in theoretical analysis because $G_3(\omega, \omega, -\omega)$ is the component which varies as excitation amplitude cubed.

The quantity most reduced and extract in seakeeping experiments in waves corresponds to the ratio of amplitude of response at frequency ω to amplitude of incident regular wave of frequency ω. Referring to Eq. (10.8), we write the equivalent RAO at the excitation frequency in the form:

$$R(\omega) = \frac{\text{output amplitude}(\omega)}{\text{input amplitude}(\omega)} = G_1(\omega) + \frac{3}{4} A_1^3 G_3(\omega,\ \omega,\ -\ \omega) \quad (10.13)$$

The first term is the well-known RAO, but the 2nd-term appears to be third-order with respect to the linear wave amplitude. In the experiment we conduct Fourier transform of the response time series to obtain the responses as function of frequency. In the above case, however, the sum will be the sum of conventional definition of RAO and 3rd-order term with respect to the linear wave amplitude as shown.

When $G_3(\omega,\ \omega,\ -\omega)$ is zero, $R(\omega)$ is $|G_1(\omega)|$ and invariant with excitation amplitude as would be expected for a system of a quadratic system. When the cubic system is present the amplitude can vary with excitation amplitude in a variety of ways, since in general both $G_1(\omega)$ and $G_3(\omega,\ \omega,\ -\omega)$ are complex and this must be taken into account of prior to performing the absolute value.

When the cubic system is excited by bichromatic wave, A_1 and A_2, ω_1 and ω_2 the equivalent RAO becomes:

$$R(\omega_1) = \frac{\text{output amplitude}}{\text{input amplitude}}$$

$$= \text{Re}\left\{ \begin{bmatrix} G_1(\omega_1) + \dfrac{3}{4} A_1^3 G_3(\omega_1,\omega_1,-\omega_1) \\ + \dfrac{3}{2} A_2^2 A_1 G_3(\omega_2,\omega_2,-\omega_1) \end{bmatrix} e^{i(\omega_1 t - \varepsilon_1)} \right\} \quad (10.14)$$

The 1st-two terms are the same as in the corresponding term of the monochromatic wave experiment as given in Eq. (10.13). The 3rd represents a cubic-order interaction between the two frequency components.

The 3rd-order terms, in Eq. (10.10) involve a frequency of new form; that is:

$$\left(2\omega_i \pm \omega_j \right) \quad (10.15)$$

and the corresponding special values of quadratic response function:

$$\left(\omega_i, \omega_i, \pm\omega_j\right) \tag{10.16}$$

Equation (10.12) indicates the response of the system to a trichromatic excitation. In this case all the types of frequency components noted in the bichromatic excitation reappear. Essentially, the results of bichromatic excitation are repeated for all the possible combinations of two of three frequencies. The last four lines of the response shown in Eq. (10.12) involve four new frequencies:

$$\begin{aligned} \left(\omega_1 + \omega_2 + \omega_3\right) \\ \left(\omega_1 + \omega_2 - \omega_3\right) \\ \left(\omega_1 - \omega_2 + \omega_3\right) \\ \left(-\omega_1 + \omega_2 + \omega_3\right) \end{aligned} \tag{10.17}$$

10.4 Simulation of Nonlinear Response

A nonlinear response of a cubic system to the Gaussian sea was simulated, and the 3rd-order equation in frequency domain was developed. Thus the LFRF, QFRF and CFRF were fixed. These frequency response functions were transformed to impulse response functions LIRF, QFRF, CIRF, which were then convolved with a single wave, two waves, and with three waves, to simulate the response in time domain, according to Eq. (10.1). Extensive data were generated and discussions were given (Dalzell, 1982). A typical example is shown in Fig. 10.1. We have already observed 1st-order responses in Chap. 3 and 2nd-order responses in Chaps. 7-9. Most interesting behavior of 2nd-order response is that the 2nd-order response becomes large in the large linear wave groups. Here in Fig. 10.1, we see the time series of random wave $x(t)$, and 1st-, 2nd-, and 3rd-order responses, $y_1(t)$, $y_2(t)$, and $y_3(t)$. The last one $y(t)$ is the resultant response. The excitation $x(t)$ has evidently two large wave groups, which create familiar groups in the 1st-order and 2nd-oder response which includes the mean. In addition to these, especially, the 3rd-order response has the form of pure wave groups at the corresponding time instants which are nearly same as the peaks of zero mean 2nd-order. It is clear from the simulation that the 3rd-order interaction creates higher nonlinearity than the linear peak wave group.

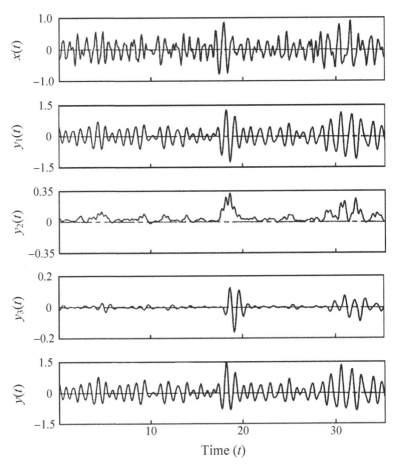

Fig. 10.1. Time series of third-order interaction of random wave (trial test by Dalzell, 1982).

10.5 3rd-Order Stokes Wave

10.5.1 Introduction

Stokes 3rd-order wave was solved by Pierson (1993) and Kim et al. (1993) applied the theory for the effect of extremely large wave on a vertical truncated cylinder. The results of the foregoing studies will be reviewed referring to the 3rd-order solutions.

For a single wave with a wavenumber k and amplitude A, the usual procedure is to use the artifice of steady motion. A current, C, is assumed to be present in the water such that the free surface is a function of x and the potential function of x and z alone. The well-known solution of the above to third-order (Lamb, 1932) is:

$$\varphi(z,x) = -AC_0 e^{kz} \sin kx + \frac{A^3 k^2 C_0}{8} e^{kz} \sin kx$$

$$\eta(x) = A\cos(kx) + \frac{1}{2}A^2 k \cos(2kx) + \frac{3}{8}A^3 k^2 \cos(3kx) \tag{10.18}$$

where

$$C_0 = \sqrt{\frac{g}{k}} \; ; \quad C = \sqrt{\frac{g}{k}(1+k^2 A^2)} \tag{10.19}$$

The phase velocity C_0 is the linear dispersion velocity. The phase velocity C depends on the wave amplitude and wavenumber which is called amplitude dispersion. The phase velocity C_0 in deep water is the same in the 1st- and 2nd-order waves and is independent of the amplitude. When the amplitude increases the wave becomes 3rd- or higher order which propagates faster than the 1st- or 2nd-order wave.

The 3rd-order Stokes irregular waves are represented (Pierson, 1993) by the sum of many 3rd-order nonlinear interaction terms. The theory is believed to be a fundamental breakthrough that overcomes the difficulties encountered by the linear theory (St. Denis and Pierson, 1953).

The 3rd-order theory has been developed starting with the Stokes 3rd-order wave equation for a single wavenumber (Lamb, 1932). Sums of such waves were assumed to be nonlinearly interacting with each other to the

3rd-order. A 3rd-order perturbation scheme was used to satisfy the free surface boundary conditions. The wavenumbers of the component waves are nonlinear with respect to the wave amplitude if the frequencies of the 1st-order components are constant. They decrease as the amplitudes of the 1st-ordere waves increase.

In order to employ the theory for representing nonlinear waves, in the model wave tank, the frequencies were held constant and used as the frequencies of the flap motion, with the associated necessary modification of the wavenumber as the amplitudes vary. Nonlinear periodic Stokes-like irregular waves were successfully generated in the model wave tank at the U.S. Naval Academy as shown in (Pierson, 1993). The theory is useful for generating numerical nonlinear interacting irregular waves in numerical wave tank.

Any number of Airy waves can be used as the 1st-order input to calculate 3rd-order interacting nonlinear waves. Three Airy waves can illustrate all of the possible 3rd-order interaction terms. It should be possible to generate a realistic practical representation for irregular waves with ten Airy waves.

At present, we use the formulas developed for four Airy waves input. With four Airy waves as input, the interaction produces a total of sixty-eight (68) sinusoids for the 3rd-order model with zero mean. The Airy wave amplitudes are determined using their frequencies and prescribed maximum slope for the sinusoidal waves.

The 1st-order solution for sums of 1st-order terms is well known. The potential function depends only on the amplitudes and frequencies of the 1st-order terms. A major difficulty with the linear representation is that the velocity field at the crests of the waveform is not correctly represented.

The 2nd-order solution for sums of 1st-order waves involves only the sums and differences of combinations of two frequencies (or wavenumbers). The wave elevation is modified by 2nd-order effects that produce higher crest height and shallow trough depths. The velocity potential has a small 2nd-order correction that depends on the differences of pairs of wavenumbers.

The Stokes 3rd-order irregular wave model used for this analysis (Kim et al., 1993; Pierson, 1993), assumes a 128 s long record that repeats exactly every 128 s. The waves are generated from one end ($x = 0$) and propagated along the wave tank without any reflection.

The properties of the kinematics of the irregular wave are independent of the location x at which the wave is measured and independent of the random phase angles of input Airy waves.

These formulas are much too complex to be given in the writing because of limitations on the length of this note. A single Stokes wave at 3rd-order is described in Lamb (1932) who used the artifice of steady motion. At 3rd-order, an important result is that the phase speed is a function of the wave steepness. The 1st-step in the derivation for sums of Stokes waves is to eliminate the artifice of steady motion and to show that the wavy free surface boundary condition can be satisfied at 3rd-order with a wave as a function of time, distance and depth. This makes it possible to consider what happens when the solutions for two, or more, different Stokes waves are added.

Stokes found the 3rd-order (and a few higher order terms) solution by replacing the exponential term in the potential function, for an Airy wave as a start, by its series expansion, substituting the equation for the free surface into it, collecting terms of the same order step by step and solving for them. The derivation follows the same technique with the necessary modifications such as using the sum of the free surfaces for appropriate orders.

As an example the perturbation expansion yields a term at 3rd-order for the Bernoulli equation that involves the product of three cosine terms of phase function θ, as in the sample solution below, which transforms to:

$$\cos(\theta_1)^2 \cos(\theta_2) = \frac{1}{2}\cos(\theta_2) + \frac{1}{4}\cos(2\theta_1 + \theta_2) + \frac{1}{4}\cos(2\theta_1 - \theta_2) \quad (10.20)$$

Similar terms occur in the kinematic boundary condition. The equations for the amplitudes of the 3rd-order terms can then be found.

The formulas for the 3rd-order nonlinear interacting irregular waves mentioned in the foregoing are described in detail in Pierson (1993).

10.5.2 Properties of 3rd-order solution referring to sample solutions

The properties of the 3rd-order oscillatory perturbation solution are illustrated by referring to the sample solutions given in Kim et al. (1993) and

Pierson (1993). These have similar sample solutions but different interests are shown.

10.5.2.1 Sample solution of 3rd-order interaction (Kim et al., 1993)

A numerical solution $\eta(x, t)$ of 3rd-order interaction of four Airy waves is listed below. The 1st-order steepness is calculated from the linear relationship for wavenumber given by $k_1 = \Omega_1^2 / g$ so that $A_1 k_1 = 0.07$. For example $\Omega_1 = 2\pi f_1$. The four frequencies were $f_1 = 107 / 128$, $f_2 = 113 / 128$, $f_3 = 137 / 128$, and $f_4 = 151 / 128$ in Hz.

The sample listing of 3rd-order wave interaction below describes a Fourier series because all of the frequencies are members of the orthogonal basis set for sinusoids that repeat exactly every 128 s.

The interactions of the four 1st-order terms with amplitudes of 2.5, 2.2, 1.5, and 1.2 cm produce all of the other terms in the sample listing. The phase functions are given by:

$$\theta_1 = \frac{2\pi}{L_1} x - \Omega_1 t$$

$$\theta_2 = \frac{2\pi}{L_2} x - \Omega_2 t$$

$$\theta_3 = \frac{2\pi}{L_3} x - \Omega_3 t \qquad (10.21)$$

$$\theta_4 = \frac{2\pi}{L_4} x - \Omega_4 t$$

where L_i denotes nonlinear wavelength and $K n_i = 2\pi / L_i$ presents nonlinear wavenumber which are given in Table 10.2 and below:

$$L_1 = 2.386, \quad L_2 = 2.139, \quad L_3 = 1.455, \quad L_4 = 1.198 \ (m) \qquad (10.22)$$

$$\eta(x,t) = 2.5\cos\theta_1 + 2.2\cos\theta_2 + 1.5\cos\theta_3 + 1.2\cos\theta_4$$
$$-1.835313\cos\theta_1 - 1.497668\cos\theta_2 - 0.7188991\cos\theta_3 - 0.4086303\cos\theta_4$$
$$+8.3503268E - 3\cos(\theta_2 - \theta_1) + 8.347571E - 3\cos(\theta_4 - \theta_3)$$
$$+2.277361E - 2\cos(\theta_3 - \theta_2) + 3.157251E - 2\cos(\theta_3 - \theta_1)$$
$$+3.0461998E - 2\cos(\theta_4 - \theta_2) + 3.917062E - 2\cos(\theta_4 - \theta_1)$$
$$+8.230275E - 2\cos(2\theta_1) + 0.1532032\cos(\theta_1 + \theta_2)$$
$$+7.108353E - 2\cos(2\theta_2) + 0.1303358\cos(\theta_1 + \theta_3)$$
$$+0.1197057\cos(\theta_2 + \theta_3) + 0.1181813\cos(\theta_1 + \theta_4)$$
$$+0.1080077\cos(\theta_2 + \theta_4) + 4.857249E - 2\cos(2\theta_3)$$
$$+8.606356E - 2\cos(\theta_3 + \theta_4) + 3.77644E - 2\cos(2\theta_4)$$
$$-6.063041E - 3\cos(2\theta_1 - \theta_4) - 2.416535E - 4\cos(\theta_1 + \theta_2 - \theta_4)$$
$$-4.391903E - 3\cos(2\theta_2 - \theta_4) - 1.414769E - 3\cos(2\theta_1 - \theta_3)$$
$$+1.319552E - 2\cos(\theta_1 + \theta_2 - \theta_3) + 9.797904E - 3\cos(2\theta_2 - \theta_3)$$
$$+1.684153E - 2\cos(\theta_1 + \theta_3 - \theta_4) + 2.481629E - 2\cos(\theta_2 + \theta_3 - \theta_4)$$
$$+0.116481\cos(2\theta_1 - \theta_2) + 0.1352193\cos(2\theta_2 - \theta_1)$$
$$+0.9084416\cos(\theta_1 - \theta_3 + \theta_4) + 4.242141E - 2\cos(2\theta_3 - \theta_4)$$
$$+0.3735828\cos(\theta_2 - \theta_3 + \theta_4) + 0.308197\cos(\theta_1 - \theta_2 + \theta_3)$$
$$+0.1870351\cos(\theta_3 + \theta_2 - \theta_1) + 0.3131976\cos(\theta_1 - \theta_2 + \theta_4)$$
$$+0.1622416\cos(\theta_4 + \theta_2 - \theta_1) + 0.0653955\cos(2\theta_3 - \theta_2)$$
$$+0.0644173E - 2\cos(2\theta_4 - \theta_2) + 5.963631E - 2\cos(2\theta_3 - \theta_1)$$
$$+8.586461E - 2\cos(\theta_4 + \theta_3 - \theta_2) + 7.422636E - 2\cos(\theta_4 + \theta_3 - \theta_1)$$
$$+4.245579E - 2\cos(2\theta_4 - \theta_2) + 4.046057E - 2\cos(2\theta_4 - \theta_1)$$
$$+4.064245E - 3\cos(3\theta_1) + 1.157016E - 2\cos(2\theta_1 + \theta_2)$$
$$+1.094932E - 2\cos(\theta_1 + 2\theta_2) + 3.445138E - 3\cos(3\theta_2)$$
$$+1.076612E - 2\cos(2\theta_1 + \theta_3) + 2.016795E - 2\cos(\theta_1 + \theta_2 + \theta_3)$$

...

$$(10.23)$$

...

$$+9.427246E - 3\cos(2\theta_2 + \theta_3) + 1.036044E - 2\cos(2\theta_1 + \theta_4)$$

$$+1.930298E - 2\cos(\theta_1 + \theta_2 + \theta_4) + 8.97688E - 3\cos(2\theta_2 + \theta_4)$$

$$+8.92869E - 3\cos(\theta_1 + 2\theta_3) + 8.286393E - 3\cos(\theta_2 + 2\theta_3)$$

$$+1.673436E - 2\cos(\theta_1 + \theta_3 + \theta_4) + 1.546864E - 2\cos(\theta_2 + \theta_3 + \theta_4)$$

$$+7.750552E - 3\cos(\theta_1 + 2\theta_4) + 2.359287E - 3\cos(3\theta_3)$$

$$+7.139751E - 3\cos(\theta_2 + 2\theta_4) + 6.502248E - 3\cos(2\theta_3 + \theta_4)$$

$$+5.920219E - 3\cos(\theta_3 + 2\theta_4) + 1.782692E - 3\cos(3\theta_4)$$

The 1st effect of these interactions is that the wavenumbers cannot be computed from the frequencies of the 1st-order terms. The wavenumbers must be decreased because a 2nd-order correction is needed at 3rd-order which is similar to the one obtained for a Stokes wave with a given amplitude and wavenumber. Table 10.2 shows the changes required at 3rd-order for the wavenumbers to be used in the phase functions. The nominal steepness 0.07 is reduced by 6% to the value 0.0656 because the waves are longer.

Table 10.2. The linear and 3rd-order values for wavenumbers, wavelengths, and phase speed. The nonlinear wavenumbers can be found dividing the linear wavenumbers by 1.0678.

Frequency	107/128	113/128	137/128	151/128
Linear wavenumber	2.812	3.136	4.610	5.600
Linear phase speed	1.868	1.7679	1.459	1.324
Linear wavelength	2.234	2.003	1.363	1.122
Nonlinear wavenumber	2.634	2.937	4.318	5.245
Nonlinear wavelength	2.386	2.139	1.455	1.198
Nonlinear phase speed	1.993	1.885	1.557	1.413

Another effect at 3rd-order is that each 1st-order term interacts with the other 1st-order terms to change the amplitude of each 1st-order term. The amplitudes of the 1st-order terms as a result of these interactions are approximately 0.655, 0.702, 0.781 and 0.791cm. This is shown by the next four terms in the listing. For example, the amplitude of the 1st-order term with the phase function θ_1 is reduced by 1.835313 cm from an original amplitude 2.5 cm.

The next group of 6 terms in the listing shows the amplitudes of the 2nd-order terms that result from the differences of two 1st-order phase functions. The frequency of the $\cos(\theta_2 - \theta_1)$ term, for example is $113/128 - 107/123 = 6/128$ Hz.

The ten terms in the next group are the 2nd-order terms of two phase functions. For example, the $\cos(2\theta_1)$ term has a frequency of $214/128$ Hz. Its amplitude is computed from $A_1^2(Kn_1)/2$. See Eq. (10.18). These bound (phase-locked) terms propagate with phase speeds like the 1st-order terms. They shape the nonlinear appearance of the Stokes waves.

The next twenty-four terms are the 3rd-order interaction terms that have frequencies comparable to the 1st-order frequencies. For example, the 1st-term in this group has a frequency given by $(2\theta_1 - \theta_4)=2(107/128) - (151/128)$ which equals $63/128$ Hz. These terms compensate for the reduction of the amplitudes of the 1st-order terms at 3rd-order.

The decrease in amplitudes of the 1st-order terms due to their interaction with each other is compensated for by the large number of 3rd-order terms that result. The frequency of $121/128$ is the 3rd-order interaction term for $(\theta_1-\theta_3 + \theta_4) = (107/128)-(137/128) + (151/128)$. This 3rd-order term, as shown in the sample listing has an amplitude of 0.90844, which is greater than the four original 1st-order terms after they have interacted nonlinearly with each other.

The last twenty terms have frequencies corresponding to three times those of the 1st-order terms. The term $\cos(3\theta_1)$ has a frequency of $321/128$ Hz and an amplitude of $3A_1^3(Kn_1)^2/8$. See Eq. (10.18). The phase relationships among the term in the solution are an important property and result in the Stokes-like waveforms that are obtained.

In the foregoing equation we see the 2nd-order terms consist of the sum, double frequency and difference frequency terms as the Volterra quadratic model. The frequency (or phase function) combinations of the 3rd-order terms in Eq. (10.23) appear identical to those of Volterra cubic-order model as shown in Eqs. (10.15), (10.16), and (10.17). This is similar to the relation between the 2nd-order wave theory and Volterra quadratic model.

10.5.2 2 Sample solution of 3rd-order interaction (Pierson, 1993)

We have similar numerical solution of 3rd-order interaction given the four Airy waves as shown in Eq. (178) on p.375 in Pierson (1993). The linear amplitudes are defined as 3.6, 3.2, 2.2 and 1.8 cm. And the nonlinear wavelengths are given below:

$$L_1 = 2.528, \quad L_2 = 2.267, \quad L_3 = 1.542, \quad L_4 = 1.269 \ m \qquad (10.24)$$

It is our interest to observe the behaviors of the 1st-, 2nd-, and 3rd-order response terms. These response terms are separately shown in Fig. 10.2.

Figure 10.2 shows responses of each order similar to Fig. 10.1. The difference is in the input of the random wave in Fig. 10.1, while input of four Airy waves to the 3rd-order interaction in Fig. 10.2. The behaviors of each response in the two figures are similar. We see it again that the 3rd-order wave groups are directly under the waves groups of linear and quadratic waves. The 3rd-order wave group is slightly larger than the 2nd-order wave group. The higher-order waves under the 1st-order wave groups are in the form of groups and the largest resultant waves appear under the largest linear wave groups. We recall similar study of the 2nd-order wave group in Sec. 9.2.3.3, that is, if we find large wave groups in the linear waves, there are 2nd-order nonlinear resultant wave groups directly under the linear wave groups.

Figure 10.3 represents the solutions evaluated at a distance of 15 m from the wavemaker where $x = 0$ (and t is an integer multiple of 128).

Figure 10.3 (top) is from 0 to 128 s, Fig. 10.3 (middle) and (bottom) are for an expanded time scale for 0 to 64 and 64 to 128 s. The solution is periodic with a period of 128 s. The time series in Fig. 10.3 may be discrete Fourier transformed and can be brought back to the original series exactly.

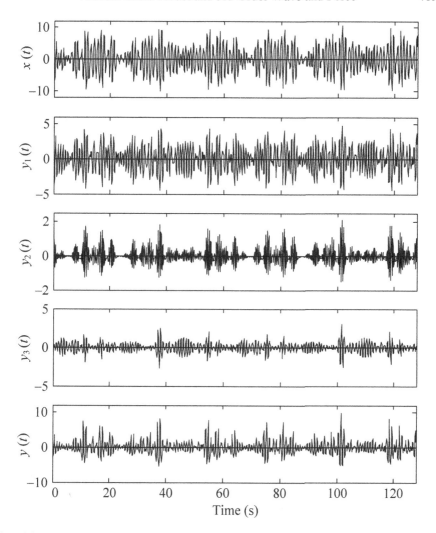

Fig. 10.2. Result of 3rd-order interaction. The top indicates the linear input and the rest are 1st-, 2nd-, 3rd-order and resultant response.

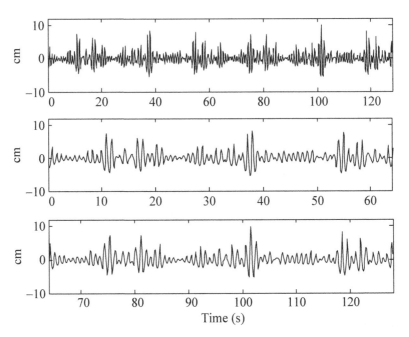

Fig. 10.3. Top fig. presents time series of 3rd-order interaction for 128 s repeating period. The 2nd presents the same as the top 0 to 64 s in an expanded scale. The 3rd presents time series from 64 to 128 s in the expanded scale.

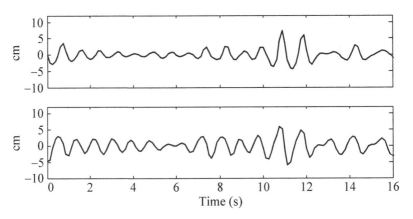

Fig. 10.4. Expanded graph of first 16 s from Fig. 10.3 of a similar calculation for 1st-order waves increased in amplitude by 10%. The higher waves arrive earlier at 15 m location as predicted.

Referring to Fig. 10.4, we consider the amplitude dispersion. We have increased 1st-order amplitudes from the previous linear amplitudes 3.6, 3.2, 2.2 and 1.8 cm by 10% so that they are evaluated 3.96, 3.52, 2.42, and 1.98 cm. The frequencies could not be changed, but the nonlinear wavelengths are changed as shown below:

$$L_1 = 2.578, \quad L_2 = 2.311, \quad L_3 = 1.572, \quad L_4 = 1.294 \ m \qquad (10.25)$$

The top graph in Fig. 10.4 is the solution for first 16 s of the theoretical time history from Eq. (178) of the paper by Pierson (1993) using Eq. (10.24) evaluated at 15 m from the wavemaker. The bottom graph is the one resulting from increasing the linear input wave amplitudes by 10%. The waves in the lower graph are higher. They have arrived at the point 15 m from the source earlier than the lower waves (in top) because of increased phase speed. Since the 3rd-order solution is only an approximation, actual measurements will differ in yet to be determined ways from the graphs.

10.5.3 3rd-order wave group

Pierson (1993) developed an analytical method in order to demonstrate the 3rd-order low frequency wave group, which is in the 3rd-order nonlinear wave group. Periodic groups of waves separated by $L = 2048$ m can be obtained by using wavenumbers $k_1 = 2\pi \times 15/2048$, $k_2 = 2\pi \times 16/2048$, and $k_3 = 2\pi \times 17/2048$ along with appropriate 1st-order amplitudes chosen so that $A_1 k_1 = A_2 k_2 = A_3 k_3 = 0.1$. The 3rd-order wave created by the three component waves is expressed in Eq. (10.26) and Figs. (10.5)(a)-(f).

Figure 10.5(a) is a complete 3rd-order solution representing a large nonlinear wave group Eq. (10.26). Decomposing the foregoing complete 3rd-order wave group we present 1st-order only in Fig. 10.5(b); the sum of 1st- plus 3rd-order amplitude change in the θ_1, θ_2, θ_3 terms in Fig. 10.5(c), the 2nd-order only in Fig. 10.5(d), low wavenumber 3rd-order terms in Fig. 10.5(e) and high wavenumber 3rd-order terms in Fig. 10.5(f). The low wavenumber 3rd-order terms are slightly higher than the 2nd-order terms. The same behaviors are found in the Volterra 3rd-order model in Fig. 10.1 and Stokes 3rd-order wave in Fig. 10.2.

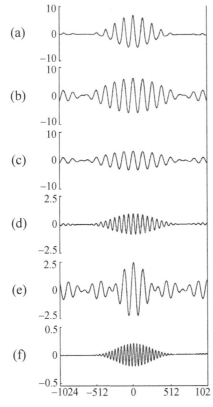

Fig. 10.5(a) Complete 3rd-order solution given in Eq. (10.26). (b) First-order only. (c) First-plus third-order term in θ_1, θ_2, θ_3. (d) 2nd-order only. (e) Low wavenumber 3rd-order terms. (f) High wave number 3rd-order term. Note the vertical scales are different (Pierson, 1993).

$$\eta(x,t) = 2.173\cos\theta_1 + 2.0371\cos\theta_2 + 1.9173\cos\theta_3$$
$$-1.080\cos\theta_1 - 0.9175\cos\theta_2 - 0.7823\cos\theta_3$$
$$+0.0067\cos(\theta_2 - \theta_1) + 0.0060\cos(\theta_3 - \theta_2)$$
$$+0.0128\cos(\theta_3 - \theta_1) + 0.1089\cos(2\theta_1) + 0.2105\cos(\theta_1 + \theta_2)$$
$$+0.1018\cos(2\theta_2) + 0.2045\cos(\theta_1 + \theta_3) + 0.1977\cos(\theta_2 + \theta_3)$$
$$+0.0957\cos(2\theta_3) + 0.1699\cos(2\theta_1 - \theta_3)$$
$$+0.2097\cos(2\theta_1 - \theta_2) + 0.3563\cos(\theta_1 + \theta_2 - \theta_3)$$
$$+0.1987\cos(2\theta_2 - \theta_3) + 0.4385\cos(\theta_3 - \theta_2 + \theta_1)$$
$$+0.2269\cos(2\theta_2 - \theta_1) + 0.2138\cos(2\theta_3 - \theta_2)$$
$$+0.4249\cos(\theta_3 + \theta_2 - \theta_1) + 0.2072\cos(2\theta_3 - \theta_1)$$

$$+ 0.0081\cos(3\theta_1) + 0.0239\cos(2\theta_1 + \theta_2)$$
$$+ 0.0234\cos(2\theta_2 + \theta_1) + 0.0235\cos(2\theta_1 + \theta_3)$$
$$+ 0.0076\cos(3\theta_2) + 0.0460\cos(\theta_1 + \theta_2 + \theta_3) \quad\quad (10.26)$$
$$+ 0.0225\cos(2\theta_2 + \theta_3) + 0.0226\cos(2\theta_3 + \theta_1)$$
$$+ 0.0220\cos(2\theta_3 + \theta_2) + 0.0072\cos(3\theta_3)$$

10.6 3rd-Order Wave Force

The development of the 3rd-order force theory was intended to simulate the ringing wave that was in a pressing need. Malenica and Molin (1995) and Faltinsen et al. (1995) formulated the wave force on the bottommounted vertical column in the water of finite depth due to a monochromatic wave. Malenica and Molin (1995):

$$F = \mathrm{Re}[(F^{(1)} + \hat{F}^{(3)})e^{-i\omega t}] + \overline{F}^{(2)}$$
$$+ \mathrm{Re}[\tilde{F}^{(2)}e^{-i2\omega t}] + \mathrm{Re}[\tilde{F}^{(3)}e^{-i3\omega t}] \quad\quad (10.27)$$

The response of the cubic-order system to a monochromatic excitation has the same form as the Volterra cubic model excited by monochromatic wave. Refer to the basic Volterra cubic model due to monochromatic excitation given in Eq. (10.8) as well as the case $p = 1$ in Table 10.1. The monochromatic wave creates additional 3rd-order response amplitude at the excitation frequency, i.e., linear term and 3rd-order term at first harmonics; mean term, double frequency term; and 3rd-order term with third 3rd-harmonics.

$F^{(1)} + \hat{F}^{(3)}e^{-i\omega t}$ represents the sum of 1st-order and 3rd-order term at the excitation frequency; $\overline{F}^{(2)} + \tilde{F}^{(2)}e^{-i2\omega t}$ represents the sum of the mean and double frequency terms; $\tilde{F}^{(3)}e^{-i3\omega t}$ expresses the 3rd-order term at the triple excitation frequency.

References

Journal references:

Pierson, W. J. (1993). Oscillatory 3rd-Order Perturbation Solutions for Sums of Interacting Long-Crested Stokes Waves on Deep Water, J. Ship Res., Vol. 37, No. 4, pp. 354–383.

Malenica, S. and Molin, B. (1995). Third-hamonic wave diffraction by a vertical cylinder, J. Fluid Mech., Vol. 302, pp. 203–229.

Faltinsen, O.M., Newman, J.N. and Vinje, T. (1995). Nonlinear wave loads on a slender vertical cylinder, J. Fluid Mech., 289, pp. 179–198.

Book references:

Lamb, Sir H. (1932) Hydrodynamics, Dover Publications, New York.

Proceedings references:

Dalzell, J.F. (1982). An Investigation of the Applicability of the Third Degree Functional Polynomial Model to Non-Linear Ship Motions Problems, Report SIT-DL-82-9-2275.

Kim, C.H., Pierson, W.J. and Tick, L.J. (1993). Extreme Nonlinear Wave Loads on a Vertical Truncated Cylinder in Nonlinear Irregular Stokes-Like Waves. Proc. 3rd. Int. Offshore Polar Eng. Conf., Vol. 3, pp. 158–166.

Chapter 11

Highly Nonlinear Waves and UNIOM

11.1 Introduction

The ultimate goal of seakeeping research would be to create an input-output model, which can simulate the extreme responses of a ship in the very high or phenomenal sea (over $H_s = 14$ m).

In Chap. 9, we reviewed springing and ringing of ISSC TLP occurring simultaneously in the very high seas. The springing can be simulated theoretically using the Volterra quadratic model in the weakly nonlinear waves, while the ringing of ISSC TLP had to be analyzed relying on the experimental data. The 3rd-order force model was proposed for TLP's ringing load by Faltinsen et al. (1995) and Marion and Molin (1995) but these were still at the infant stage.

The extremely nonlinear seas may contain strongly nonlinear asymmetries as shown in Fig. 11.1 (Longuet-Higgins and Cokelet, 1976, Funke and Mansard, 1982, Myrhaug and Kjeldsen, 1984). Because no theories are available for such strongly asymmetric waves and loads, we had to use scaled model test data for the analysis of ringing of ISSC TLP (Zou et al., 1998). We tried to simulate ringing of ISSC TLP in the replica of Heidrun ringing wave, but the ringing in Heidrun ringing wave was not found. This indicates that Heidrun ringing wave does not necessarily produce ringing in the other TLP such as ISSC TLP. Thus we have made a decision to generate a storm sea that could cause ringing for ISSC TLP.

The Heidrun ringing wave was a replica (Stansberg et al., 1995; Statoil-MARINTEK, 1996), which has 3 hrs long record of JONSWAP spectrum:

H_s = 15.4 m, T_m = 17.8 s, and γ = 1.7 and vertical asymmetry λ = 1.50. The storm sea generated for ISSC TLP at TAMU wave tank was the wave of JONSWAP spectrum: H_s = 9.88 m, T_m = 11.02, γ = 6.5 and the vertical asymmetry λ = 2.15. It was measured 20 s in each run, a total of 8 runs due to the limited length of the wave tank, resulting in 27 min ensemble (length scale is 1/100 , thus it is equivalent to 4.5 hrs in real scale). At each run the random seed number was changed. The foregoing TAMU storm wave showed significantly stronger nonlinearity than the Heidrun ringing wave.

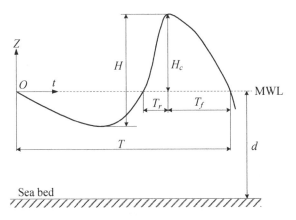

Fig. 11.1. Description of strongly asymmetric transient wave.

In the research of springing and ringing we have developed two basic input-output models; ULSM-kinematics and ULSM-diffraction. The former model is for determining wave kinematics given the wave elevation, with which one can determine the wave load by Morison's equation. The latter model can determine the diffraction load, given the wave elevation at the center of the structure. ULSM has later been renamed as UNIOM (universal nonlinear input-output model) by Kumar and Kim (2002A; 2002B).

UNIOM is a quasi-Volterra higher-order model that can accept highly nonlinear (non-Gaussian) input and it is allowed to assume the system as a linear, quadratic, or linear plus quadratic. Thus, the simulation by UNIOM has to be verified by comparing with the measured response force.

In view of the above remarks, the research needs to generate both weakly and highly nonlinear waves in the wave tank. In particular, a strongly nonlinear wave is needed both as an isolated transient and transient in the random seaway. These experiments include the measurement of wave elevations, weakly and highly nonlinear horizontal particles kinematics and non-impact and impact loads on the structure.

11.2 Background of UNIOM

As mentioned in Sec. 11.1, we made a decision to use UNIOM instead ULSM. Thus hereafter we use UNIOM and ULSM. The model is to predict the wave kinematics and force on a structure given the wave elevation time series at the center of the structure.

Kim et al. (1992) generated strongly asymmetric wave or extreme 2-D transient wave in the TAMU wave tank and measured the kinematics using LDA (Laser Doppler Anemometer). It was necessary to simulate the wave kinematics by Wheeler's stretching method (Wheeler, 1970) in order to compare the theory and experiment. The simulation was much weaker than the measured transient wave kinematics as shown in Fig. 11.8. It was our goal to develop a model to simulate the foregoing transient wave kinematics as well as nonlinear random waves so that eventually we could predict the wave load on a fixed vertical column.

Kim and Zou (1995) proposed ULSM-kinematics with a different stretch form from Wheeler's and delta stretching by Rodenbusch and Forristall (1986), respectively to simulate nonlinear wave kinematics. The same ULSM-kinematics was applied in Kim et al. (1996). However, the stretching formula was not adequate to match the laboratory generated asymmetric irregular waves with the highly nonlinear wave kinematics.

In the simulation of ISSC TLP response, Zou et al. (1997) proposed a 2nd-version of stretching formula to match with strongly asymmetric wave kinematics. Further the 2nd-version of the ULSM-kinematics was employed in the simulation of the strongly asymmetric transient wave kinematics in the work Kim et al. (1997) to investigate the impact loads on a series of columns varying radius and transient waves. It was found that weak and strong asymmetric waves produced non-impact and impact and springing and ringing, respectively.

Another approach was developed by Kim and Zou (1997). It is ULSM-diffraction for diffraction force on a large vertical truncated column in the replica of Heidrun ringing wave (Stansberg et al., 1995; MARINTEK-Statoil, 1996). The model was validated by comparing the experiment and simulation.

11.2.1 Review of springing and ringing

A review of the progress and development of springing and ringing (Kim, 1998) was on the observations of Heidrun ringing waves:

Natvig (1994) observed that ringing occurs in all extreme seas with modal periods, typically from 10 s and above. The ringing wave is single height, but not breaking, with a steep front or steep back or both.

Davies et al. (1994) observed that the ringing wave events usually include largest crest in the sea state. Such crests propagate through the structures with much less change in the form than would be expected from linear dispersion. The ringing waves exhibit strong asymmetries identified by Kjeldsen and Vinje (1980) and associated with breaking waves. Such waves bear little resemblance to Stokes wave forms. In fact these events much more closely resemble the transient wave conditions described by Kim et al. (1990), who compared 3rd-order Stokes and 2-D transient wave profiles. These authors also observed that such events exhibit very different, and much larger, horizontal velocity fields than the perturbation based model and that such approaches are not suitable for describing these wave conditions. Ringing events tend to be initiated by waves on the forward portion of the wave group, when such a crest encounters the leading column of the structure. A spilling breaker that maintains form as it passes through the structure is probably the worst case.

Jefferys and Rainey (1994) observed that to a fair approximation that large waves travel without distortion over distances typical of TLP structural dimensions. Also the capacity to create ringing changes very little over distances typical of TLP dimensions, although ringing is sensitive to kinematics.

11.3 UNIOM-Kinematics

11.3.1 Wheeler's method

Kim et al. (1992) simulated the measured transient wave and peak particles velocities employing Wheeler's method.

The Fourier transform of the measured transient wave elevation η (x, t) (Fig. 2, Kim et al., 1992) is given by:

$$\eta(x,t)= \text{Re}\left[\sum_{j=1}^{n} A_j\, e^{i(k_j x - \omega_j t + \phi_j)} \right] \quad (11.1)$$

where A_j and ϕ_j represent the amplitudes and phases, respectively. And Eq. (11.1) is the precise reproduction (simulation) of the measured wave in the same time-interval. The horizontal velocity at $(0, z, t)$ is:

$$u(0,z,t)= \text{Re}\left[\sum_{j=1}^{n} A_j \cdot G_u(z,\omega_j) e^{i(k_j x - \omega_j t + \phi_j)} \right], \quad z \leq 0 \quad (11.2)$$

with

$$G_u(z,\omega_j) = \frac{gk_j}{\omega_j} \frac{\cosh k_j (z + h)}{\cosh k_j h} \quad (11.3)$$

where G_u is a LTF representing the horizontal velocity per unit wave amplitude at a point z. In order to have kinematics in the crest zone, one has to stretch the velocity profile (Wheeler, 1970) by replacing z by z_e.

$$z_e = (z_a + h)\frac{h}{h + \eta} - h \quad (11.4)$$

where z_e is the effective coordinate in the range $-h \leq z_e \leq 0$; z_a is the actual vertical coordinate in the range of $-h \leq z_a \leq \eta$; η is the instantaneous free surface, positive upward and zero at MWL. Equation (11.4) represents mapping between the effective and actual coordinate.

Wheeler-stretch simulation for the peak velocities shows a large difference from the experiment as shown in Fig. 11.8 in Sec. 11.6. The Wheeler stretch gives larger value than that of the transient wave in the region below MWL, whereas, it is much less than the measured transient wave in the crest region. The maximum of the simulated velocities is 61.4% of the maximum measured value. This represents a significant difference in the kinematics of the simulated and the measured transient wave.

11.3.2 1st-version of UNIOM-kinematics

A new stretching model was proposed by Kim and Zou (1995) and Kim et al. (1996), which may be called 1st-version of UNIOM-kinematics. The measured wave elevation is expressed in terms of Fourier complex amplitude spectrum:

$$\eta(x,t) = \mathrm{Re}\left[\sum_{j=1}^{n} A_j\, e^{i(k_j x - \omega_j t + \phi_j)}\right] \tag{11.5}$$

The linear wave theory gives the horizontal velocity below the measured wave in the form:

$$u(0,z,t) = \mathrm{Re}\left[\sum_{j=1}^{n} A_j \cdot G_u(z,\omega_j)\, e^{i(k_j x - \omega_j t + \phi_j)}\right], \quad z \le 0 \tag{11.6}$$

$$G_u(z,\omega_j) = \frac{gk_j}{\omega_j} \frac{\cosh k_j(z+h)}{\cosh k_j h} \tag{11.7}$$

The stretching is formulated as:

$$z_e = \left(h - \frac{H_t H_c}{H_t + H_c}\right)\frac{z_a}{h} - \frac{H_t H_c}{H_t + H_c} \quad \text{for } -h \le z_a \le 0$$

$$z_e = \left(\frac{H_t}{H_t + H_c}\right)z_a - \frac{H_t H_c}{H_t + H_c} \quad \text{for } 0 \le z_a \le H_c \tag{11.8}$$

where H_c and H_t are the crest height and trough depth whose crest indicates the time instant at which the velocity u is measured. z_e is the effective vertical coordinate ($-h \le z_a \le 0$) whereas z_a the actual coordinate ($-h \le z_a \le H_c$). Equation (11.8) represents mapping between z_e and z_a in the region below and above the MWL, respectively.

The 1st-version was applied to simulate particles velocities in the three typical nonlinear irregular waves; transient wave, Stokes higher-order-like irregular wave and nonlinear irregular wave. Both measured and predicted particles velocities were in excellent agreement, except a slight over prediction of the particle velocity on the MWL.

11.3.3 2nd-version of UNIOM-kinematics

Here we consider the 2nd-version of UNIOM-kinematics (Zou et al., 1997) for the simulation of wave kinematics of strongly asymmetric waves as shown in Figs. 11.1, 11.5, and 11.9. The 1st-version was less adequate in simulating a strongly asymmetric wave.

The measured wave elevation is expressed in terms of Fourier complex amplitude spectrum:

$$\eta(x,t) = \mathrm{Re}\left[\sum_{j=1}^{n} A_j\, e^{i(k_j x - \omega_j t + \phi_j)} \right] \tag{11.9}$$

The linear wave theory gives the horizontal velocity below the measured wave in the form:

$$u(0,z,t) = \mathrm{Re}\left[\sum_{j=1}^{n} A_j \cdot G_u(z,\omega_j)\, e^{i(k_j x - \omega_j t + \phi_j)} \right], \quad z \le 0 \tag{11.10}$$

$$G_u(z,\omega_j) = \frac{gk_j}{\omega_j} \frac{\cosh k_j(z+h)}{\cosh k_j h} \tag{11.11}$$

$$z_e = az_a^3 + bz_a^2 + cz_a + d \quad \text{for } -h \le z_a \le H_c \tag{11.12}$$

with a new stretch formula:

$$a = \frac{(-h + H_c) + k(h + H_c)}{(h + H_c)^3}$$

$$b = \frac{-2(h^2 - hH_c + H_c^2) - k(h + H_c)(H_c - 2h)}{(h + H_c)^3}$$

$$c = \frac{H_c(H_c^2 - H_c h + 4h^2) + kh(h + H_c)(h - 2H_c)}{(h + H_c)^3}$$

$$d = -\frac{h^2 H_c [k(h + H_c) + 2H_c]}{(h + H_c)^3}$$

$$k = (2.00 - \lambda)\frac{H_t}{H}$$

$$\lambda = \frac{T_f}{T_r}$$

(11.13)

If $H_c/H_t \le 1.0$, then $\lambda = 1.0$, and if $\lambda > 1.95$, then $\lambda = 1.95$. z_e is the effective vertical coordinate ($-h \le z_e \le 0$), whereas z_a is the actual vertical coordinate $-h \le z_a \le H_c$. Equations (11.12) and (11.13) are the formulas mapping z_e to z_a. Substituting the effective coordinate into z in Eq. (11.11), one can obtain the LTF G_u as a function of the actual coordinate. The definitions of the symbols used in Eqs. (11.12) and (11.13) are given in Fig. 11.9.

11.4 UNIOM-Diffraction

11.4.1 UNIOM-linear diffraction

Assume we have the wave elevation time series at the center of the structure which is expressed in terms of complex amplitude spectrum. Then we assume the response diffraction force is given by:

$$y(t) = \text{Re}\left[\sum_{j=1}^{n} A_j (LTF)_j e^{-i\omega_j t} \right]$$

(11.14)

where A_j denotes the complex amplitude of the measured wave, indicating both amplitude and phase angle. The above relation is named UNIOM-linear diffraction.

11.4.2 UNIOM-linear and quadratic diffraction

Consider we have the wave elevation time series at the center of the structure which is expressed in terms of complex amplitude spectrum. Then we assume the response diffraction force is given by:

$$
\begin{aligned}
y(t) = \text{Re}\left[\sum_{j=1}^{n} A_j (LTF)_j e^{-i\omega_j t}\right] \\
+ \text{Re}\frac{1}{2}\left[\sum_{j=1}^{n}\sum_{k=1}^{n} A_j A_k (QTF)_{jk}^{\pm} e^{-i(\omega_j \pm \omega_k)t}\right]
\end{aligned}
\tag{11.15}
$$

where A_j denotes the complex amplitude of the measured wave, indicating amplitude and phase angle. This is not the familiar random phase angle as used in the Volterra quadratic model. Equations (11.14) and (11.15) transmit precisely the effect of the nonlinearity of the wave onto the response at each time step. We see that Eq. (11.15) has the same form of Volterra quadratic model as described in Chaps. 7, 8, and 9. The difference of the UNIOM-diffraction from Volterra quadratic model is that the UNIOM-diffraction accepts non-Gaussian measured input and assumes arbitrary system characteristics. Equation (11.15) is called UNIOM-linear and quadratic diffraction.

11.4.3 UNIOM-linear and quadratic system

We may extend the foregoing models for response forces to any kind of response through the linear and quadratic system. For instance we want to have vertical response motion of structure in the nonlinear seas. If the system is linear we may need LTFs. If the system is quadratic, then we may need QTFs of the system. The formulas for the foregoing responses have the same form as Eqs. (11.14) and (11.15). The QTFs for the vertical motion of structure had not been developed yet. Under the circumstances one may determine the QTFs by conducting the cross-bi-spectral analysis of the measured input and output data in the random seas.

11.5 Generation of Nonlinear Waves in Wave Tank

11.5.1 Introduction

We can obtain a variety of nonlinear waves in wave tank. Such laboratory waves may vary from weak to highly nonlinear. Highly nonlinear single wave was numerically generated by Longuet-Higgins and Cokelet (1976) which has the shape as shown in Fig. 11.1. The typical laboratory-generated highly nonlinear transient wave is identical to the highly nonlinear wave as shown in Fig. 11.1. The wave is also frequently called strongly asymmetric wave.

The weakly nonlinear Stokes 5th-order wave (large regular wave) and the highly nonlinear transient wave are usually compared to demonstrate the significant difference of the weakly and strongly nonlinear wave.

The generation of the isolated strongly asymmetric wave and the same wave in the random seas will be discussed later.

The degree of nonlinearity of the phenomenal random sea such as the ringing wave of Heidrun TLP was statistically investigated by Kumar et al. (2002C), by comparing the crest heights of the experimental wave with the Rayleigh distribution as illustrated in Fig. 9.8(c). When the sea severity is lower than $H_s = 9$ m, we could assume the wave is weakly nonlinear. If the sea severity is phenomenal $H_s = 15.4$ m as Heidrun ringing wave it may be called highly nonlinear.

11.5.1.1 Generation of Stokes 5th-order wave

The Stokes 5th-order wave is weakly nonlinear and a large regular wave that has commonly been used as a design wave of the highest load on offshore structures (API, 1987).

Given the water depth and period, one determines the fundamental frequency and five amplitudes of higher harmonics using tables (Skjelbreia and Hendrickson, 1960). The Stokes 5th-order wave profile y is given in the form:

$$\beta y(t) = \lambda \cos \omega t + (\lambda^2 B_{22} + \lambda^4 B_{24}) \cos 2\omega t$$
$$+ (\lambda^3 B_{33} + \lambda^5 B_{35}) \cos 3\omega t + (\lambda^4 B_{44}) \cos 4\omega t \qquad (11.16)$$
$$+ (\lambda^5 B_{55}) \cos 5\omega t$$

Given the wave height H and wave period T or wavelength L, and water depth d, then β and λ and B_{ij} are determined as function of d/L from the tables of Stokes 5th-order theory, where ω is the fundamental frequency derived from the period T. Equation (11.16) may be written in the form similar to Fourier series:

$$y(t) = A_1 \cos \omega t + A_2 \cos 2\omega t + A_3 \cos 3\omega t$$
$$+ A_4 \cos 4\omega t + A_5 \cos 5\omega t \qquad (11.17)$$

The Stokes 5th-order wave was approximately reproduced in wave flume, and it was confirmed that the theory and measurements were almost identical (Alex and Kim, 2000). The horizontal particle velocity field along the vertical axis passing the crest point was measured by LDA, and compared with the theory as shown in Fig. 11.2. It is interesting to see that the experimental horizontal maximum velocity behaves similar to the transient wave in Fig. 11.8. The theoretical value below MWL is larger than the measurement. But in the region above MWL it is in the trend to be less than the measurement.

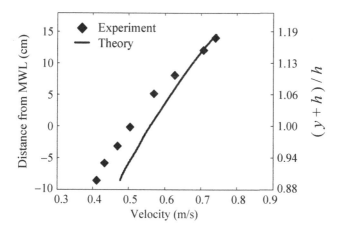

Fig. 11.2. Horizontal velocity of wave # 09.

11.5.1.2 Generation of transient wave

We can vary the laboratory waves from being weakly to strongly nonlinear. However, all the techniques to generate such waves depend basically on the use of linear input-output model. The LTFs of wave-flap is used in linking a design wave amplitude spectrum and the voltage amplitude spectrum. Given the LTFs of wave-flap and design wave amplitude spectrum, the voltage amplitude spectrum is determined, which will provide the voltage signal to derive the wave-flap. This kind of linear technique is generally applicable to generate a wave given the time series of the wave elevation at a design location along the wave tank. The effect of distance between the wave-flap and design wave location on the transfer function can be treated by assuming linear dispersion. Though we use linear model, as such, the generated waves may be linear and weakly or strongly nonlinear depending on the design wave spectrum.

The generation techniques of transient waves were developed by Davis and Zarnick (1964), Takezawa and Hirayama (1977), Mansard and Funke (1982), Kjeldsen (1982), Longuett-Higgins (1974), Rapp (1986), Kim et al. (1990 and 1992), Clauss and Kuehnlein (1994), and Chaplin (1996), and others.

The generation of extremely large 2-D transient waves at TAMU tank was based on the works of Takezawa and Hirayama (1977) and Mansard and Funke (1982) with some modifications (Kim et al., 1990 and 1992).

The amplitude spectrum Fig. 11.3(a) was designed by keeping constant slope and constant phase angle for all the wave components at the design location in the wave tank. The transfer function of the wave-flap Fig. 11.3(b) should be already known, usually by experiment.

$$A_v = \frac{A_h(\omega)}{A_{hv}(\omega)} \tag{11.18}$$

where A_h, A_{hv} and A_v represent the design amplitude spectrum, transfer function of wavemaker and voltage amplitude spectrum, respectively. Fourier transform of the voltage amplitude spectrum gives the voltage signal as shown in Fig. 11.3(d):

$$v(t) = \frac{1}{\pi} \int_0^\infty A_v(\omega) \cos \omega t \, d\omega \qquad (11.19)$$

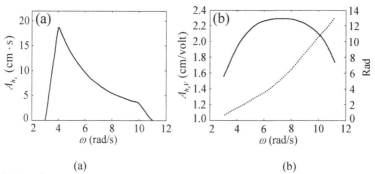

(a)　　　　　　　　　　　　　　(b)

Fig. 11.3(a) Design amplitude spectrum of constant phase. (b) Wavemaker transfer function.

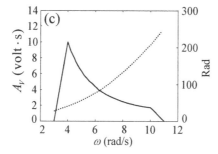

Fig. 11.3(c) Voltage amplitude spectrum.

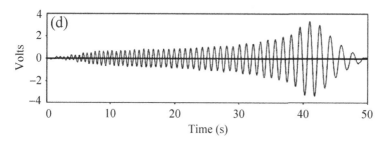

Fig. 11.3(d) Voltage time series.

Wavemaker transfer function:

The unit of wave-flap transfer function [Fig. 11.3(b)], for instance, as was measured at TAMU, is the wave amplitude (cm) measured 1 m down the tank from the wave-flap [Fig. 11.3(e)] divided by the input voltage amplitude (volt). These were measured at each frequency from 3.0 to 12.0 rad/s. Since the data points were scattered in the high frequency region, they are fitted to a smooth line by applying the cubic spline fitting method and 200 data points were created. Thus many points were required at other frequencies. These data points are built into the computer program to generate the digital voltage signals. Another software built in the program is used to convert the digital data into analog data by a D/A converter (Krafft and Kim, 1987).

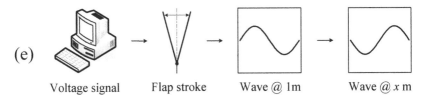

(e)   Voltage signal   Flap stroke   Wave @ 1m   Wave @ x m

Fig. 11.3(e) Wavemaker system.

Voltage signal:

It is interesting to observe the voltage signal in Fig. 11.3(d). The amplitude is initially small and the frequency is high. As the time increasing, the amplitude gradually increases while frequency decreases, reaching the highest voltage group. This will produce the largest wave group making a large water transient wave. In the wave tank, the progressing wave follows linear dispersion rule approximately. Thus shorter waves move slower and longer waves move faster. It was specifically designed as to create a large transient wave at a given location in the wave tank by keeping both phase angles and wave slopes constant for all the wave components in the transient.

11.5.2 Definition of strongly asymmetric wave

The highly nonlinear wave or strongly asymmetric wave is defined in the time domain (Mansard and Funke, 1982; Myrhaug and Kjeldsen, 1980) as shown in Fig. 11.1. Using zero-down-crossing wave time series of period T, we can produce employing UNIOM the horizontal wave kinematics along the vertical axis passing the crest point, which will be later necessary to determine impact on a fixed column using Morison's equation. To describe the strongly asymmetric wave, we define two asymmetry factors as follows: Vertical asymmetry factor,

$$\lambda \equiv \frac{T_f}{T_r} \qquad (11.20)$$

Horizontal asymmetry factor,

$$\mu \equiv \frac{H_c}{H} \qquad (11.21)$$

where T_f and T_r, denote the falling and rising times of wave crest, H_c and H wave crest height and total wave height, respectively.

A study was conducted to find the distribution of strongly asymmetric waves in the field (Myrhaug and Kjeldsen, 1984). The largest wave was strongly asymmetric both, about the vertical axis passing the crest of the wave and about the MWL.

The geometry was defined by T_a and T_b by the foregoing authors. However, Kim et al. (1997) changed the symbols; T_a by T_r and T_b by T_f, because transient wave and impacting force simultaneously rise and fall, where r and f stand for the rising and falling of the wave and force, respectively.

One may regard a random wave containing a strongly asymmetric wave as a freak (rogue) wave. Definition of the freak wave is also given in Chap. 1. Such wave can be generated in the wave tank approximately as will be shown later.

11.5.2.1 Generation of strongly asymmetric wave in random seas

The generation of freak (strongly asymmetric) wave in the wave tank is similar to the generation of random seaway containing more than one transient wave. Zou and Kim (2000) developed the method for generation of strongly asymmetric wave in the random seaway, employing distortion method by Funke and Mansard (1982). However, the techniques proposed by

Kriebel and Alsina (2000) and Clauss and Steinhagen (2000) for generation of freak wave in the random seaway are quite different from Zou and Kim (2000).

11.5.2.2 Distortion of a large wave

Funke and Mansard (1982) proposed a distortion technique to generate the asymmetries in random seas in a wave tank. The technique employs three steps to control wave asymmetries in random waves. They are amplitude distortion, time distortion and crest distortion.

Before distortion, we must numerically simulate random wave time series from the given wave spectrum and then find the largest wave. Zou and Kim (2000) applied the formula by Isherwood (1986) to easily incorporate a large peakedness parameter γ in the spectrum in order to obtain the largest amplitude wave in the simulation. After this simulation the largest zero-down crossing wave was isolated before distortion as shown in Fig. 11.4. This simulated largest wave was distorted as described below.

The basic procedure is to increase the crest height and reduce the trough depth by keeping the input and output amplitude spectra identical and changing the phase spectra. Application of this procedure resulted in decreasing the average steepness by about 10%.

Time distortion is utilized to make the duration of trough longer and that of crest shorter but the total period remains same. Refer to Fig. 11.1.

Crest distortion is employed to move the location of the highest wave crest to the left in Fig. 11.1 and thus the front steepness in space will be increased. Time distortion and crest distortion seem to be essential. The distortion of the entire time series used by Funke and Mansard (1982) is not appropriate at TAMU tank because of the limitation of tank length.

Zou and Kim (2000) utilized the time distortion and crest distortion of the highest wave simulated previously. We contained the largest wave in the short truncated record of about 20.0 s. All the other segments of the time series were not distorted, but connected to each other to make the reasonably long resultant wave elevation time series.

Time Distortion:

The wave time series of the numerically simulated irregular wave (dotted line), containing the zero-down crossing wave with the largest crest is first fixed as shown in Fig. 11.4. One isolates the zero-down-crossing wave (dotted line), then, deforms the trough and crest duration using the following relations:

$$T_t' = (1 + \delta)T_t \tag{11.22}$$

$$T_c' = T_c - \delta T_t \tag{11.23}$$

where δ denotes the time distortion factor which is usually taken in the range of 0.2–0.5 (Funke and Mansard, 1982). T_c' and T_t' represent the new crest and trough duration, respectively. During the distortion process, one has to satisfy the condition that the area in the trough and crest zone remains unaltered. That is, increasing T_t implies reducing the trough depth and reducing T_c means increasing the crest.

Crest Distortion:

Myrhaug and Kjeldsen (1984) analyzed the field data and found that the vertical asymmetric parameter λ varied in the interval from 1.2 to 2.1 with a mean value of 1.5. Due to the lack of correlation of vertical asymmetric parameter with wave spectral moment, it is difficult to create a formula to determine λ. But in general, one may take any value from 1.2 to 2.1:

$$\lambda = \frac{T_f'}{T_r'} \tag{11.24}$$

$$T_r' = \frac{T_c - \delta T_t}{1 + \lambda} \tag{11.25}$$

$$T_f' = \lambda \frac{T_c - \delta T_t}{1 + \lambda} \tag{11.26}$$

where T_r' and T_f' are the rising and falling time duration of the crest, respectively, which are obtained after the distortion. One compares the two wave profiles of the truncated time series before and after distortion and additionally the measured profile after the distortion as shown in Fig. 11.4.

The distorted wave profile is used to generate such wave at the design location in the wave tank. The above generated wave is highly nonlinear due to the nonlinear wave-wave interaction that cannot be analytically treated.

The basic governing condition in applying distortion is to keep the input power spectrum unaltered. Both spectra, before and after distortion, are different in the high frequency region; but their magnitudes are negligibly small. The phase spectra before and after distortion are different. Thus the modified wave approximately satisfies the required condition set to keep the input power spectrum unaltered.

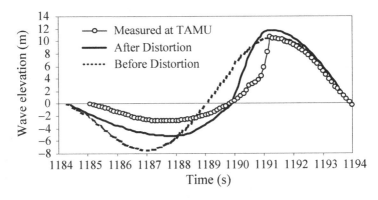

Fig. 11.4. Wave elevations before and after distortion and elevation measured after distortion.

11.5.2.3 Computer program for generation of freak wave

A computer program has been developed for generating strongly asymmetric wave (freak wave) in the given random seas in the wave tank. Such program is essential for generating distortion since each work is slightly different. Thus a successful result can be obtained later by a systematic comparison of many computer generations.

A computer program was produced to do the distortion as shown below, but it was set to use Isherwood's formula given in Sec. 2.3.3.5, at the input stage which is appropriate for the present purpose.

After the distortion we may have a new set of wave time series. Employing the new time series we generate the new wave. The wave elevation is then measured and the local part of the elevation showing isolated strongly asymmetric wave is obtained (see Fig. 11.5).

Given the zero-down-crossing wave, as shown in Fig. 11.5, one may determine the horizontal kinematics along the vertical axis through the crest by using the 2nd version of UNIOM-kinematics as will be shown in an example later.

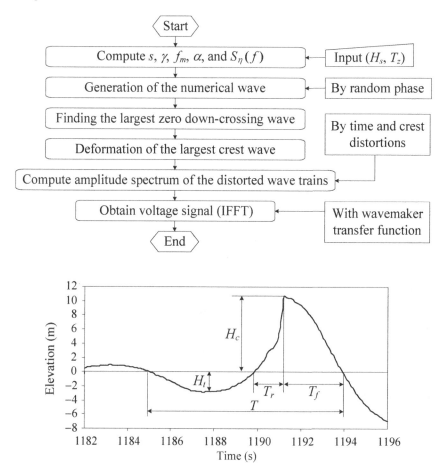

Fig. 11.5. Local view of measured highly nonlinear wave (real scale).

11.5.2.4 Generation of Draupner freak wave

Draupner freak wave data (Chap. 1) was reproduced in wave tank (Kim and Kim, 2003A). The approach was to determine the wave amplitude spectrum of the given field freak wave time series assuming the freak wave is located at the design location x, where the voltage amplitude spectrum and driving voltage signal for wave-flap are computed using the formulas Eqs. (11.18) and (11.19). The voltage signal drives the wave-flap to generate the freak wave at the design location x.

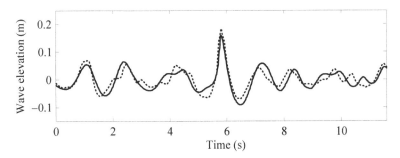

Fig. 11.6. Reproduction (solid) of Draupner freak wave (dotted) in wave tank (model scale).

The reproduction of Draupner freak wave approximately represents the original but only near the highest wave as shown in Fig. 11.6. Since our interest is to investigate the wave kinematics and impacting force due to the strongly asymmetric localized wave as shown in Fig. 11.6, this approximate reproduction seems to be adequate for the practical purpose.

We can also generate any size and shape of freak wave which satisfies the freak wave criterion (Chap. 1) by a trial and error.

11.6 Wave Kinematics

The data of horizontal particles velocity field is essential in determining the horizontal force acting on the fixed column. Since highly nonlinear wave kinematics cannot be simulated theoretically, one has to measure the kinematics. An empirical approach to simulate such measured wave kinematics is UNIOM-kinematics of 2nd version. Given the transient wave elevation time series of zero-down-crossing period, UNIOM-kinematics will provide the horizontal velocity field that are in good agreement consistently. Use of this kinematics will provide the force including impact by applying

Morison's equation. Ordinary forces that belong to the diffraction range can be determined by diffraction theory or Morison's equation. However, the impact load can only be determined by Morison's equation with a simulated wave kinematics. The simulation of kinematics can be validated by comparing with the measured kinematics.

Thus the measurement of kinematics of transient wave or strongly asymmetric nonlinear wave is one of the most fundamental works for determining the wave forces including the impact. Figure 11.7 shows the kinematics in the crest of transient wave measured by LDA (Kim et al., 1992). The highest speed is greater than 1 m/s with a length scale of 1/100. These highest speed particles are considered to play major role for producing impact on the cylinder surface. The transient wave elevation time series had the form similar to that in Fig. 11.1 or 11.5.

11.6.1 Measurement of transient wave kinematics

The positive maximum horizontal velocity field is presented in Fig. 11.8. The measured horizontal particle velocity under crest peak of a transient wave is compared to those of similar height regular wave, Stokes 3rd-order theory, and simulated transient wave with stretching. The horizontal velocity is normalized by the phase velocity. The phase velocity of Stokes 3rd-order wave and the transient are estimated to be 2.18 m/s and 2.29 m/s, respectively.

If the horizontal particles velocity reaches unity, the wave particles will move forward with the wave surface. The transient wave kinematics is featured by the velocity smaller than the simulation in the region below MWL, while it becomes much larger than the simulation in the region above MWL. The similar characteristics are found in the Stokes 5th-order wave as shown in Fig. 11.2.

Wheeler stretching model was used for simulation of the transient wave kinematics employing the transient wave elevation data such as given in Fig. 11.5. However, the kinematics of Wheeler stretching is far less than the real measured data in the region above the position $(y + h)/h = 1.05$, while the measured transient wave kinematics is less than the simulation in the range below MWL. Wheeler stretching does not work well.

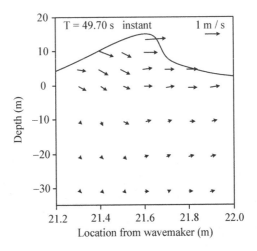

Fig. 11.7. Velocity under crest of 25 cm extreme transient wave measured in wave tank of 1/100 scale (Kim et al., 1992).

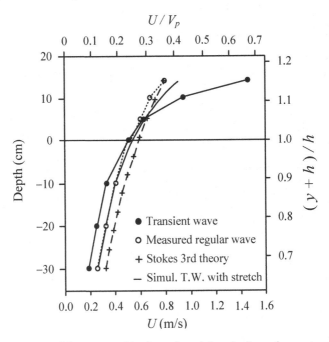

Fig. 11.8. Comparison of the measured horizontal particle velocity under crest peak of 25 cm transient wave, similar height regular wave, Stokes 3rd-order theory, and simulated transient wave with stretching.

11.6.2 Hybrid wave model for transient wave by perturbation

A hybrid wave model was proposed by Spell et al. (1996) to simulate the unidirectional irregular wave kinematics, assuming weakly nonlinear wave. The work conducted a steep transient wave generation and measurement. The theory was in excellent agreement with the measurement. However, the maximum height of the particles kinematics distribution is about the half of the real measured transient wave crest height. The Hybrid model is in excellent agreement with the LDA measurement. The satisfactory agreement between the numerical results of the hybrid wave model and the related experimental measurements indicates that the hybrid wave model, especially near steep wave crests, can predict irregular wave elevation and velocities more accurately and more reliably than those methods commonly used by the offshore industry. The basics of hybrid wave model are given in Zhang et al. (1992).

The foregoing study is intended to simulate the horizontal kinematics of the unidirectional irregular waves by implementing the 2nd-oder wave interaction of the dual waves. Accordingly it is expected that there is similar limitation as the application of the 2nd-order wave theory to simulate the strongly asymmetric wave in the wave tank.

11.6.3 UNIOM-kinematics for strongly asymmetric wave kinematics

For the strongly asymmetric wave as shown in Figs. 11.1, 11.5, and 11.9, UNIOM-kinematics of 2nd version has been developed in Sec. 11.3.3. In employing UNIOM-kinematics one has to use the symbols as illustrated in Fig. 11.9. Additional comparison can be seen for the measured kinematics of a regular wave and the Stokes 3rd-order wave in Fig. 11.8. The regular wave is a large one with equivalent height of the transient wave. The Stokes 3rd-order wave is approximately close to the Stokes 5th-order wave as shown in Fig. 11.2.

The Wheeler stretching cannot simulate the kinematics of the transient wave as shown in Fig. 11.8. The simulation method of such strongly nonlinear wave kinematics was developed by Kim et al. (1997) and the technique has been applied in a variety of cases such as the kinematics of

Draupner freak wave in Figs. 11.9 and 11.10. We find in Fig. 11.9 two neighboring zero-down-crossing waves; the first strongly asymmetric wave is contained between two zero-down-crossing points at $t = 4.68$ s and 6.08 s, while the second ordinary wave is contained between two time instants 6.08 s and 7.7 s.

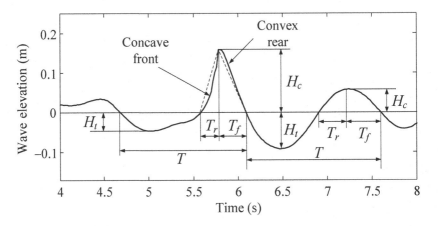

Fig. 11.9. Transient wave with definitions of symbols given in Eq. (11.9).

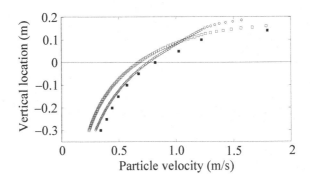

Fig. 11.10. Comparison of UNIOM simulated velocities of both laboratory freak wave (□) and field freak wave (○), with laboratory measured velocity (■), (Kim and Kim, 2003A; 2003B).

Figure 11.10 presents comparison of the measured particles velocity field of Draupner freak wave in the wave tank and simulations of the laboratory freak wave (LFW) and field freak wave (FFW), using 2nd version of

UNIOM-kinematics. The simulations are generally less than the measured data. The measured horizontal velocity profile is similar to those of the transients (Kim et al., 1992; Kim et al., 1997).

11.7 Wave Impact

In the previous research of large transient wave and kinematics measurement (Kim et al., 1992) the authors had proposed to develop a technique to predict the wave kinematics and horizontal impact load by the transient wave on a fixed vertical column. For this purpose, a series of research on the wave impact load have been conducted experimentally (Zou and Kim, 1995; Zou and Kim 1996; Kim et al., 1997; Kim and Kim, 2003A; 2003B)

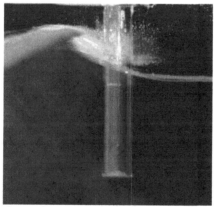

Fig. 11.11. Photograph of strongly asymmetric freak wave impacting on the vertical column (diameter=10 cm, draft=30 cm) (Kim and Kim, 2003A; 2003B).

11.7.1 Wave impacts for series of cylinders

As noted earlier, Kim et al. (1997) conducted a systematic and an in-depth study of horizontal impact load by varying transient waves and cylinder sizes. The research has developed a prediction technique of the strongly asymmetric wave kinematics by employing UNIOM-kinematics of 2nd-version, and applying Morison's equation to simulate the horizontal impact load. The new freak wave impact data (Kim and Kim, 2003A, 2003B) were added to the result of the previous impact research work (Kim et al., 1997) as shown in Fig. 11.12.

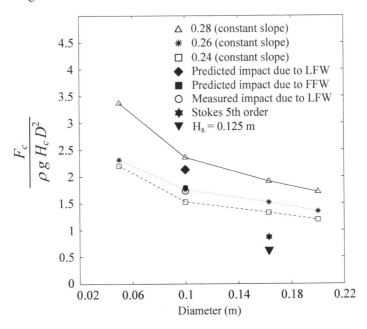

Fig. 11.12. Laboratory impact loads on cylinder due to transients, Stokes 5th-order, random wave of H_s of 0.125 m, and simulated and measured impact loads due to LFW and FFW.

Kim and Kim (2003B) applied the UNIOM-kinematics of the 2nd-version to the freak waves as shown in Fig. 11.12, which contains the result of experimental impact of LFW. These data fall in the same range of the previously measured transient wave loads (Kim et al., 1997). The important finding is that the weakly nonlinear Stokes 5th-order wave force is about one half of the equivalent height transient wave or strongly nonlinear wave force. It is well-known that the Stokes wave force is the design wave load recommended by API (1987).

There are similar researches on the impact behavior by Chan et al. (1991) and Oumerachi et al. (1992).

11.7.2 UNIOM-kinematics or diffraction for impact

The impact due to freak wave in the wave tank is shown in Fig. 11.13. The figure shows that the diffraction theory prediction is in good agreement, except at the special instantaneous point of impact. This indicates that diffraction theory cannot predict impact. In other words, diffraction is not in the category of impact, because the diffraction theory is based on the sinusoidal wave, which cannot give such an infinitesimal time effect on the load. However, Morison's equation can provide such a short time effect on the load (Kim et al., 1997).

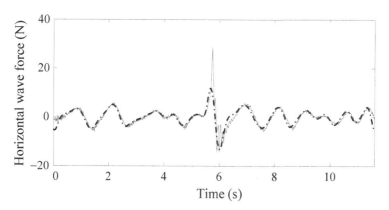

Fig. 11.13. Instant of impacting peak load around at 5.7 s; comparison of measured freak wave force (solid line) and simulation by UNIOM-diffraction (chain-dot).

11.7.3 Morison equation for impact

The simulated kinematics using the 2nd-version of UNION-kinematics for the freak wave is used in Morison's equation for the impact force:

$$F(t) = \rho C_M \frac{\pi}{4} D^2 \int_{-d}^{\eta} \frac{\partial u}{\partial t} dz + \frac{1}{2} \rho C_D D \int_{-d}^{\eta} u |u| dz \qquad (11.27)$$

where C_M and C_D are assumed to be uniformly distributed along the cylinder axis and the values are 1.95 and 0.95 respectively (Kim et al., 1997).

The impacting load due to laboratory freak wave and field freak wave are simulated applying Morison's equation Eq. (11.27). The comparison of the simulations with the measured force is shown in Fig. 11.14. The simulated impact load due to the laboratory freak wave is in phase with the measured impact load because the simulation used the same laboratory freak wave. The simulated impact load due to field freak wave is slightly out of phase from the laboratory impact load. The simulated impact loads due to field freak wave and laboratory freak wave are 31.6 N and 33.7 N, respectively, while the laboratory experimental impact load is 28.4 N. The simulated impact loads are higher than the experimental data. This may be partly due to the experimental procedure that truncates off the high frequency components.

The simulated impact load due to field freak wave is lower than the laboratory freak wave. This may be attributed to the local acceleration of the field freak wave being lower than laboratory freak wave.

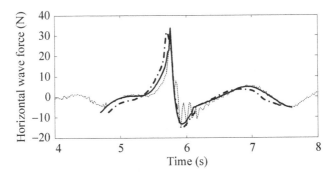

Fig. 11.14. Comparison of laboratory impact load (dotted line) and simulated impact loads due to LFW (solid line) and FFW (dashed dot line).

It is to be noted that the asymmetry of impact time history (Fig. 11.14) is similar to the wave crest in Fig. 11.9. The impact force by transient wave of nominal constant slope 0.24 in the generation of transient wave has very small concave in front and zero convex in rear, while the transient wave impact force of nominal constant slope 0.28 has a clear concave in front and zero convex in rear as shown in Fig. 11.9 (Kim et al., 1997).

All of the simulated freak wave impact loads have concave in front, but measured and simulated impact due to the laboratory freak wave have nearly zero convex in rear, while simulated impact load due to field freak wave has convex in rear.

Comparison of the asymmetry of the impact time series of the transients and freak waves indicates that the crest forms of the measured impact loads in laboratory freak wave and simulated impact load due to laboratory freak wave behave similar to the transient waves of nominal constant slope 0.24 and 0.28.

For the asymmetry parameter of the impact load, we use F_c/T, in the unit of N/s, where F_c and T represent the impact load and period between the zero-down-crossings of the impact load. The asymmetry parameters F_c/T of transient waves of constant slopes of 0.24, 0.26, and 0.28 are 52.6 N/s, 76.3 N/s, and 106.7 N/s, respectively (Kim et al., 1997). The laboratory and simulated impact loads due to laboratory freak wave are 61.7 N/s and 56.3 N/s, both falling in between 52.6 N/s and 76.3 N/s of the foregoing transient waves.

The above comparison indicates that the asymmetry characteristics of laboratory freak wave impact load and simulated impact load due to laboratory freak wave are nearly same as those of the transient waves of nominal constant slope 0.24 and 0.26.

11.7.4 Analysis of impact frequency by wavelet

It is of our interest to know how high the frequency of the impact load was at the time of impacting instant. Such high frequency can be searched applying a wavelet transform technique as shown in Figs. 11.14 and 11.15.

It is known that fast Fourier transform cannot analyze such large amplitudes and higher frequency components contained in such Draupner freak wave impact load (Fig. 11.13). Wavelet transform can accurately assess the extreme waves and associated impact loads on offshore structures and detect the large and high frequency components of the freak wave (Jacobsen et al., 2001).

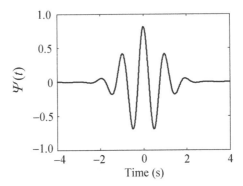

Fig. 11.15. Gabor wavelet.

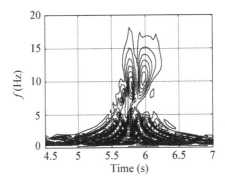

Fig. 11.16. Isograms of magnitude of CWT of laboratory freak wave impact load in time-frequency domain.

The Gabor wavelet in Fig. 11.15 is used here since it is known to be effective in providing a smaller area of time-frequency window than any other functions according to the uncertainty principle (e.g., Chui, 1992).

The continuous wavelet transform (CWT) by Gabor wavelet is shown in Fig. 11.16. The highest values of CWT of the force are contained during the time of impacting, from 5.5 s to 6.5 s, and these CWT are distributed in the range of model frequencies 0.0 to 18 Hz. A closer look of impacting instant in Fig. 11.13 may be made in the isograms in Fig. 11.16.

The result may be interpreted using the familiar wave energy spectrum in real scale; the maximum frequency of wave energy spectrum is usually about 0.3 Hz, while the peak frequency of the impact force is 1.8 Hz in real scale.

Hence the peak frequency contained in the impact force in real scale is about 6 times the maximum frequency of the wave energy spectrum. The rest of the entire force time series contains frequency components below 0.25 Hz, which fall in the ordinary frequency range of the wave energy spectra.

Since the impact force is due to the freak wave, it is necessary to observe the CWT of the freak wave in the time-frequency domain. The highest CWT and the high frequency of the freak wave are contained in the same time interval between 5.5 s and 6.5 s as in the impact force Fig. 11.13. The maximum frequency of the wave is 0.65 Hz in real-scale. During the time of non-impacting the frequency is in the range between 0.0 Hz to 0.2 Hz, which is usually observed in the ordinary nonlinear irregular waves. Thus the peak frequency of impact load is about 3 times the freak wave. This may be attributed to the strong nonlinear dynamic impact load of high speed water masses on the hard surface-piercing column. The freak wave and its impact load can be also analyzed by a wavelet transform based coherence analysis (Kwon et al., 2005). The coherence analysis revealed that some high-frequency components were highly correlated with the impact forces. The coherence analysis can also be used to locate exact time of the maximum coherence of the wave and its impact forces.

11.7.5 Kinematics of strongly asymmetric freak wave

The strongly asymmetric freak waves are similar in appearances to the extreme transient waves (Kim et al., 1992; Kim et al., 1997). The nominal constant wave slopes of the transients that were tested were 0.28, 0.26, and 0.24. The crest shapes of the transient wave had concave in front and convex in rear, and the ratios of crest height to period H_c/T (cm/s) are 18.3, 17.4, and 14.8 as shown in Fig. 11.9.

The laboratory freak wave crest has concave in front and convex in rear and $H_c/T = 14.0$ cm/s. Thus the crest shape of laboratory freak wave is similar to the transient wave of nominal constant slope of 0.24. On the other hand field freak wave has nearly zero concave in front and convex in rear and $H_c/T = 11.4$ cm/s. Thus the crest form of field freak wave is found to be less asymmetrical than transient wave of slope 0.24.

11.8 Application of UNIOM-Diffraction

11.8.1 Load on large column in Heidrun ringing wave

Kim and Wang (1999) applied UNIOM-diffraction for simulation of horizontal load on a single column of large diameter tested by Stansberg et al. (1995) in the replica of Heidrun ringing wave. Figure 11.17 illustrates the incident wave and Fig. 11.18 shows the comparison of the simulation and measurement. The simulation was made using UNIOM-diffraction (L + Q), where L and Q denote the LTFs and QTFs of the horizontal force on the column, respectively.

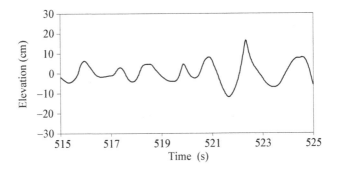

Fig. 11.17. Wave elevation measured at the center of the vertical truncated column.

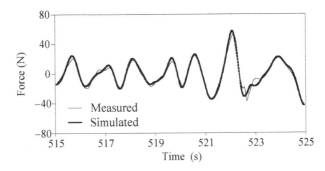

Fig. 11.18. Comparison of the measured and simulated horizontal force on the vertical truncated column.

Based on the foregoing work, Wang and Kim (2001) applied UNIOM to a different structure ISSC TLP in the replica of the Heidrun ringing wave. It was found that the ringing of ISSC TLP was not observed in the Heidrun ringing wave. This work was later extended by Kumar and Kim (2002A; 2002B) to simulate the ringing of Heidrun TLP in the Heidrun ringing wave as follows.

11.8.2 Uniom-diffraction for Heidrun ringing

UNIOM-diffraction-(L + Q) was applied by Kumar and Kim (2002A; 2002B) to simulate ringing of Heidrun TLP in the replica of Heidrun ringing wave. Figure 11.19 illustrates the replica of Heidrun ringing wave and Fig. 11.20 illustrates an example of the QTFs computed for the Heidrun TLP using HOBEM (Liu et al., 1995).

The normalized high frequency tension is presented in the time series as shown in Fig. 11.21 and probability of exceedence is shown in Fig. 11.22.

The simulation work extracted the high frequency tension (HFT) from the total tension amplitude spectrum and brought it in the time domain in the normalized form by the std as shown in Fig. 11.21. The statistics of total frequency tension (TFT), high frequency tension (HFT), low frequency tension (LFT), and wave frequency tension (WFT) are shown in Table 11.1. The kurtosis of HFT is 18.21, which is much higher than the required criterion 5 for ringing (Jefferys and Rainey, 1994).

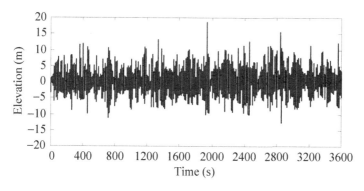

Fig. 11.19. Replica of Heidrun ringing wave (Stansberg et al., 1995; MARINTEK-Statoil, 1996).

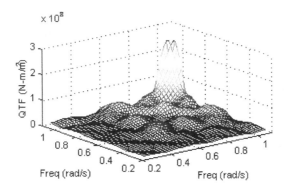

Fig. 11.20. QTFs distribution for Heidrun TLP in the bi-frequency domain.

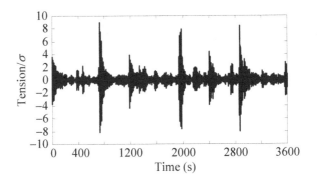

Fig. 11.21. Normalized high frequency tension time series of Heidrun TLP.

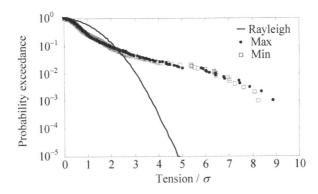

Fig. 11.22. Probability of exceedence of normalized high frequency peak tension of Heidrun TLP.

Table 11.1. Statistics of tensions of Heidrun TLP tendons

	TFT	HFT	LFT	WFT
Mean (N)	6.84E7	1.14E2	-1.43E2	2.62E2
Max (N)	1.56E8	5.12E7	6.66E6	5.13E7
Min (N)	-1.28E7	-4.74E7	-1.02E7	-5.45E7
σ (Std, N)	1.69E7	5.75E6	1.9E6	1.58E7
Skewness	-0.003	8.27E-3	-0.529	1.23E-2
Kurtosis	3.99	18.21	7.59	2.95

The other criterion is that the normalized peak HFT has to be greater than 7 as Figs. 11.21 and 11.22 illustrate the peak tensions greater than 7.

The result of this study shows good agreement with the typical normalized HFT time series and the corresponding distribution of ringing factor obtained from the typical experiment as shown in Figs. 1(a) and 1(b) on p.590 of the work by Jefferys and Rainey (1994).

The springing is observed from Figs. 11.21 and 11.22, where the ringing factors are less than those ringing criteria. Springing analysis of a TLP was discussed in Sec. 9.7, which was solved applying the Volterra quadratic model for determining the wave-exciting forces in the time domain and the sea severity was $H_s = 8$ m which is below the limitedness of Volterra quadratic model.

We may easily conclude that the ringing is only due to highly nonlinear impacting wave as discussed in Sec. 11.5. Ringing of ISSC TLP in the strongly asymmetric wave was observed by Zou et al. (1998).

However, here the Heidrun ringing wave was a phenomenal sea $H_s = 15.4$ m, that does not contain such strongly asymmetric wave as that generated at TAMU tank. All we know is that it does not contain the highly nonlinear wave such as freak wave in it. Thus we may conclude that ringing of Heidrun TLP was not due to highly nonlinear impact load but due to the action of large nonlinear waves of the phenomenal sea in association with the design characteristics of Heidrun TLP. We have found that ringing was not due to the impact load.

References

Journal references:

Chan, E.S., Tan, B.C. and Cheong, H.F. (1991). Variability of Plunging Wave Pressures on Vertical Cylinders, Int. J. Offshore Polar Eng., ISOPE, Vol. 1, No. 2, pp. 94–100.

Chaplin, J.R. (1996). On Frequency-Focussing Unidirectional Waves, Int. J. Offshore Polar Eng. ISOPE. Vol. 6, No. 2, pp. 131–137.

Faltinsen, O.M., Newman, J.N. and Vinje, T. (1995). Nonlinear wave loads on a slender vertical cylinder, J. Fluid Mech., Vol. 289, pp. 179–198.

Isherwood, R.M. (1987). Technical Note: A revised parameterization of the JONSWAP spectrum, Appl. Ocean Res., Vol. 9, No.1, pp. 47–50.

Kim, N.S. and Kim, C.H. (2003A). Investigation of Dynamic Property of Draupner Freak Wave, Int. J Offshore Polar Eng., Vol. 13, No. 1, pp. 38–42.

Kim, N.S. and Kim, C.H. (2003B). Simulation of Draupner Freak Wave Impact on a Vertical Truncated Cylinder, Int. J. Offshore Polar Eng., Vol. 13, No. 3, pp. 20–265.

Kim, C.H., Randall, R.E., Boo, S.Y. and Krafft, M.J. (1992). Kinematics of 2-D Transient Water Waves Using Laser Doppler Anemometry, J. Waterways Port Coastal Ocean Eng., ASCE, Vol. 118, No. 2, pp. 147–165.

Kim C.H. and Zou, J. (1995). A Universal Linear System Model for Kinematics and Forces Affected by Nonlinear Irregular Waves, Int. J. Offshore Polar Eng., Vol. 5, No. 3, pp. 166–170.

Kim, C.H., Xu, Y. and Zou, J. (1997). Impact and Non-Impact on Vertical Truncated Cylinder Due to Strong and Weak Asymmetric Wave, Int. J. Offshore Polar Eng., ISOPE, Vol. 7, No. 3, pp. 161–167.

Kim, C.H., Zhao, C.T., Zou, J. and Xu, Y. (1997). Springing and Ringing Due to Laboratory-Generated Asymmetric Waves, Int. J. Offshore Polar Eng., Vol. 7, No. 1, pp. 30–35.

Kim, C.H. and Wang, Z.M. (1999). Universal Linear System Model for Storm Sea Loads on Column, Int. J. Offshore Polar Eng., Vol. 9, No. 1, pp. 49–54.

Kumar, A. and Kim, C.H. (2002B). Ringing of Heidrun TLP in High and Steep Random Waves, Int. J. Offshore Polar Eng., Vol.12, No. 3, pp. 189–195.

Kumar, A., Kim, C.H. and Zou, J. (2002C). Limitedness of 2nd-order Theories for Laboratory High and Steep Sea Waves and Forces on Structures, Int. J. Offshore Polar Eng., Vol. 12, No. 4, pp. 243–248.

Kwon, S.H. and Lee, S.H. and Kim, C.H. (2005). Wavelet transform based coherence analysis of freak wave and its impact, Ocean Eng., Vol. 32, pp. 1572–1589.

Liu, Y.H., Kim, M.H. and Kim, C.H. (1995). The Computation of Second-Order Mean and Double Frequency Wave Loads on a Compliant TLP by HOBEM, Int. J. Offshore and Polar Eng., Vol. 5, No. 2, pp. 111–119.

Longuet-Higgins, M.S. and Cokelet, E.D. (1976). The Deformation of Steep Surface Waves on Water-I, A Numerical Method of Computation, Proc. Roy. Soc. London, A350, pp. 1–26.

Malenica, S. and Molin, B. (1995). Third-hamonic wave diffraction by a vertical cylinder, J. Fluid Mech., Vol. 302, pp. 203–229.

Myrhaug, D. and Kjeldsen, S.P. (1984). Parametric Modeling of Joint Probability Density Distribution for Steepness and Asymmetry in Deep Water Waves, Appl. Ocean Res., Vol. 6, No. 4, pp. 207–220.

Spell, C.A., Zhang, J. and Randall, R.E. (1996). Hybrid wave model for unidirectional irregular waves — Part II. Comparison with laboratory measurements, Appl. Ocean Res., 18, pp. 93–110.

Takezawa, S. and Hirayama, T. (1977). Advanced Experimental Techniques for Testing Ship Models in Transient Water Waves — Part 2. The Controlled Transient Water Waves for Using in Ship Motion Tests, Inst. of Mech. Eng., pp. 37–54.

Wheeler, J.D. (1970). Method for calculating forces produced by irregular waves, J. Petroleum Tech., pp. 359–367.

Zhang, J., Chen, L., Ye, M. and Randall, R.E. (1996). Hybrid wave model for unidirectional irregular waves — Part I. Theory and Numerical Scheme, Appl. Ocean Res., 18, pp. 77–92.

Zhang , J., Randall, R.E. and Spell, C.A. (1992). Wave Kinematics, J. Waterway Port Coastal Ocean Eng., Vol. 118, No. 4, pp. 401–415.

Zou, J. and Kim, C.H. (1996). Experimental Study of Impacting Wave Force on Vertical Truncated Cylinder, Int. J. Offshore Polar Eng., ISOPE, Vol. 6, No. 4, pp. 291–293.

Zou, J., Xu, Y., Kim, C.H. and Zhao, C.T. (1998). Ringing of ISSC TLP Due to Laboratory Storm Seas, Int. J. Offshore Polar Eng., Vol. 8, No. 2, pp. 81–89.

Book references:

Chui, C. K. (1992) An Introduction to Wavelets, Academic Press, San Diego.

Proceedings references:

Alex, H. and Kim, C.H. (2000). Laboratory Stokes 5th-Order Waves and Forces on a Vertical Truncated Cylinder, Proc 10th Int. Offshore Polar Eng Conf., ISOPE, Vol. 6, pp. 131–137.

API RP 2T (1987). Recommended Practice for Planning, Designing and Constructing Tension Leg Platforms, The API Recommended Practice 2T (RP 2T), American Petroleum Institute, Washington, D.C.

Clauss, G.F. and Kuehnlein, W.L. (1994). Seakeeping Tests of Marine Structures with Deterministic Wave Groups and Tank Side Wall Wave Absorbers, Proc. 7th Int. Conf. Behavior Offshore Structures, Vol. 2, pp. 769–785.

Clauss, G.F. and Steinhagen, U. (2000). Optimization of Transient Design Waves in Random Seas, Proc 10th Int. Offshore Polar Eng. Conf., Vol. 3, pp. 229–236.

Davis, M.C. and Zarnick, E.E. (1964). Testing Ship Models in Transient Waves, Proc. 5th Symp. Naval Hydrodynamics, Bergen, Norway.

Funke, E.R. and Mansard, E.P.D. (1982). The Control of Wave Asymmetries in Random Waves, Proc. 18th Int. Conf. Coastal Eng., Cape Town, South Africa, pp. 725–744.

Davis, M.C. and Zarnick, E.E. (1964). Testing Ship Models in Transient Waves, Proc. 5th Symp. Naval Hydrodynamics, Bergen, Norway.

Funke, E.R. and Mansard, E.P.D. (1982). The Control of Wave Asymmetries in Random Waves, Proc. 18th Int. Conf. Coastal Eng., Cape Town, South Africa, pp. 725–744.

Jakobsen, J.B., Haver, S. and Odegard, J.E. (2001). Study of Freak Waves by Use of Wavelet Transform, Proc.11th Int. Offshore Mechanics Polar Eng. Conf., Vol. 3, pp. 59–64.

Jefferys, E.R. and Rainey, R.C.T. (1994). Slender Body Models of TLP and GBS 'Ringing', Proc. Int. Conf. Behavior Offshore Structures, Vol. 2, pp. 587–605.

Kim, C.H., Randall, R.E., Krafft, M.J. and Boo, S.Y. (1990). Experimental Study of Kinematics of Large Transient Wave in 2-D Wave Tank, Proc. 22nd Ann. Offshore Tech. Conf., Houston, Texas, OTC 6364, pp. 195–202.

Kim, C.H., Zhao, C.T. and Zou, J. (1995). Springing and Ringing Due to Nonlinear Waves on a Coupled TLP, Proc. 5th Int. Offshore and Polar Eng. Conf., ISOPE, The Hague, Netherlands, Vol. 1, pp. 83–89.

Kim, C.H., Xu, Y., Zou, J. and Won, Y.S. (1996). A Model for Weak Impacting Force on Vertical Truncated Cylinder Due to Steep Asymmetric Wave, Proc. 6th Int. Offshore Polar Eng. Conf., Los Angeles, ISOPE, Vol. 3, pp. 215–220.

Kim, C.H. (1998). Recent Progress in Springing and Ringing Research — A Review, Proc. 8th Int. Offshore Polar Eng. Conf., Montreal, Canada, Vol. 1, pp. 20–28.

Kjeldsen, S.P. and Vinje, T. (1980). Kinematics of Deep Water Breaking Waves, Proc. Offshore Tech. Conf., Houston, TX, OTC 3714.

Krafft, M.J. and Kim, C.H. (1987). Extreme transient water waves generation at Texas A & M University, COE Report 294.

Kriebel, D.L. and Alsina, M.V. (2000). Simulation of Extreme Waves in a Background Random Sea, Proc. 10th Int. Offshore Polar Eng. Conf., Vol. 3, pp. 31–37.

Kumar, A. and Kim, C.H. (2002A). Time Domain Simulation of Ringing of Heidrun TLP, Proc. 12th Int. Offshore Polar Eng. Conf., Kitakyushu, Japan, Vol. 1, pp. 176–183.

Mansard, E.P.D. and Funke, E.R. (1982). A New Approach to Transient Wave Generation, Proc. 18th Int. Conf. Coastal Eng., Cape Town, South Africa, pp. 710–724.

Natvig, B.J. (1994). A Proposed Ringing Analysis Model for Higher Order Tether Response, Proc. 4th Int. Offshore Polar Eng. Conf., Osaka, ISOPE, Vol. 1, pp. 40–51.

Oumerachi, H., Partenseky, H.W. and Tautenhain, E. (1992). Breaking Wave Loads on Vertical Gravity Strucures, Proc. 2nd Int. Offshore Polar Eng. Conf., ISOPE, Vol. 3, pp. 532–539.

Rapp, R.J. (1986). Laboratory measurements of deepwater breaking waves, Dissertation for Ph.D., MIT, Cambridge, Massachusetts.

Rodenbusch, G. and Forristall, G.Z. (1986). An empirical model for random directional wave kinematics near the free surface, Proc. 18th Annual Offshore Tech. Conf., Houston, Texas.

Skjelbreia, L. and Hendrickson, J. (1960). Fifth Order Gravity Wave Theory, Proc. 7th Coastal Eng. Conf., The Hague, pp. 184–196.

Stansberg, CT, Huse, E, Krogstad, J.R. and Lehn, E. (1995). Experimental study of non-linear loads on vertical cylinders in steep random waves, Proc. 5th Int. Offshore Polar Eng. Conf., ISOPE, Vol. 1, pp. 75–82.

Statoil (1996). Single Column Test Data Produced at MARINTEK, Stavanger, Norway.

Wang, Z.M. and Kim, C.H. (2001). Nonlinear Response of ISSC TLP in High and Steep Random Waves, Proc. 11th Int. Offshore Polar Eng. Conf., Vol.3, pp. 440–446.

Zou, J. and Kim, C.H. (1995). Extreme Wave Kinematics and Impact Loads on a Fixed Truncated Circular Cylinder, Proc. 5th Int. Offshore Polar Eng. Conf., The Hague, Netherlands, Vol. 3, pp. 216–225.

Zou, J., Xu, Y. and Kim, C.H. (1997). Time Domain Simulation of a TLP's Response Due to Laboratory-Generated Asymmetric Irregular Waves, Proc. 7th Int. Offshore Polar Eng. Conf., Honolulu, ISOPE, Vol. 1, pp. 193–200.

Zou, J, Huang, E.W., and Kim, C.H. (1999). Nonlinear and Non-Gaussian Effects on TLP Tether Responses, Proc. 9th Int. Offshore Polar Eng. Conf., Brest, France, Vol. 1, pp. 315–324.

Zou, J., Xu, Y. and Kim, C.H. (1997). Time Domain Simulation of a TLP's Response Due to Laboratory-Generated Asymmetric Irregular Waves, Proc. 7th Int. Offshore Polar Eng. Conf., Honolulu, ISOPE, Vol. 1, pp. 193–200.

Zou, J. and Kim, C.H. (2000). Generation of Strongly Asymmetric Wave in Random Seaway, Proc. 11th Int. Offshore Polar Eng. Conf., Vol. 3, pp. 95–102.

Chapter 12

Numerical Wave Tank

12.1 Introduction

Engineers and designers are commonly faced with problems of fluid/structure interaction, not only in naval architecture and ocean engineering, but also in various other industrial fields, such as car design, aeronautics, spacecraft design, internal combustion engines, turbo-machinery, etc. To increase the performance of their products, they generally resort to simulation on models to validate their solutions before bringing their project to the production stage. Depending on the kind of problem being considered, the choice may be open between physical models or numerical models. Historically there were not many choices left and the former experimental approach was obviously the only solution to visualize the fluid flowing around bodies. In aeronautics, for instance, wind tunnels were systematically used for the design of aircraft till sixties, and the knowledge of the phenomena involved was essentially acquired through experiments. The equivalent of wind tunnel in naval and offshore applications is the wave basin or towing tank.

As in the case of aeronautics where computer program codes have progressively replaced the wind tunnels, researchers are developing numerical wave tanks (NWT) intended to reproduce as closely as possible the flow around marine structures like ship or offshore rigs as they are excited by ocean waves and currents. For historical and technical reasons, these computer codes are less advanced in hydrodynamics than they are in aerodynamics. The presence of the moving free surface, which adds one complexity level to the air/body interaction problem, is in large part responsible for this.

In the design and analysis of marine structures, the random wave loading is usually assumed to be a Gaussian process and the response of the structure is assumed to be linear. These assumptions assure designers they will obtain simple and complete response characteristics of the system through spectral analysis. However, the above assumptions remain valid as long as the waves are not very rough (H_s = 4 to 6 m). The recent 2nd-oder nonlinear diffraction theory with the aid of Volterra model for the 2nd-order wave force may be valid for the high seas (H_s = 6 to 9 m), though it remains to be extensively investigated before coming to a conclusion. Very high seas (H_s = 9 to 14 m) and phenomenal seas (H_s over 14 m) usually contain nonlinear components higher than 5th order. Thus the design estimates based on the linear and 2nd-order theory need to be modified by a factor of safety to compensate for the unknown higher order wave effects. As these modification still cannot ensure final safety, one resort to the conducting model test in EWT. It is believed that the nonlinear forces will eventually be simulated in NWT.

Many nonlinear hydrodynamic problems were usually observed in EWT during the industrial design work. MARINTEK found ringing phenomena on the Heidrun TLP during the experiment in a wave tank. Ringing is a dynamic transient stage or burst in a steady state springing signal (Natvig, 1994). The bursting tendon vibration was due to the impact of waves that were asymmetric about the vertical axis (Davies et al., 1994). The nonlinear hydrodynamic problems in ship motions were usually investigated with the model tests in EWT. Green water shipping is a highly nonlinear and enormously difficult problem due to the breaking of waves in both EWT and NWT. The bow submergence of the destroyer model was highly over-predicted by linear theory with the Gaussian sea assumption (Cummins, 1973). A ship capsizing in a roll had been studied in both theory and experiment, mainly looking at the effect of the nonlinear restoring moment with a negative metacentric height. The effect of a highly nonlinear asymmetric wave could be another cause of ship capsizing in a roll.

The main part of this chapter is obtained from the work coauthored with Drs. Clément and Tanizawa on the recent research and development of NWT (Kim et al., 1999).

12.1.1 Highly and weakly nonlinear wave

Mathematical waves are usually classified as linear, weakly nonlinear or highly (or strongly) nonlinear. A sinusoidal wave is linear and so is the sum of many sinusoids with random phase angles are linear. The theoretical Gaussian packet and Cauchy-Poisson waves are linear transient waves. Given a number of sinusoidal waves, one can create, based on perturbation scheme, 2nd- or 3rd-order nonlinear interacting irregular waves (Longuet-Higgins, 1963; Pierson, 1993). These are weakly nonlinear higher-order waves. Stokes 2nd-, 3rd-, 4th-, and 5th-order waves are weakly nonlinear self-interacting waves.

Nonlinear waves are generally asymmetric about the horizontal MWL axis as well as about vertical axes through crests and troughs, whereas the sinusoidal wave is symmetric about both axes. Longuet-Higgins and Cokelet (1976) numerically generated a highly nonlinear wave whose crest has a distinct convex front and a concave rear, and has a strong vertical as well as horizontal asymmetry. We may call the above highly nonlinear wave a strongly asymmetric wave (about the vertical axis). The vertical asymmetry is expressed by asymmetry parameter, i.e., the ratio of the falling to rising period of the crest, with field data varying from 1.2 to 2.1 (Myrhaug and Kjeldsen, 1984). All the other waves without the above property are weakly asymmetric or weakly nonlinear. Hereafter these new terms will be used without further explanation.

Conventional EWT can generate steep regular waves (roughly similar to Stokes higher-order-like waves) and strong asymmetric transient waves, e.g., Kim et al. (1992). Zou and Kim (2000) have generated strongly asymmetric waves in JONSWAP seaway, applying local distortion (Funke and Mansard, 1982) to the largest wave in the wave tank.

12.2 Numerical Wave Tank

The ideal fluid NWT employs the boundary integral equation (BIE) to determine the velocity and acceleration field in the fluid domain. A typical formulation is given below to provide a basis for the following discussions about the time marching solution, numerical implementation, and application of NWT for simulation of nonlinear waves, forces and motions.

12.2.1 Typical formulation in potential theory-fixed body

A rectangular wave tank (Fig. 12.1) of constant depth equipped with a wavemaker at one end and a wave absorbing condition at the other is considered. A surface-piercing structure is fixed in the basin. We assume that an ideal fluid is set in motion from the initial state of rest. Then there exists a velocity potential satisfying the Laplace's equation in the fluid domain:

$$\nabla^2 \phi = 0 \qquad (12.1)$$

Given the boundary conditions over the entire boundary surfaces, the velocity potential can be determined by solving the direct or indirect BIE. The modified Green's third identity (Chap. 6) is a direct BIE as given by:

$$\frac{1}{4\pi} \int_S \left[G(P;P_0) \frac{\partial \phi(P_0)}{\partial n(P_0)} - \phi(P_0) \frac{\partial G(P;P_0)}{\partial n(P_0)} \right] d\sigma(P_0)\; C(P)\; \phi(P), \; P \in S$$

$$(12.2)$$

where P and P_0 represent the fixed and moving points as defined in Sec. 4.1.6. S denote the boundary surfaces with suffices f, s, and r standing for the free surface, solid boundary, and radiation or open boundary. Green's function $G(P;P_0)$ is defined in Sec. 4.1.6 or in 5.3.3. $C(P)$ is the normalized solid angle α at a point on the boundary surface in the form $C(P) = \alpha/4\pi$. The unit normal vector on the boundary is directed outward from the fluid domain.

The fully nonlinear kinematic and dynamic boundary conditions on the free surface are written either in Lagrangian or semi-Lagrangian manner, assuming unit water density:
The Lagrangian manner:

$$\frac{D\mathbf{X}}{Dt} = \nabla \phi(\mathbf{X},t) \quad \text{and} \quad \frac{D\phi}{Dt} = -gZ + \frac{1}{2}(\nabla\phi)^2 \quad \text{on } S_f \qquad (12.3)$$

Semi-Lagrangian manner:

$$\frac{\partial Z}{\partial t} = \frac{\partial \phi}{\partial z} \quad \text{and} \quad \frac{\partial \phi}{\partial t} = -gZ - \frac{1}{2}(\nabla \phi)^2, \quad \text{on } S_f \qquad (12.4)$$

where D/Dt indicates material derivative (differentiation following the moving particle), Z the wave elevation at (X, Y, t) and the $\mathbf{X}$ position vector of a moving particle with respect to the space fixed coordinate system O-XYZ, Z upward from the mean water surface and X toward the beach in the longitudinal direction of the tank.

The boundary condition on the solid surface is:

$$\frac{\partial \phi}{\partial n} = V_n \quad \text{on } S \qquad (12.5)$$

where V_n is a prescribed normal velocity of a point on S consisting of the surfaces of the tank sides and tank ends and bottom, wavemaker and fixed structure.

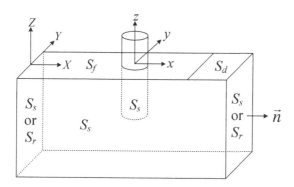

Fig. 12.1. Schematic diagram of NWT.

If sinusoidal waves are to be generated at the up-tank by a wavemaker, the user will prescribe $V_n(t)$ the sinusoidal normal velocity in the same way as in EWT. It is necessary to include a ramp or modulation function to the wavemaker motion in order to ensure a smooth start and avoid numerical breaking at the beginning of the run. The Neumann condition Eq. (12.5) is therefore known explicitly.

Absorption of the outgoing waves at the down-tank wall is a serious problem in NWT. No perfect absorption condition, local in time, exists for that purpose. In the frequency domain where one has to deal with a single frequency wave, the well known Sommerfeld condition applies exactly:

$$\frac{\partial \phi}{\partial t} + C \frac{\partial \phi}{\partial n} = 0 \quad \text{on} \quad S_r \qquad (12.6)$$

where $C(\omega)$ is the phase velocity while ω the wave frequency. The term-to-term transposition of this condition in the time domain is usually referred to as the Orlanski condition (1976). It necessitates the on-line evaluation of the velocity coefficient $C(t)$. Furthermore, it is not well fitted to broad banded irregular incident waves.

A pragmatic solution, often referred to as the numerical beach (Cointe, 1989), or sponge layer (Israeli and Orszag, 1981; Romate, 1992) was proposed by Baker et al. (1981), Cointe et al. (1990) and others. It consists of some extra dissipative terms to the kinematic and dynamic boundary conditions, in their Lagrangian or semi-Lagrangian form.

$$\frac{D\mathbf{X}}{Dt} = \nabla \phi(\mathbf{X}, t) - v(X)\,(\mathbf{X} - \mathbf{X_e})$$

$$\frac{D\phi}{Dt} = -gZ + \tfrac{1}{2}(\nabla \phi)^2 - v(X)(\phi - \phi_e) \quad \text{on} \ S_d \qquad (12.7)$$

$$\frac{\partial Z}{\partial t} = \frac{\partial \phi}{\partial z} - v(X)(Z - Z_e)$$

or $\qquad\qquad\qquad\qquad\qquad\qquad\qquad\qquad\qquad\qquad\qquad\qquad (12.8)$

$$\frac{\partial \phi}{\partial t} = -gZ - \frac{1}{2}(\nabla \phi)^2 - v(X)\,(\phi - \phi_e) \quad \text{on} \ S_d$$

with

$$v(X) = \begin{cases} \alpha\omega\left(\dfrac{k}{2\pi}X - \beta\right)^2 & \text{in } X_o \leq X \leq X_1 = X_o + \dfrac{2\pi\beta}{k} \\ 0 \ \text{except in } X_o \leq X \leq X_1 \end{cases} \qquad (12.9)$$

where S_d denotes the absorbing zone of the free surface.

The nature of the extra terms in Eqs. (12.7) and (12.8) is arbitrary and their form may vary from one author to another. A detailed discussion is given below under Absorbing Beach and Wavemaker Absorber and Active Absorption of Waves.

12.2.2 Marching solution of free surface-MEL procedure

Let us consider the fluid at rest up to $t = 0$, being set in motion by the prescribed motion of a wavemaker. Then the initial boundary conditions are:

$$\phi(\mathbf{X},t) = 0, \quad Z = 0$$
$$\text{with specified } (X,Y), \quad t = 0 \text{ on } S_f + S_d \tag{12.10}$$

$$\frac{\partial \phi}{\partial n} = V_n \text{ at } t = 0 \quad \text{on } S_s \tag{12.11}$$

and

$$\frac{\partial \phi}{\partial t} + C \frac{\partial \phi}{\partial n} = 0 \quad \text{on } S_r \tag{12.12}$$

From the above initial boundary conditions, Eq. (12.2) may be solved for $\partial \phi / \partial n$ on S_f and S_d and ϕ on S_s at time t by a BEM. After that, the potential and its normal derivative are known everywhere over the whole fluid domain boundary. Thus they can be used to determine the particle velocity contained in Eq. (12.3) or (12.4), and Eq. (12.7) or (12.8). The particle velocity, if a higher order BEM (e.g., HOBEM) is used, can be easily determined by the following formula (Lee, 1992, Xü and Yue, 1992, and Boo, 1993):

$$\begin{pmatrix} \dfrac{\partial \phi}{\partial x} \\[2mm] \dfrac{\partial \phi}{\partial y} \\[2mm] \dfrac{\partial \phi}{\partial z} \end{pmatrix} = \begin{pmatrix} \dfrac{\partial x}{\partial \xi} & \dfrac{\partial y}{\partial \xi} & \dfrac{\partial z}{\partial \xi} \\[2mm] \dfrac{\partial x}{\partial \eta} & \dfrac{\partial y}{\partial \eta} & \dfrac{\partial z}{\partial \eta} \\[2mm] n_x & n_y & n_z \end{pmatrix} \begin{pmatrix} \dfrac{\partial \phi}{\partial \xi} \\[2mm] \dfrac{\partial \phi}{\partial \eta} \\[2mm] \dfrac{\partial \phi}{\partial n} \end{pmatrix} \tag{12.13}$$

where ξ, η are the coordinates of the master element. If a low-order BEM is used, one employs other methods, such as those given under Time-Stepping Integration.

Time integration of Eq. (12.3) or (12.4) and Eq. (12.7) or (12.8) on the free surface and damping zone yields the new position of the fluid particle and velocity potential on $S_f + S_d$. Given these updated initial conditions, the above time-stepping process can then be iterated.

In the original formulation (Longuet-Higgins and Cokelet, 1976), the present BIE is solved in Eulerian frame, while the time integration of free surface boundary condition is performed in Lagrangian manner, following the moving fluid particles. For this reason, it was called mixed-Eulerian-Lagrangian (MEL) procedure.

12.2.3 Typical formulation in potential theory — floating body motion

Let us now consider the case where a moving floating body is introduced in the NWT. The space coordinate system O-XYZ was defined in above. Now we add the body coordinate system o-xyz to deal with floating body motion. The kinematic boundary condition for the velocity field due to body motion:

$$V_n = \mathbf{n} \cdot (\mathbf{v_0} + \boldsymbol{\omega} \times \mathbf{x}) \qquad (12.14)$$

where $\mathbf{n}$, v_0 and $\boldsymbol{\omega}$ denote the unit normal, translation and angular velocity of the body. $\mathbf{x}$ is the position vector of the moving body surface referring to the body fixed frame.

Given a forced motion of the body, the normal velocity in Eq. (12.14) is known a priori. Then, V_n can be substituted into Eq. (12.5) in the foregoing section to solve the boundary value problem for the velocity field due to the body motion.

Now suppose a freely floating body moves in response to the hydrodynamic force excitation and affects the hydrodynamic force vice versa. Then one needs to determine the acceleration field and body acceleration by taking into account the coupling of fluid and body motions. Once the acceleration field is solved and body acceleration obtained, the time integral of the body acceleration gives the body velocity, and the velocity field can be solved in the same way as the case of prescribed body motion. Figure 12.2 shows the relation between the values in the acceleration and velocity field. As the body acceleration is not known prior to solving the BIE of ϕ_t, one has to solve the loop of the acceleration field implicitly or iteratively (Fig. 12.2). Tanizawa (1996) introduced the nonlinear acceleration potential:

$$\Phi = \frac{\partial \phi}{\partial t} + \frac{1}{2}(\nabla \phi)^2 = -p - gZ \tag{12.15}$$

Similar to the velocity potential, the gradient of the nonlinear acceleration potential gives fluid acceleration. Using Φ, the Neumann boundary condition for ϕ_t can be systematically derived:

$$\frac{\partial \Phi}{\partial n} = \mathbf{n} \cdot (\dot{\mathbf{v}}_0 + \dot{\boldsymbol{\omega}} \times \mathbf{x}) + q = \mathbf{N} \cdot \boldsymbol{\alpha} + q \tag{12.16}$$

where $\boldsymbol{\alpha} = (\dot{\mathbf{v}}_0, \dot{\boldsymbol{\omega}})$ is the generalized acceleration of floating body, q is the additional normal acceleration that can be determined with the solution of the velocity field and $\mathbf{N} = (\mathbf{n}, \mathbf{x} \times \mathbf{n})$ is the generalized unit normal on the body surface.

The first term of right hand side of Eq. (12.16) is the normal acceleration of fluid due to body acceleration and the second term q is the normal acceleration due to body velocity and fluid flow, which can be explicitly determined from the solution of the velocity field:

$$q = -k_n (\nabla \phi - \mathbf{v}_0 - \boldsymbol{\omega} \times \mathbf{x})^2 + \mathbf{n} \cdot \boldsymbol{\omega} \times (\boldsymbol{\omega} \times \mathbf{x})$$
$$+ \mathbf{n} \cdot 2\boldsymbol{\omega} \times (\nabla \phi - \mathbf{v}_0 - \boldsymbol{\omega} \times \mathbf{x}) - \frac{\partial}{\partial n}\left(\frac{1}{2}(\nabla \phi)^2\right) \tag{12.17}$$

where k_n denotes the normal curvature of the body surface. The existence of the term q is due to the difference between the velocity and acceleration field. Even if the body is not accelerated, the body and fluid velocity accelerates the fluid itself. Thus there exists a normal component of fluid acceleration on the body surface. The term q is indispensable for a fully nonlinear solution of the acceleration field. The body motion equation is:

$$M\boldsymbol{\alpha} + \boldsymbol{\beta} = \int_S \rho(-\phi_t - \frac{1}{2}(\nabla \phi)^2 gZ)\mathbf{N}dS + \mathbf{F}_g \tag{12.18}$$

where M is the generalized inertia tensor, β generalized gyroscopic moment, $\mathbf{F}_g$ the gravity force of the body.

By eliminating α from Eqs. (12.16) and (12.18) we obtain the implicit boundary condition for the ϕ_t:

$$
\begin{aligned}
\frac{\partial \phi_t}{\partial n} &= \mathbf{N}M^{-1} \cdot \left(\rho \int_S -\phi_t \mathbf{N} dS \right) + \mathbf{N}M^{-1} \\
&\cdot \left\{ \rho \int_S \left(-gZ - \frac{1}{2}(\nabla \phi)^2 \right) \mathbf{N} dS \right\} \\
&+ \mathbf{N}M^{-1} \cdot \left(\mathbf{F}_g - \beta \right) + q - \frac{\partial}{\partial n} \left(\frac{1}{2}(\nabla \phi)^2 \right)
\end{aligned}
\tag{12.19}
$$

The free surface boundary condition is obtained from Eq. (12.15):

$$
\phi_t = -gZ - \frac{1}{2}(\nabla \phi)^2
\tag{12.20}
$$

which is semi-Lagrangian for marching integration. In solving the free surface boundary condition one may choose either Lagrangian or semi-Lagrangian, as shown in Eq. (12.7) or (12.8).

For the numerical evaluation of the term $\partial / \partial n (1/2(\nabla \phi)^2)$ in Eq. (12.17), the following formula is useful (Tanizawa, 1996):

$$
\frac{\partial}{\partial n} \left(\frac{1}{2}(\nabla \phi)^2 \right) = -k(\nabla \phi)^2 + \frac{\partial \phi}{\partial n} \left(-\frac{\partial^2 \phi}{\partial s^2} \right) + \frac{\partial \phi}{\partial s} \frac{\partial}{\partial s} \left(\frac{\partial \phi}{\partial n} \right)
\tag{12.21}
$$

where k is the curvature of the body surface.

The simulation is performed through five steps:
1. determine the velocity field ;
2. determine ϕ_t : by solving implicit boundary condition Eq. (12.19);
3. determine the pressure distribution on the body surface, hydraulic force and acceleration of the body using the solution of ϕ_t and acceleration field Eq.(12.15);
4. estimate the new position and velocity of the body at the next time step by integrating the velocity and acceleration of the body;
5. utilize the MEL procedure for the renewal of the free surface.

Acceleration Field

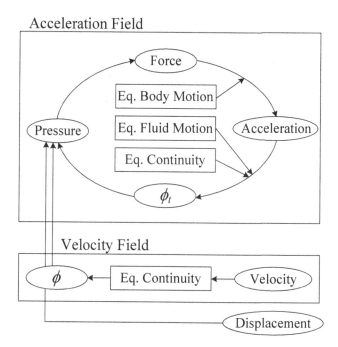

Fig. 12.2. Schematic diagram of velocity and acceleration field.

12.3 Numerical Implementations

The numerical implementation includes the boundary element method (BEM), discretization tools, treatment of corners and edges, efficient matrix equation solvers, time-stepping integration, control of saw-tooth stability, choosing the time increment, radiation boundary, absorbing beach and wavemaker absorber, active absorption of waves, accuracy test and incident wave generation. We will now focus on the techniques developed to handle these problems.

12.3.1 Boundary element method

The BIE Eq. (12.2) is numerically solved by discretizing the boundary surface with BEM and approximating the unknown functions by basic shape functions on the elements. (Brebbia et al., 1984).

In the constant panel method CPM (zeroth-order BEM. Hess and Smith, 1964), unknown sources are considered to be constant over each plane panel, and collocation points are located at the centroid of the facet. CPM can be used for discretization of both direct and indirect BIE.

Application of the higher order boundary element method (HOBEM) to the direct BIE Eq. (12.2) is relatively new (Liu et al., 1990). The boundary surface is discretized by higher-order isoparametric quadrilateral elements. The body surface, velocity potential and its normal derivative are expressed by the same higher-order shape functions on each element. The details of HOBEM are referred to Sec. 6.2.

As described above, a BIE may be discretized by a BEM. The original CPM was improved by implementing the effect of curvature, which is called a higher-order CPM (Hess, 1979).

In CPM, quadrilateral or triangular facets replace the 3-D boundary surface. Each facet has one node (control point) at the center, and supports the source and/or dipole distribution of constant strength. The panels are chosen such that the control points contact or nearly contact the actual body surface from outside or inside of the fluid domain. For curved surfaces, the quadrilateral faceted surface becomes discontinuous as all four of corner points generally do not coincide with those of the neighboring panels. Webster (1975) partly overcame this difficulty. The method was further improved by Beck et al. (1993) and is known as desingularized method. In this method, the solution is constructed by integrating a distribution of fundamental singularities over a surface located outside the fluid domain. For instance, Xu and Mori (1989), Romate (1989), Broeze (1992) and Yang et al. (1992) used CPM. Ferrant (1994) avoided this problem of discontinuity in his 3-D NWT.

The 3-D HOBEM was applied by Chau and Eatock Taylor (1988) and Liu (1988) for problems in the frequency domain. Liu et al. (1990) conducted a comparison study of HOBEM and CPM from which it was concluded that HOBEM can provide more complete and precise model than CPM. Liu et al. (1993) computed the vertical mean drift force on a vertical truncated column. Xü and Yue (1992), Lee et al. (1994), Boo (1993) and Boo and Kim (1994) recently employed 3-D HOBEM in NWT simulation. Grilli et al. (1989) and Kashiwagi et al. (1998) used 2-D HOBEM.

As mentioned above, CPM had to use enormous number of panels to make accurate computation. However, the difficulty was overcome by vectorization and parallelization. Another breakthrough was made recently by Korsmeyer et al. (1996). They developed a fast algorithm for CPM computations. In contrast to the standard CPM of order N^2 for cpu time and order N^2 for memory, the pre-corrected-FFT has order $N \log N$ for cpu time and order N for memory. Thus, it can employ enormous amount of panels for discretization of boundary surface in a relatively short time period.

Another improvement was proposed by Wang et al. (1993). Their Longtank code is based on a multi-subdomain approach. The initial fluid domain is divided into adjacent subdomains in which a boundary integral equation has to be solved. The coupling between subdomains is provided by continuity of the potential and normal velocity. This results in a matrix which is larger than in the original method, but which is now block-diagonal instead of being full non-symmetric. The cpu time is drastically reduced compared to the standard full domain method. A variant of this multidomain method was proposed by De Haas and Zandbergen (1996).

12.3.2 Control of corners and edges

CPM fails to converge at the corner/edge of rectangular barge bottom or bottom of truncated cylinder. Lin et al. (1984, 2-D) addressed the difficulty caused by the singularity at the intersection of wavemaker and the free surface. They implemented new algorithm, based on Cauchy theorem, with complex potential. Cointe (1989, 2-D) investigated the foregoing same problem and concluded that the former work (Lin et al., 1984) has not provided definitive answer on the singularity problem along the section between the free surface and wavemaker surface.

Dommermuth and Yue (1987, 3-D) computed the section flows using CPM with quadratic and cubic splines.

Grilli and Svenson (1989, 2-D) employed the Berthod-Zaborowsky quadrature formula. They used a HOBEM to accurately integrate the rapidly varying velocity potential in the edge region when the wavemaker is moving.

Clément (1991, 2-D) used a double-node technique at the corner. Lee (1992, 3-D) avoided the singularities at the corners and edges of rectangular shape NWT by employing multi-node technique, while Boo (1993, 3-D) used the discontinuous element (Brebbia 1984). Sung et al. (1997, 3-D) used also HOBEM with a discontinuous element. Kashiwagi et al. (1998, 2-D), using HOBEM, placed double nodes at the intersections between the free surface and a freely floating body surface as well as the wavemaker.

12.3.3 Efficient matrix equation solvers

The influence matrix arising from the mixed BIE is full and asymmetric. A direct solution of the resulting system of linear equations requires an effort of $O(N^3)$ and the number of unknowns for 3-D problems is astronomical. Xü and Yue (1992) investigated a number of existing iterative methods for full and asymmetric influence matrix. They found that the generalized minimum residual (GMRES) iterative scheme (Saad and Schultz, 1986) with the symmetric successive overrelaxation (SSOR) preconditioner was effective. Thus, the computational effort of quadratic order boundary element method (QBEM) for solving the BIE was reduced to $O(N^2)$ and the CPU time βN^2. The coefficient β for the fully vectorized implementation of QBEM was less than twice that of vectorized CPM for a flow past a sphere. Thus they estimated $N \approx 150$ for QBEM and $N \approx 2000$ for CPM. Therefore, QBEM was approximately two orders of magnitude more efficient than CPM for the flow past a sphere. Ferrant (1994) also investigated a number of algorithms for his LBEM. Finally he adopted GMRES with a diagonal preconditioner. Another study of efficient solver (Prasad and Kane, 1994) for boundary element analysis of elastic structure concludes that GMRES with diagonal preconditioner out performs the other solvers.

Zou and Kim (1996) investigated the effectiveness of GMRES for computation of hydrodynamic force using HOBEM in the frequency domain. The advantage of GMRES in the above case was negligible. Sung et al. (1997) developed 3-D HOBEM and GMRES and effectively treated free surface problem intersecting with vertical boundaries.

12.3.4 Time-stepping integration

The free surface geometry and potential are updated by integrating Eq. (12.3) or (12.4) and Eq. (12.7) or (12.8) in time. A 4th-order Runge-Kutta scheme (RK4) is generally used for this purpose, but some authors also implemented the 4th-order Adams-Bashforth-Moulton method (ABM4) with success. Longuet-Higgins and Cokelet (1976, 2-D) numerically treated time stepping integration of Lagrangian formulation of the free-surface boundary conditions using both ABM4 and RK4 scheme. On the solid surface, the normal velocity at the next time-step is either changed from user prescribed wavemaker velocity or may be derived from the absorption scheme chosen at the down-tank region.

Dold and Peregrine (1984, 1986, 2-D) developed and used an accurate explicit time-stepping method by using a higher-order Taylor expansion in time for the free surface position in a Lagrangian formulation of 2-D flow. Dommermuth and Yue (1987), Grilli et al. (1989, 2-D), Nakayama (1990, 2-D) and Tanaka and Nakamura (1994, 2-D) adopted the same method.

For the time-stepping integrations, one needs to compute $\nabla \phi$ (ϕ_n, ϕ_s) on the free surface. A low-order BEM calculates ϕ_s by numerical differentiation of the potential over the free surface using a second-order finite-difference formula (Dommermuth and Yue, 1987, 3-D), weighted Arctang technique Clément (1996, 2-D) or cubic splines (Sen et al.,1989). HOBEM employs Eq. (12.13), as discussed previously.

Because the position and velocity potential of the free surface continuously change in time, the computation of influence coefficients is repeated, demanding enormous computational effort in 3-D problem. Therefore, Lee (1992) and Lee et al. (1994), Boo (1993) and Boo and Kim (1994), Ferrant (1994) and others used frozen coefficient approach. The influence coefficients were computed once during the first evaluation and fixed for the remaining three sub-time-steps. The time step was reduced as small as possible to keep it accurate. Ferrant (1994) used the method with sufficiently small time steps in his computations and confirmed that there was no noticeable effect on the accuracy of numerical results.

12.3.5 Control of saw-tooth instability

In nearly all computations, the wave profile, after a sufficiently long time, develops a saw-toothed appearance, in which the computed positions of the particles lie alternately above and below a smooth curve. Longuet-Higgins and Cokelet (1976) interpreted that it does not simply develop from rounding error but partly from some physical reason. They applied a smoothing procedure, such as 5 point smoothing, which has been frequently used by other researchers.

Dold and Peregrine (1984; 1986) developed an accurate and efficient time-stepping technique, and tested the technique for a number of cases. They found that there were no signs of the "saw-tooth" numerical instability.

Dommermuth and Yue (1987) performed a von Neumann stability analysis for RK4, with linearized free surface conditions, and obtained Courant condition:

$$\Delta t^2 \le \frac{8}{\pi} \frac{\Delta x}{g} \qquad (12.22)$$

where Δt is the time step, and Δx the local grid spacing. At least Eq. (12.22) should be satisfied for stability. Dommermuth and Yue (1987, 2-D) performed linearized von Neumann stability analysis for using ABM4 scheme and found that ABM4 is weakly unstable with a growth rate proportional to $\Delta t^6/\Delta x^3$.

Dommermuth and Yue (1987) proposed, a regridding idea to remove instability instead of smoothing by using CPM. They highlighted advantages of regridding: (1) it can potentially lower the energy of the system, (2) the arbitrariness in the choice of a smoothing formula is avoided, (3) smoothing cannot be easily applied at the intersections, and (4) the difficulties associated with the loss or gain of Lagrangian points as they cross the matching boundary are completely avoided. Application of 3-D CPM generally necessitates the use of automatic regridding for updating the moving boundary surface at each time step. Romate (1989) and Broeze (1992) employed "adaptive grid generation method" in their calculations. Dold (1992) proposes a high-order smoothing technique to remove saw-tooth instabilities. Tanizawa and Yue (1991; 1992, 2-D) developed a robust regridding method based on smoothing cubic spline and mesh function for their simulation of a plunging wave and its impact on a vertical wall.

Kashiwagi et al. (1998, 2-D) employed HOBEM for solving a new indirect method for the acceleration field, where fourth-order Runge-Kutta-Gill method was adopted at the time marching scheme. At each time step, rearranging the nodes on the free surface is conducted to avoid dense or sparse distribution of nodes, which is necessary for stable and accurate simulations for a long time.

12.3.6 Choosing time increment

In the majority of work, the size of Δt is usually constant during simulation. Nakayama (1990, 2-D) developed a variable time-stepping technique, i.e. the size of time increment is calculated every time step by controlling the remainder of truncated Taylor series. The Δt was calculated in such a way that the remainder is equal to a pre-assigned error limit. Clément and Mas (1995) implemented a dynamic control of their RK scheme by assigning a limit value to the maximum local Courant number around the domain boundary; the time-step is then continuously adapted to the higher local fluid velocity in the NWT.

Broeze (1992) found that stability restrictions based on linear theory (Dommermuth and Yue, 1987, 2-D) are not suitable in computations on highly nonlinear wave propagation problems. He considered that Nakayama's condition (1990) is rather arbitrary and lacks a theoretical basis. A straightforward condition was derived for the time increment from a perturbation analysis of the solution around the nonlinear solution.

12.3.7 Radiation (open) boundary

Longuet-Higgins and Cokelet (1976, 2-D) used spatial periodic lateral boundary conditions on the inflow and outflow boundary. Thus, no radiation boundary conditions were necessary. Many other papers employed the same approach. For instance, Xü and Yue (1992, 3-D) used spatial periodic boundary conditions both in the x- and y-directions.

Isaacson (1982, 2-D) employed a reflection condition on the outflow boundary because it was a transient problem. Similarly applied reflection boundary on the down-tank wall in order to simulate an asymmetric transient wave.

Dommermuth and Yue (1987) employed matched boundary conditions for the outer boundary in a circular wave tank. The inner solution is nonlinear while external solution is a linear transient wave on the radiation boundary.

Xu and Mori (1989) and Lee (1992) applied Sommerfeld/Orlanski radiation boundary for rectangular tank and matched the inner-solution to the linear potential in outer region. Yang and Ertekin (1992) extended Sommerfeld radiation condition to be applied to all the vertical walls of a rectangular tank and implemented the numerical radiation condition. Boo (1993) and Boo et al. (1994) used Sommerfeld/Orlanski boundary condition by taking the radiation as Dirichlet boundary. Since the wave was nonlinear and irregular an average phase speed was used in Sommerfeld/Orlanski condition. In addition, they used matching condition to a linear stretching potential as an outer solution.

Romate (1992) reviewed the existing methods of radiation-boundary conditions for four linear and nonlinear waves, and introduced the first- and second-order partial differential equation as the radiation-boundary condition for a linearized model. Romate (1992) performed a numerical implementation and stability analysis of the foregoing conditions for simulating the free surface waves.

Sierevogel and Hermans (1996, 2-D) developed a kind of Dirichlet to Neumann condition well suited to simultaneous wave/current simulations.

12.3.8 Absorbing beach and wavemaker absorber

Cointe et al. (1990, 2-D) following Baker et al. (1981), proposed a pragmatic technique similar to that used for an experiment. The free surface boundary condition Eq. (12.3) or (12.4) was modified by adding a damping term as shown in Eq. (12.7) or (12.8). The principle of this damping zone is to absorb the incident wave energy before it can reach the down-tank wall. If the absorption is too weak, part of incident energy will reach the wall and be reflected. Inversely, if the absorption is too strong, part of the energy will be reflected by the damping zone. The damping coefficient in Eq. (12.9) is preferably continuous and differentiable in the damping zone and it can be tuned to a characteristic wave frequency and characteristic wavenumber. The same idea is used for wavemaker absorber in front of the wavemaker. The energy corresponding to the reflected wave from the body can be similarly damped. The reflected wave is defined as the difference between the actual wave and the free incident wave. The damping is then applied to the difference. Cointe et al. found that the reflection coefficient is less than 2% with the parameters $\alpha = 1$ and $\beta = 1$ in Eq. (12.9) and that a damping zone of given characteristics absorbs reasonably well over a large range of frequencies. The pragmatic model of numerical beach does not involve any hypothesis concerning the steepness of the outgoing waves. The choice of $\alpha = 1$ and $\beta = 1$ seems to unfavorable behavior of the beach for low frequency waves.

All the damping coefficients employed by other researchers do not have the frequency and wave number as used by Cointe et al. (1990) above. Nakos et al. (1993) proposed Newtonian cooling terms for absorbing beach. Cao et al. (1993, 3-D) and Clément (1996) added a damping coefficient proportional to normal velocity in the dynamic free surface condition. This ensures analytically the positiveness of the energy flux from the fluid domain to outside, which is not always true with other kinds of extra dissipative terms. Grilli and Horillo (1997) similarly proposed to adapt the damping coefficient to the energy flux entering the beach. Boo and Kim (1995) added a damping coefficient multiplied with the velocity potential to the kinematic free surface boundary condition, in which the damping coefficient begins to increase gradually from zero at the front to unity at the end of the beach. Tanizawa and Naito (1997, 1998) used piston-type wavemaker, absorbing beach and wavemaker absorber identical to those of Cointe et al. (1990). Kashiwagi et al. (1998) employed the Newtonian cooling term similar to Nakos et al. (1993).

12.3.9 Active absorption of waves

Active absorbers were primarily developed for physical wave basins EWT (Milgram, 1970). The waves impinging on a steady wavemaker are totally reflected by it. The principle of active absorption is to move the wavemaker in such a way that radiated waves and reflected waves cancel each other at some distance from the paddle (Clément, 1988, 2-D). The instantaneous knowledge of the incident wave to be absorbed is necessary to determine, in real time, the driving velocity or position signal. This feedback can be provided by either a wave height signal taken on the paddle itself (Schoeffer et al., 1994) or at some distance from it (Milgram, 1970), or by the hydrodynamic force upon it (Salter 1976; Clément and Maisondieu, 1993). The common basis of these studies is the linear potential theory in frequency domain. The methods developed and optimized in this framework are then transposed to nonlinear time-domain applications.

Active absorption in NWT naturally provides us with an alternative method of numerical wave absorption. The control of the wave absorber is easier in numerical application than the experimental, due to the lack of noise and the availability of nonphysical quantities such as velocity potential and individual pressure components. Sophisticated control laws were proposed and are still in development (Skourup and Schoeffer, 1997, 2-D; Chatry et al., 1998). Active absorption is now extended to 3-D NWT (Skourup and Schoeffer, 1998, 3-D), where arrays of 2-D wavemakers are independently controlled by a two-dimensional filter.

Active wave absorbers are ideally fitted in the low frequency range and give excellent results. Numerical beaches as described above, feature an opposite behavior, i.e., they work well for short waves and fail to absorb long waves. The combined use of both methods, from these considerations was shown to give excellent results in the whole frequency range of interest (Ohyama and Nadaoka 1991, Clément, 1996, and Grilli and Horillo, 1997).

12.3.10 Accuracy test

The method of accuracy test used by Longuet-Higgins and Cokelet (1976) has been widely employed by others. NWT codes based on Laplace's equation structurally satisfy mass conservation; Nevertheless, testing it may be useful for detection of mistakes or weakness in programming. Energy and momentum conservation are more sensitive to the method parameter, which can be tuned by checking the accuracy of these global quantities (Tanizawa, 1996, 2-D).

12.4 Method of Wave Generation in NWT

To generate wave in the NWT, one is offered five possibilities; (1) space-periodic wave in tank; (2) an incident wave due to wavemaker motion; (3) numerical velocity or potential input on inflow boundary; (4) a prescribed incident wave in a wave tank; and (5) the Discrete Internal Singularities method.

12.4.1 Space periodic waves

The evolution of steady progressing waves is treated by imposing space-periodic boundary conditions of the wave on the tank walls in x- or x- and y-directions, depending on 2-D or 3-D waves. No open boundary condition is necessary for Longuet-Higgins and Cokelet (1976, 2-D). Similar works are given, for instance, in Vinje and Brevig (1981a, 2-D), Dold and Peregrine (1986, 2-D), and Xü and Yue (1992, 3-D).

12.4.2 Wavemaker driven waves

This mode of wave generation in NWT is very close to the EWT technique; waves are generated by user-prescribed motions of one of the tank end walls. Piston-type wavemaker was studied by Lin et al. (1984), which had a singularity problem at the intersection of the flap and free surface. Grilli et al. (1989) showed improved simulation avoiding the singularity applying HOBEM. Cointe and Boudet (1991), Clément and Mas (1995), and Tanizawa and Naito (1997; 1998) used piston-type wavemaker. She et al. (1992) employed flap-type wavemaker. Kashiwagi et al. (1998) used a plunger-type wavemaker. In these methods, the singularity at the intersections must be treated with care.

12.4.3 Feeding velocity on inflow boundary

The fluid in the wave tank is initially at rest and the incident wave is generated by feeding numerical velocity or velocity potential of Stokes 2nd-order or solitary waves on the up-tank boundary (Yang and Ertekin, 1992, 3-D). To remove the impulse effect, a modulation equation is used. Boo et al. (1994) used the numerical input data of Stokes 2nd-and 3rd-order irregular waves (Longuet-Higgins, 1963 and Pierson, 1993).

12.4.4 Prescription of incident wave

Incident wave field is prescribed in NWT and time marching simulation is started smoothly; a ramp function is applied to the boundary condition, which appears progressively by Ferrant (1994; 1995, 3-D), De Haas et al. (1995, 3-D), and Sung et al. (1998, 3-D).

12.4.5 Discrete internal singularities method

A discrete set of singularity is introduced in the NWT. The characteristics (strength, position) are imposed by the user. Brorsen and Larsen (1987, 2-D) initiated this technique using fixed sources on a vertical line. Grilli and Svendsen (1989), Ohyama and Nadaoka (1991, 3-D) and Tanaka and Nakamura (1994) adopted the same scheme. As a short illustration, Grilli and Svensen termed it: the "transparent" wavemaker. Recently, Clément (1999a) introduced the self explanatory "spinning dipole" technique which has the property to generate an unsymmetrical one-sided wave field. With this method, the transparent wavemaker generates waves forward, to the opposite tank end, but no waves backward, which is highly beneficial in this context.

12.5 Application of NWT

A variety of NWT technique was developed for simulation of waves, forces and motions of floating structures. We will review highly and weakly nonlinear waves, wave-wave interaction, wave forces on fixed body, motion-induced force, force due to an advancing submerged body in calm water, floating body motion due to waves.

12.5.1 Highly nonlinear waves

In their pioneering paper, Longuet-Higgins and Cokelet (1976, 2-D) simulated asymmetric overturning wave working in a conformally mapped space. They applied forcing surface pressure distribution on the free surface with MEL scheme to make the wave overturn. There is similar but improved research, for instance, Vinje and Brevig (1981a, 2-D) treated the same problem in the physical domain; Dold and Peregrine (1984, 2-D) computed overturning wave using a higher-order Taylor expansion in the time variable for the free surface position in a Lagrangian formulation.

Lin et al. (1984, 2-D) simulated a transient breaking wave due to large-stroke. Grilli et al. (1989, 2-D) simulated the asymmetric wave using HOBEM and a flap-type wavemaker.

An overturning deepwater wave was simulated by Xü and Yue (1992, 3-D). The technique is similar to the treatment by Longuet-Higgins and Cokelet (1976). QBEM (HOBEM) was used with an efficient matrix equation solver GMRES. Broeze (1992, 3-D) computed pre-breaking wave due to a submerged bottom structure using CPM.

Dommermuth et al. (1988, 2-D) experimentally generated plunging breakwater in a wave flume, simultaneously measuring the horizontal velocity of the piston-type wavemaker. This paper documents the Fourier coefficients of the piston velocity. Using the test data as input normal velocity on the inflow boundary, they simulated the plunging breakers. Excellent agreement was observed between the simulation and experiment. Cointe (1989, 2-D) and Beck et al. (1993, 2-D) simulated the same plunging breakers inputting the same normal velocity on the inflow boundary as given above.

Dold and Peregrine (1986, 2-D) generated breaking waves by computing evolution of modulated wave train in a wave tank with space-periodic boundary conditions on the up- and down-tank walls.

Skourup (1994, 2-D) also computed a similar evolution of modulated wave. He compared the result with Stokes 5th-order wave kinematics and found the result was closer to experiment than the theory. The method can simulate freak wave, which cannot be modeled by mathematical theory.

Cointe and Boudet (1991, 2-D) simulated steep asymmetric wave starting with bichromatic modulated incident wave, which was in excellent agreement with experiments. They also investigated the characteristics of third-order Schrödinger analytical model, and found that the waves were weakly nonlinear, whereas the fully nonlinear numerical and physical evolution of bichromatic waves created highly nonlinear waves.

Wang et al. (1993, 2-D) simulated breaking wave due to evolution of an incident modulated wave in a long tank. They created NWT code "LONGTANK" employing a multi-sub-domain approach.

12.5.2 Simulation of weakly nonlinear waves

Weakly nonlinear waves are obtainable in principle by perturbation based theory such as Stokes higher-order interacting waves by Longuet-Higgins (1963) and Pierson (1993).

She et al. (1992, 2-D) produced weakly nonlinear numerical waves which agree with Stokes wave theory. These authors employed a flap-type wavemaker, a damping beach and RK4 for time-stepping with a special care for the intersection flow between the flap and wave surface.

Yang and Ertekin (1992, 3-D), using CPM, simulated nonlinear waves by feeding numerical data of solitary waves and Stokes second-order regular waves on the inflow boundary.

Boo et al. (1994, 3-D), employing HOBEM, computed fully nonlinear irregular waves by feeding theoretical Stokes 3rd-order-like nonlinear interacting wave (Pierson, 1993), at the inflow boundary, for the repeating period of 128 s. This paper uses full-nonlinear free surface condition with semi-Lagrangian marching scheme.

Contento and Casole (1995, 2-D) generated weakly nonlinear waves with a flap-type wavemaker. With large amplitude of sinusoidal strokes in shallow wave tank, they could generate nonlinear regular waves of steepness 1/15 to 1/10. Orlanski open boundary was installed. Marching procedure employs semi-Lagrangian with RK4 algorithm. The free surface was regridded following Dommermuth and Yue (1987) with a special care using interpolation-extrapolation scheme. Comparison of simulations by Lagrangian with a second-order time stepping and semi-Lagrangian with RK4 algorithm is given. Both simulations are in good agreement. Since the former used smoothing while the latter did not, the agreement between the two indicates that the filtering does not alter the energy of the system. More detailed works and discussions were given, including the effect of grid size on the resolution and energy conservation.

Clément and Gil (1997) numerically simulated head to head collision of monochromatic wave packets. The interaction between two colliding monochromatic wave packets in water of finite depth is investigated through these numerical experiments. Each end of this flume was a piston acting as a wavemaker, and could be switched to be a wave absorber when the wave packet emitted from the other reached it. By doing so waves are not reflected. The main effect of the nonlinear interaction during head-on collision consists in a phase lag of the wave train due to a decrease of the phase velocity. The paper investigated the influence of the frequency, amplitude and wave number of the two incident waves on this phase lag by numerical experiments. The previous analytical results by Longuet-Higgins and Phillips (1962) related to the infinite water depth case were recovered and their validity was extended to finite water depth.

12.6 Wave Force

12.6.1 Wave force on fixed body

Vinje and Brevig (1981b, 2-D) computed wave forces on a fixed submerged circular cylinder under the action of breaking highly nonlinear waves.

Clément and Mas (1995, 2-D) simulated force on a submerged fixed body due to a solitary wave. The NWT is featured with the coupled piston-beach wave absorber (Clément, 1996). Motion of the piston-type wavemaker was derived from the solution of Boussinesq equation. The procedure employs LBEM, MEL and RK4 algorithm for time-stepping. An improved method was used for computing tangential velocity on the free surface, in such a way that no saw-tooth instabilities were observed. Regridding was performed on the vertical boundaries where the wavemaker and free surface make intersections. They investigated interactions between the solitary wave and the horizontal force of a submerged circular cylinder. The result agrees with other experimental and numerical simulation. Using the same scheme, Clément (1991) investigated the interaction of solitary waves with a fixed free-surface piercing cylinder.

Yang and Ertekin (1992, 3-D) computed forces on a surface-piercing bottom-mounted cylinder in solitary wave and Stokes 2nd-order waves. They used a rectangular wave tank with CPM and Sommerfeld radiation boundaries for all the tank walls. The Stokes wave force was compared with experimental data and 2nd-order diffraction theory against fundamental frequency. NWT result agreed with experiment better than the diffraction theory.

Boo and Kim (1995, 3-D) simulated Yang and Ertekin's problem. Rectangular NWT boundaries were discretized by HOBEM. Semi-Lagrangian marching with ABM4 and RK4 was used. An absorbing beach was implemented in addition to open boundary condition and matched with outflow velocity field outside. The linear boundary condition was imposed on the free surface, while satisfying an exact fully nonlinear boundary condition on the intersection flow between the wave surface and body boundary. The diffraction forces were found to contain higher-order harmonics to 3rd-order and the horizontal force coefficient was larger than the linear solution.

Boo and Kim (1996, 3-D) simulated Yang and Ertekin's Problem (1992, 3-D) again and found that the maximum force is generally greater than the linear force. Boo and Kim (1997, 3-D) simulated diffraction force of a vertical truncated column in the linear and Stokes 2nd-order nonlinear irregular waves. The NWT technique was practically the same as Boo and

Kim (1996, 3-D) above, except for imposing the linear and fully nonlinear free surface boundary conditions for the linear and Stokes 2nd-order wave. The saw-tooth instability was removed by applying the method of Longuet-Higgins and Cokelet (1976). The accuracy test found a small error. However, these simulations could not be verified because of the lack of experimental data.

Ferrant (1994, 3-D) computed forces on a submerged sphere. It employs triangular LBEM, MEL, and RK4 algorithm with 'frozen' coefficients for time-stepping, a GMRES matrix solver with diagonal pre-conditioning and annular absorbing beach for the absorption of diffracted waves. The incident wave is given by the stream function theory of Rienecker and Fenton (1981), and used to describe the initial conditions as well as the time dependent boundary conditions on the outer surface of the computational domain.

Using above computational procedure, Ferrant (1995, 3-D) simulated Yang and Ertekin's problem described in the foregoing. Simulation was conducted to compute wave run-up and horizontal force.

De Haas et al. (1995, 3-D) also simulated Yang and Ertekin's problem using CPM and MEL procedure. Near the cylinder the procedure employs Eulerian grid description, while far from the cylinder the collocation points moved in a Lagrangian fashion. An exact periodic steady wave (Rienecker and Fenton theory) was used for the incident wave. The paper computed the maximum run-up and maximum horizontal force and compared with other numerical and experimental data.

Lalli et al. (1996, 3-D) simulated 3-D diffraction by a bottom-mounted column in both submerged and surface-piercing columns. The velocity potential was split into the Stokes wave potential and the scattered term to avoid the need for generating the incident wave train. They employed CPM for indirect BIE with a desingularized scheme (Cao et al., 1991), time marching in semi-Lagrangian with RK algorithm on the free surface, and a damping beach suggested by Baker et al. (1989). The results were compared with the theoretical second-order run-up by Kriebel (1990; 1992) and experimental force by Hogben and Standing (1975).

Ferrant (1998a, 3-D) applied fully nonlinear BIE for the simulation of wave-body-current interaction in the time domain. The potential is separated into explicitly known and unknown part. Nonlinear incident regular waves due to Rienecker and Fenton (1981), LBIE, an RK4 algorithm for time-stepping and an absorbing beach were employed. The effect of nonlinear waves was investigated varying the wave steepness and Froude number. The same problem was treated by Cheung et al. (1996) and others as shown in the following section. Cheung et al. (1996) employed the formulas of 1st- and 2nd-order forces based on the perturbation expansion of boundary conditions with respect to wave steepness and current, and simulated the interaction in the time domain.

Ferrant (1998b) applied the above method to predict the force and run-up in the time and frequency domain. He compared a harmonic analysis of his fully nonlinear NWT results with the EWT data from Huseby and Grue (1998) for the bottom-mounted cylinder in regular nonlinear waves. Overall fair and good agreements were found.

Sung et al. (1998, 3-D) simulated wave force and wave elevations on the wall of the bottom-mounted column. They used 3-D HOBEM and Rienecker-Fenton theory (1981) as the incident wave and GMRS solver. Differently from Ferrant (1998a), they did not separate the velocity potential. RK4 for marching integration and 5-points smoothing algorithm for saw-tooth instability was employed.

Celebi et al. (1998, 3-D) investigated diffraction of bottom-mounted and truncated column in monochromatic waves. They employed DBIEM (Beck et al., 1993), the MEL procedure, feeding the wave particle velocity on the inflow boundary, implementing the absorbing beach similar to that of Israeli and Orszg (1981), and regridding similar to that of Dommermuth and Yue (1987). The pressure and run-up were compared more favorably with experimental data than the 2nd-order diffraction theory.

12.6.2 Hybrid NWT for wave force on fixed body

Isaacson and Cheung (1990, 2-D) computed the 1st-and 2nd-order force of a submerged circular cylinder in the time domain and compared with other theories and experiments. They formulated 1st- and 2nd-order free surface boundary conditions and corresponding BIEs based on a perturbation scheme, which were solved using CPM NWT. The incident wave and scatter wave potentials were treated separately to develop a 1st-order radiation condition on both the up-wave and down-wave control surface.

Isaacson and Cheung (1992, 3-D) computed the 1st- and 2nd-order forces on a bottom-mounted column in the time domain and compared them with other theoretical data. The basic methodology is similar to the above. Both free surface boundary conditions and radiation conditions are satisfied to 2nd-order. The solution was separated into a known incident potential corresponding to a Stokes 2nd-order wave field and a scattered potential. The radiation condition applied to the scattered potential was modified to include a time dependent celerity in order to account for the motions of forced and free waves at 2nd-order. The initial condition corresponds to an undisturbed Stokes 2nd-order wave field in the time domain, and the scattered potential was developed in time and space through the gradual imposition of the body surface boundary condition over one wave period. 3-D CPM NWT was used.

Cheung et al. (1996, 3-D) computed the 1st-order oscillatory force and 2nd-order mean drift force on a vertical bottom-mounted column in the linear sinusoidal wave and current. The model was formulated to 1st- and 2nd-order in wave slope, both for linear force and 2nd-order mean force, and computed in NWT similar to the work of Isaacson and Cheung (1992, 3-D). The current was shown to significantly affect the steady drift force and run-up.

Buechmann et al. (1997, 3-D) simulated wave run-up due to the interaction of wave, current and bottom-mounted column. The model was formulated to 2nd-order in wave steepness and 1st-order in current. Active wave absorbers were applied at the radiation boundaries. The numerical results agree with the previous analytical and numerical data. It was shown that the 2nd-order terms can be significantly important for run-up calculation.

Kim and Kim (1997) calculated the 1st-order wave force and 2nd-order mean drift force, wave drift damping, wave field and wave run-up of a bottom-mounted column in 1st-order wave and small current. The model was formulated to 1st-order in wave steepness and in the current speed, while the mean drift force being to 2nd-order in steepness was treated independently. The 3-D NWT and HOBEM were used. The current effects were more significant in the drift force and wave run-up than in the 1st-order wave force.

Ferrant (1998b, 3-D) simulated forces on a bottom-mounted column due to a 2nd-order wave packet, which is close to EWT data. He employed 2nd-order perturbation solution similar to Isaacson and Cheung (1992, 3-D). The 1st-order problem is solved using the 1st-order wave from the packet and consequently the 2nd-order problem is solved using the 1st-order solution and the 2nd-order wave from the packet. Unlike the others, he used the total velocity potential, and a numerical wave instead of theoretical Stokes 2nd-order wave. The NWT result agrees with test data. It employs 3-D LBEM, time-marching with RK4 and absorbing beach.

12.6.3 Force due to forced heaving body

Dommermuth and Yue (1987) simulated 3-D nonlinear flow due to a forced heaving motion of an axisymmetric body starting from rest. CPM and Rankine ring source with an MEL scheme were used. Regridding was applied to avoid saw-tooth instability, and a nonlinear solution inside was matched with linear transient wave-field outside.

Forced heaving of a vertical truncated circular cylinder was also simulated by Lin et al. (1984) and the same was done by Yang et al. (1989, 3-D). The former used Rankine ring source, while the latter used Rankine sources with suitable images. Yang et al. (1989) applied a combined use of LBEM and time-marching finite difference to improve the numerical technique. They concluded that the nonlinear wave force could be accurately evaluated with less computer time than the use of LBEM alone.

The nonlinear forced heaving motion of 2-D ship shape body was simulated by Kashiwagi (1996). MEL with RK4-Gill algorithm was applied, while the BIE for the velocity and acceleration fields were solved by applying QBEM. Stable simulations over a large number of periods were made using an efficient damping beach similar to Nakos et al. (1993), and regridding was performed to avoid the cluster of nodal points on the free surface. The results were compared with other theory and experiment. Good agreement was obtained.

12.6.4 Submerged body advancing in calm water

Xu and Mori (1989, 3-D) investigated the waves and forces due to a submerged sphere advancing uniformly with their CPM NWT. Lee et al. (1994, 3-D) simulated the waves and forces due to a submerged sphere advancing at transient and constant speed employing 3-D HOBEM, a semi-Lagrangian procedure, and Sommerfeld/Orlanski radiation boundaries at the inflow and outflow boundaries of the rectangular NWT. Saw-tooth instabilities were smoothed applying the Longuet-Higgins and Cokelet method (1976, 2-D). The result was compared with a linear analytical solution by Bessho (1957) and the result of Xu and Mori (1989). Lee and Fang (1997, 3-D) applied the above technique to the motion of spheroids and compared with other analytical solution.

12.7 Floating Body Motion

Isaacson (1982, 3-D) simulated the interaction of nonlinear waves with a surface-piercing structure of 3-D arbitrary shape. The interaction was treated as a transient problem with known initial conditions. Reflection condition was used on the radiation boundary because the wave was transient. The intersection between the free surface and structure was not treated.

As given under typical formulation in potential theory-floating body motion in Sec. 12.2.3, one needs to evaluate the ϕ_t. The simplest way is to use a backward finite difference scheme in time. However, the scheme makes a solution inherently unstable. An alternative is to solve the boundary value problem for the ϕ_t which was first proposed by Vinje and Brevig (1981C, 2-D). They decomposed the acceleration field into 4 modes corresponding to the unit acceleration of 3 body motions (heave, sway and roll) and centripetal-like acceleration coming from the velocity field, then they solved the boundary value problem corresponding to each mode. The solutions of each mode were used with the equations of the floating body motions to determine the 3 modes of acceleration.

Tanizawa and Sawada (1990, 2-D) proposed a more rational method to solve simultaneous equations in the acceleration fields. The authors introduced an "implicit boundary condition" derived from the kinematic boundary condition of the body and the equation of body motions, and showed that simultaneous equations of ideal fluid and floating body motions could be solved without decomposition. Tanizawa (1995, and 1996, 2-D) introduced the nonlinear acceleration potential and derived the exact body boundary condition for the acceleration field. Using the exact body surface boundary condition, he simulated free heave motion in a closed domain and confirmed the simulated wave field well satisfies conservation laws of volume, momentum and energy. He extended the above formulation to the multiple fluid domain and fluid-body interaction problem including numerical solution. It also adds an alternative formulation for numerical methods of the implicit boundary condition.

Cointe et al. (1990, 2-D) computed linear and nonlinear motion of rectangular barge in a NWT. The wave tank had an absorbing beach and a wavemaker absorber. The authors damped the reflected wave from the floating structure before the waves reach the wavemaker.

Van Daalen (1993, 2-D) also formulated nonlinear wave-body interaction in the acceleration field and computed nonlinear responses of circular and rectangular cylinder using implicit boundary condition.

Sen (1993, 2-D) simulated ship motion in waves in a fully nonlinear way. He employed LBEM, semi-Lagrangian scheme, imposing the particle velocity of Airy wave on the inflow boundary or piston-type wavemaker, and Orlanski radiation boundary. He encountered numerical instabilities in the section between the free surface and wavemaker and between the wave and body surfaces and saw-tooth instability. He found that the use of backward finite difference scheme for the temporal derivative of the velocity potential ϕ_t was the source of error for divergence, and tried to overcome the instabilities applying simple techniques.

Tanaka and Nakamura (1994, 2-D) simulated the slow drift motion of a floating moored structure. The method employs MEL, LBEM, submerged pulsating source wavemamker and Sommerfeld radiation boundary.

Contento (1996, 2-D) simulated ship's heave-sway-roll in NWT. This work is a continuation from the previous research (Contento and Casole, 1995) for generation of progressing waves in 2-D wave tank. The procedure employed a flap-type wavemaker, the Orlanski open boundary condition, a semi-Lagrangian scheme, and a long wave tank to avoid undesirable reflection. The ϕ_t was derived from the computation of the material derivative of the velocity potential. Regular waves with a high steepness of 1/16 were generated. The fundamental frequencies of the waves were set to 2.0, 2.5 and 3.0 times the natural roll frequency. He found sub-harmonic or parametric behavior consistently in the rolling motions, whereas the ship was steadily oscillating in the sway and heave mode.

Tanizawa and Naito (1997, 2-D) studied parametric roll motion of a ship in NWT developed by Tanizawa (1995 and 1996), employing LBEM, the MEL procedure, a piston-type wavemaker, absorbing beach and wavemaker absorber. The steepness of the generated regular waves varied from 0.0018 to 0.004. The paper did not have the saw-tooth instability. The volume convergence error was about 0.016%. Following the experiment, they found qualitative and quantitative agreements. Tanizawa and Naito (1998) simulated chaotic roll motion of a 2-D unstable floating body with small negative metacentric height using the above NWT.

Wu and Eatock Taylor (1996) proposed a new technique to solve the velocity and acceleration field simultaneously instead of computing velocity and acceleration field separately. The merit of the method is to compute motion faster than that of Tanizawa, while the disadvantage is that it cannot compute the pressure distribution on the body surface unless it solves another boundary value problem for the auxiliary function. Wu and Eatock Taylor (1996), Kashiwagi et al. (1998, 2-D) solved the 2-D floating body motions in regular waves. It employed MEL with the RK4 algorithm, a plunger-type wavemaker, an absorbing beach, HOBEM and regridding. No saw-tooth instability was found. The model test was conducted in a small EWT, which was simulated in an NWT. The wave reflections were repeated between a floating body and the wavemaker. The overall agreement between NWT and EWT was very good.

Berkvens (1998, 3-D) developed a 3-D NWT for nonlinear simulation of floating body motions and simulated free motions of a floating sphere, and checked the accuracy of the method. He applied the implicit boundary condition to solve for the acceleration field.

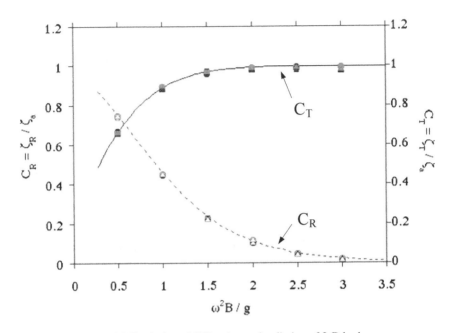

Fig. 12.3. Benchmark NWT solution of diffraction and radiation of 2-D body.

12.8 Benchmark Test

ISOPE NWT Group invited interested researchers to participate in conducting a benchmark test of numerical wave absorption (1997–1998), the results of which were discussed at the 1st Workshop of ISOPE NWTG (1998), and presented at ISOPE's 1999 Conference by Clément (1999b). The NWTG further conducted a benchmark test of 2-D numerical radiation (1998–1999), the results of which were discussed at the 2nd Workshop of ISOPE NWT in Brest, France (1999). The result is in the proceedings of ISOPE 2000 Conference. The NWTG conducted a new bench mark test of 2-D numerical diffraction (1999–2000), that was discussed at the 3rd Workshop of ISOPE-NWTG in 2000 (Seattle). Again the above results were presented in the proceedings of ISOPE 2001 Conference. In this manner, the benchmark test approach established milestones steadily. Figure 12.3 illustrates a result of benchmark solution for the freely floating 2-D body.

References

Journal references:

Baker, G.R., Meiron, D.J. and Orszag, S.A. (1989). Generalized Vortex Methods for Free Surface Flow Problems. Ii: Radiating Waves, J. Sci. Comp., Vol. 4, pp. 237–259.

Berkvens, R.J.F. (1998). Floating Bodies Interacting with Water Waves, Ph.D. Thesis, University of Twente.

Bessho, M. (1957). On the Wave Resistance Theory of a Submerged Body, 60th Anniversary. Series, Soc. Naval Arch. Japan, Vol. 2, pp. 135–172.

Boo, S.Y. (1993). Application of Higher Order Boundary Element Method to Steady Ship Wave Problem and Time Domain Simulation of Nonlinear Gravity Waves, Ph.D. Dissertation, Ocean Engineering Program, Texas A & M University.

Boo, S.Y., Kim, C.H. and Kim, M.H. (1994). A Numerical Wave Tank for Nonlinear Irregular Waves by 3-D Higher Order Boundary Element Method, Int. J. Offshore Polar Eng., ISOPE, Vol. 4, No. 4, pp. 265–272.

Broeze, J. and Romate, J.E. (1992). Absorbing Boundary Conditions for Free Surface Wave Simulation with a Panel Method, J. Comp. Phys., Vol. 99, pp. 146–158.

Brorsen, M. and Larsen, J. (1987). Source Generation of Nonlinear Gravity Waves with the Boundary Integral Equation Method, Coastal Eng., 11, pp. 93–113.

Cao, Y., Schultz, W.W. and Beck, R.F. (1991). Three-Dimensional Desingularized Boundary Integral Methods for Potential Problems, Int. J. Num. Meth. Fluids., Vol. 12, pp. 785–803.

Celebi, M.S., Kim, M.H. and Beck, R.F. (1998). Fully Nonlinear 3-D Numerical Wave Tank Simulation, J. Ship Res., Vol. 42, No. 1, pp. 33–45.

Cheung, K.F., Isaacson, M. and Lee, J.W. (1996). Wave Diffraction around a Three-Dimensional Body in a Current, ASME J. Offshore Mech. Arc. Eng., Vol. 118, No. 4, pp. 247–252.

Clément, A.H. (1996). Coupling of Two Absorbing Boundary Conditions for 2-D Time-Domain Simulations of Free Surface Gravity Waves, J. Comp. Phys., Vol. 126, pp. 139–151.

De Haas P.C.A.and Zandbergen P.J. (1996). The Application of Domain Decomposition to Time-Domain Computations of Nonlinear Water Waves with a Panel Method, J. Comp. Phys., 129, pp. 332–344.

Dold, J.W. (1992). An Efficient Surface-Integral Algorithm Applied to Unsteady Gravity Waves, J. Comp. Phys., 103, pp. 90–115.

Dommermuth, D.G. and Yue, D.K.P. (1987). Numerical Simulation of Nonlinear Axisymmetric Flows with a Free Surface, J. Fluid Mech., Vol. 178, pp. 195–219.

Dommermuth, D.G., Yue, D.K.P., Lin, W.M., Rapp, R.J., Chan, E.S. and Melville, W.K. (1988). Deep-Water Plunging Breakers a Comparison Between Potential Theory and Experiments, J. Fluid Mech., Vol. 189, pp. 423–442.

Grilli, S.T., Skourup, J. and Svendsen, I.A. (1989). An Efficient Boundary Element Method for Nonlinear Water Waves, Engineering Analysis with Boundary Elements., Vol. 6, No. 2, pp. 97–107.

Grilli, S.T. and Horrillo J. (1997). Numerical Generation and Absorption of Fully Nonlinear Periodic Waves, J. Eng. Mech., ASCE. Vol. 123, No. 10, pp. 1060–1069.

Hess, J.L. and Smith, A.M.O. (1964). Calculation of Non-Lifting Potential Flow About Arbitrary Three-Dimensional Bodies, J. Ship Res., Vol. 8, pp. 22–44.

Hess, J.L. (1979). A Higher Order Panel Method for Three Dimensional Potential Flow, Douglas Report N62269-77-C-0437.

Isaacson, M. (1982). Nonlinear Wave Effects on Fixed and Floating Bodies, J. Fluid Mech., Vol. 120, pp. 267–281.

Isaacson, M. and Cheung, K.F. (1990). Time Domain Second-Order Diffraction in Three Dimensions, J. Waterways Port Coastal Ocean Eng., ASCE, Vol. 116, No. 2, pp. 191–210.

Isaacson, M. and Cheung, K.F. (1992). Time-Domain Solution for Second-Order Diffraction, J. Waterways Port Coastal Ocean Eng., ASCE, Vol. 116, No. 2, pp. 496–516.

Israeli, M. and Orszag, S.A, (1981). Approximation of Radiation Boundary Conditions, J. Comp. Phys., Vol. 41, pp. 115–195.

Kashiwagi, M. (1996). Full-Nonlinear Simulations of Hydrodynamic Forces on a Heaving Two-Dimensional Body, J. Soc. Naval Arch., Japan, Vol. 180, pp. 373–381.

Kashiwagi, M., Momoda, T. and Inada, M. (1998). A Time-Domain Nonlinear Simulation Method for Wave-Induced Motions of a Floating Body, J. Soc. Naval Arch., Japan, Vol. 84, pp. 143–152.

Kim, C.H., Randall, R.E., Boo, S.Y. and Krafft, M.J. (1992). Kinematics of 2-D Transient Water Waves Using Laser Doppler Anemometry, J. Waterway Port Coastal Ocean Eng., ASCE , Vol. 118, No. 2, pp 147–165.

Kim, C.H., Clement, A.H. and Tanizawa, K. (1999). Recent Research and Development of Numerical Wave Tanks — A Review, Int. J. Offshore Polar Eng., Vol. 9, No. 4, pp. 241–256.

Kriebel, D.L. (1990). Nonlinear Wave Interaction with a Vertical Circular Cylinder. Part I: Diffraction Theory, Ocean Eng., Vol. 17, No. 1, pp. 345–377.

Kriebel, D.L. (1992). Nonlinear Wave Interaction with a Vertical Circular Cylinder. Part II: Wave Run-up, Ocean Eng., Vol. 19, No. 1, pp. 75–99.

Lalli, F., Di Mascio, A. and Landrini, M. (1996). Nonlinear Diffraction Effects Around a Surface-Piercing Structure, Int. J. Offshore Polar Eng., ISOPE, Vol. 6, No. 2, pp. 104–111.

Lee, C.C. (1992). Higher Order Boundary Element Method for Three Dimensional Fully Nonlinear Wave Generated by a Submerged Moving Body in Time Domain Simulation, Ph.D. Dissertation, Ocean Engineering Program, Texas A & M University.

Lee, C.C., Liu, Y.H. and Kim, C.H. (1994). Simulation of Nonlinear Waves and Forces Due to Transient and Steady Motion of Submerged Sphere, Int. J. Offshore Polar Eng., ISOPE, Vol. 4, No. 3, pp. 174–182.

Lee, C.C. and Fang, M.C. (1997). Numerical Simulation for the Wave Resistance of a Submerged Spheroid, Proc. 7th Int. Offshore Polar Eng Conf., Honolulu, Vol. 3, pp. 116–121.

Liu, Y.H. (1988). Analysis of Fluid-Structure Interaction by Using Higher Order Boundary Elements in Potential Problems and Its Application in Coupling Vibrations of Bending and Torsion of Ships, Ph.D. Thesis, Shanghai Jiao Tong University.

Liu, Y.H., Kim, C.H. and Kim, M.H. (1993). The Computation of Mean Drift Forces and Wave Run-up by Higher-Order Boundary Element Method, Int. J. Offshore Polar Eng., Vol. 3, No. 2, pp. 101–106.

Longuet-Higgins, M.S. and Phillips, O.M. (1962). Phase Velocity Effects in Tertiary Wave Interaction, J. Fluid Mech., Vol. 12, pp. 321–332.

Longuet-Higgins, M.S. (1963). The Effect of Nonlinearities on the Statistical Distributions in the Theory of Sea Waves, J. Fluid Mech., Vol. 17, pp. 459–480.

Longuet-Higgins, M.S. and Cokelet, E.D. (1976). The Deformation of Steep Surface Waves on Water — Part 1. A Numerical Method of Computation, Proc. Roy. Soc. London., A, 350, pp. 1–26.

Milgram, J. (1970). Active Water-Waves Absorbers, J. Fluid Mech., Vol. 43, No. 4, pp. 845–859.

Myrhaug, D. and Kjeldsen, S.P. (1984). Parametric Modeling of Joint Probability Density Distribution for Steepness and Asymmetry in Deep Water Waves, Appl. Ocean Res., Vol. 6, No. 4, pp. 207–220.

Nakayama, T. (1990). A Computational Method for Simulating Transient Motions of an Incompressible Inviscid Fluid with a Free Surface, Int. J. Num. Meth. in Fluids., Vol. 10, pp. 683–695.

Newman, J.N. (1989). The Numerical Towing Tank — Fact or Fiction?, Schiffstechnik Bd. 36, pp. 155–168.

Orlanski, I. (1976). A Simple Boundary Condition for Unbounded Hyperbolic Flows, J. Comp. Phys., Vol. 72, pp. 373–421.

Ohyama, T. and Nadaoka, K. (1991). Development of a Numerical Wave Tank for Analysis of Nonlinear and Irregular Wave Field, Fluid Dyn. Res., Vol. 8, pp. 231–251.

Pierson, W.J. (1993). Oscillatory Third Order Perturbation Solutions for Sums of Interacting Long Crested Stokes Waves on Deep Water, J. Ship Res., Vol. 37, No. 4, pp. 354–383.

Prasad, K.G.and Kane, J.H. (1994). Preconditioned Krylov Solvers for BEA, Int. J. Num. Meth. in Eng., Vol. 37, pp. 1651–1672.

Rienecker, M.M. and Fenton, J.D. (1981). A Fourier Approximation Method for Steady Water Waves, J. Fluid Mech., Vol. 104, pp. 119–137.

Romate, J.E. (1989). The Numerical Simulation of Nonlinear Gravity Waves in Three Dimensions Using a Higher Order PANEL Method, Ph.D. Thesis, Twente University, The Netherlands.

Romate, J.E. (1992). Absorbing Boundary Conditions for Free Surface Wave, J. Comp Phys., Vol. 99, pp. 135–145.

Saad, Y. and Schultz, M.H. (1986). GMRES: A Generalized Minimal Residual Algorithm for Solving Nonsymmetric Linear Systems, SIAM J. Sci. Stat. Comp., Vol. 7, pp. 856–869.

Sen, D. (1993). Numerical Simulation of Motions of Two-Dimensional Floating Bodies, J. Ship Res., Vol. 37, No. 4, pp. 307–330.

Sierevogel, L.M. and Herman, A.J. (1996). Absorbing Boundary Condition for Floating Two-Dimensional Objects in Current and Waves, J. Eng. Math., 30, pp. 573–586.

Tanizawa, K. and Sawada, H. (1990). A Numerical Method for Nonlinear Simulation of 2-D Body Motions in Waves by means of BEM, J. Soc. Nav. Arch. Japan, Vol. 168, pp. 223–228.

Tanizawa, K. (1995). A Nonlinear Simulation Method of 3-D Body Motions, J. Soc. Nav. Arch. Japan, Vol. 178, pp. 179–191.

Webster, W.G. (1975). The Flow About Arbitrary Three-Dimensional Bodies, J Ship Res., Vol. 19, pp. 206–218.

Xu, Q. and Mori, K.H. (1989). Numerical Simulation of 3-D Nonlinear Water Waves by Boundary Element Method, J. Soc. Nav. Arch. Japan, Vol. 165, pp. 9–15.

Yang, C. and Ertekin, R.C. (1992). Numerical Simulation of Nonlinear Wave Diffraction by a Vertical Cylinder, J. Offshore Mech. Arctic Eng., Vol. 114, pp. 36–44.

Zou, J., Xu, Y. and Kim, C.H. (1998). Ringing of ISSC TLP Due to Laboratory Storm Seas, Int. J. Offshore Polar Eng., Vol. 8, No. 2, pp. 81–89.

Book references:

Brebbia, C.A., Telles, J.C.F. and Wrobel, L.C. (1984) Topics in Boundary Element Research, Vol 1: Basic Principles and Applications, Springer-Verlag, Berlin.

Clément, A.H. (1988) Dynamic Absorption of Outgoing Waves in the Computation of 2-D Unsteady Hydrodynamics, Computer Modeling in Ocean Engineering, Schrefler & Zienkiewicz Eds pp.127–134.

Proceedings references:

Baker, G.R., Merion, D.J. and Orszag, S.A. (1981). Applications of a Generalized Vortex Method to Nonlinear Free-Surface Flows, 3rd Int. Conf. Num. Ship Hydrodynamics, Paris, pp. 179–191.

Beck, R.F., Cao, Y. and Lee, T.H. (1993). Fully Nonlinear Water Wave Computations Using the Desingularized Method, Proc. 6th Int. Conf. Num. Hydrodynamics, pp. 3–20.

Boo, S.Y. and Kim, C.H. (1994). Simulation of Fully Nonlinear Irregular Waves in a 3-D Numerical Wave Tank, Proc. 4th Int. Offshore Polar Eng. Conf., Osaka, ISOPE, Vol. 3, pp. 17–24.

Boo, S.Y. and Kim, C.H. (1995). Weakly Nonlinear Diffraction Due to Vertical Cylinders in a 3-D Numerical Wave Tank, Proc. 5th Int. Offshore Polar Eng. Conf., The Hague, ISOPE, Vol. 3, pp. 19–25

Boo, S.Y. and Kim, C.H. (1996). Fully Nonlinear Diffraction Due to a Vertical Circular Cylinder in a 3-D HOBEM Numerical Wave Tank, Proc. 6th Int. Offshore Polar Eng. Conf., Los Angels, ISOPE, Vol. 3, pp. 23–30.

Boo, S.Y. and Kim, C.H. (1997). Nonlinear Irregular Waves and Forces on Truncated Vertical Cylinder in a Numerical Wave Tank, Proc. 7th Int. Offshore Polar Eng. Conf., Honolulu, ISOPE, Vol. 3, pp. 76–84.

Broeze, J. (1992). Computation of Breaking Waves with a Panel Method, Proc. 23rd Int. Coastal Eng. Conf., Vol. 1, pp. 89–102.

Büchmann, B., Skourup, J. and Cheung, K.F. (1997). Run-up on a Structure Due to Waves and Currents, Proc. 7th Offshore Polar Eng. Conf., Honolulu, ISOPE, Vol. 3, pp. 48–55.

Cao, Y., Beck, R. and Schultz, W.W. (1993). An Absorbing Beach for Numerical Simulations of Nonlinear Waves in a Wave Tank, Proc. 8th Int. Workshop Water Waves Floating Bodies, Oita, pp. 17–20.

Chatry, G., Clément, A.H. and Gouraud, T. (1998). Self-Adaptive Control of a Piston Wave-Absorber, Proc. 8th Int. Offshore Polar Eng. Conf., Montréal, ISOPE, Vol. 1, pp. 127–133.

Chau, F.P. and Eatock Taylor, R. (1988). Second-Order Velocity Potential for Arbitrary Bodies in Waves, Proc. 3rd Int. Workshop Water Waves Floating Bodies.

Clément, A.H. (1991). The Diffraction of a Solitary Wave by a Free-Surface Piercing Cylinder, Proc. 6th Int. Workshop Water Waves Floating Bodies, Woods Hole, pp. 37–41.

Clément, A.H. and Maisondieu, C. (1993). Comparison of Time Domain Control Laws for a Piston Wave Absorber, Proc. Eur. Wave Energy Symp., pp. 117–122, Edinburgh, Elliot & Caratti Eds, NEL Publishers, Scotland.

Clément, A.H. and Mas, S. (1995). Hydrodynamic Forces Induced by Solitary Wave on a Submerged Circular Cylinder, Proc. 5th Int. Offshore Polar Eng. Conf., The Hague, ISOPE, Vol. 3, pp. 339–347.

Clément, A.H. and Gil, L. (1997). Numerical Simulation of Short Wave-Wave Interaction, Proc. 7th Int. Offshore Polar Eng. Conf., Honolulu, ISOPE, Vol. 3, pp. 92–97.

Clément, A.H.(1999a). The Spinning Dipole: An Efficient Unsymmetrical Numerical Wavemaker, 14th Int. Workshop Water Waves Floating Bodies, Port Huron, Michigan.

Clément A.H. (1999b). Benchmark Test Cases for Numerical Wave Absorption: 1st Workshop ISOPE Numerical Wave Tank Group, Montreal, Proc. 9th Int. Offshore Polar Eng. Conf., Brest, Vol. 3, pp. 266–296.

Cointe, R. (1989). Nonlinear Simulation of Transient Free Surface Flows, Proc. 5th Int. Conf. Num. Ship Hydrodynamics, pp. 239–250.

Cointe, R., Geyer, P., King, B., Molin, B. and Tramoni, M. (1990). Nonlinear and Linear Motions of a Rectangular Barge in a Perfect Fluid, Proc. 18th Symp. Naval Hydrodynamics, pp. 85–99.

Cointe, R. and Boudet, L. (1991). Nonlinear and Breaking Waves in Bichromatic Wave-Trains: Experiments and Numerical Simulations, Proc. 1st Int. Offshore Polar Eng. Conf., ISOPE, Vol. 3, pp. 517–522.

Contento, G. and Casole, S. (1995). On the Generation and Propagation of Waves in 2-D Numerical Wave Tanks, Proc. 5th Int. Offshore Polar Eng. Conf., The Hague, ISOPE, Vol. 3, pp. 10–18.

Contento, G. (1996). Nonlinear Phenomena in the Motions of Unrestrained Bodies in a Numerical Wave Tank, Proc. 6th Int. Offshore Polar Eng. Conf., Los Angeles, ISOPE, Vol. 3, pp. 18–22.

Cummins, W.E. (1973). Pathologies of the Transfer Functions, The Seakeeping Symposium Commemorating the 20th Anniversary of the St. Denis-Pierson Paper.

Davies, D.B., Leveretti, S.J. and Spillane, M.W. (1994). Ringing Response of TLP and GBS Platforms, Proc. BOSS, 94, pp. 569–585.

Dold, J.W. and Peregrine, D.H. (1984). Steep Unsteady Water Waves: An Efficient Computational Scheme, Proc. 19th Int. Coastal Eng. Conf., Houston, TX, Vol. 1, pp. 955–967.

Dold, J.W. and Peregrine, D.H. (1986). Water Wave Modulation, Proc. 20th Int. Coastal Eng. Conf., Vol. 1, pp. 163–175.

Ferrant, P. (1994). Radiation and Diffraction of Nonlinear Waves in Three Dimensions, Proc. Int. Conf. Behavior Offshore Structures, MIT, Cambridge, Mass.

Ferrant, P. (1995). Time Domain Computation of Nonlinear Diffraction Loads Upon Three Dimensional Floating Bodies, Proc. 5th Int. Offshore Polar Eng. Conf., The Hague, ISOPE, Vol. 3, pp. 280–288.

Ferrant, P. (1998a). Run-up on a Cylinder Due to Waves and Currents: Potential Flow Solution with Fully Nonlinear Boundary Conditions, Proc. 8th Int. Offshore Polar Eng. Conf., Montreal, ISOPE, Vol. 3, pp. 332–339.

Ferrant, P. (1998b). Interaction of 2nd-Order Wave Packets with a Vertical Cylinder, Proc. 8th Int. Offshore Polar Eng. Conf., Montreal, Vol. 3, pp. 340–347.

Funke, E.R. and Mansard, E.P.D. (1982). The Control of Wave Asymmetries in Random Waves, Proc. 18th Int. Conf. Coastal Eng., Cape Town, South Africa, pp. 725–744.

Grilli, S.T. and Horrillo J. (1998). Periodic Wave Shoaling Over Barred-Beaches in a Fully Nonlinear Numerical Wave Tank, Proc. 8th Int. Offshore Polar Eng. Conf., Montreal, ISOPE, Vol. 3, pp. 294–300.

Grilli, S. and Svendsen, I.A. (1989). The Modeling of Highly Nonlinear Waves: Some Improvements to the Numerical Wave Tank, Proc. 11th Conf. Bound Elem., Cambridge, Mass.

Halsey,N.D.(1977). Potential Flow Analysis of Multiple Bodies Using Conformal Mapping, Ph.D. Thesis, California State University.

De Haas, P.C.A., Berkvens, P.J.F., Broeze, J., van Daalen, E.F.G. and Zandbergen, P.J. (1995). Computation of Hydrodynamic Loads on a Bottom-Mounted Surface Piercing Cylinder, Proc. 5th Int. Offshore Polar Eng. Conf., The Hague, ISOPE, Vol. 3, pp. 304–307.

Hogben, N. and Standing, R.G. (1975). Experience in Computing Wave Loads on Large Bodies, Proc. Offshore Tech. Conf., Houston, Vol.1, pp. 413–431.

Huseby, M. and Grue, J. (1998). An Experimental Investigation of Higher Harmonic Forces on a Vertical Cylinder in Long Waves, Proc. 13th Int. Workshop Water Waves Floating Bodies, Aalphen an den Rijn.

Kim, D.J. and Kim, M.H. (1997). Wave-Current-Body Interaction by a Time-Domain High-Order Boundary Element Method, Proc. 7th Int. Offshore Polar Eng. Conf., Honolulu, ISOPE, Vol. 3, pp. 107–115.

Korsmeyer, T., Phillips, J. and White, J. (1996). A Pre-corrected-FFT Algorithm for Accelerating Surface Wave Problems, Proc. 11th Int. Workshop Water Waves Floating Bodies, Hamburg.

Lin, W.M., Newman, J.N. and Yue, D.K.P. (1984). Nonlinear Forced Motions of Floating Bodies, Proc. 15th Symp. Naval Hydrodynamics, Hamburg, Germany.

Liu, Y.H., Kim, C.H. and Lu, X.S. (1990). Comparison of Higher-Order Boundary Element and Constant Panel Methods for Hydrodynamic Loadings, Proc. 1st Eur. Offshore Mech. Symp, pp. 54–63.

Maisondieu, C. and Clément, A.H. (1993). A Realizable Force Feedback-Feedforward Control Loop for a Piston Wave-Absorber, Proc. 8th Int. Workshop Water Waves Floating Bodies, St John's Newfoundland, pp. 79–82.

Natvig, B.J. (1994). A Proposed Ringing Analysis for Higher Tether Response, Proc. 4th Int. Offshore Polar Eng. Conf., Vol. 1, pp. 40–51.

Nakos, D.E., Kring, D. and Sclavounos, P.D. (1993). Rankine Panel Methods for Transient Free Surface Flows, Proc. 5th Int. Conf. Num. Ship Hydrodynamics, Iowa, pp. 29–48.

Salter. S. (1976). Improvements Relating to Apparatus for Creating Surface Waves in a Body of Liquid, British Patent n°1553206.

Sen, D., Pawlowski, J.S., Lever, J. and Hinchey, M.J. (1989). Two-Dimensional Numerical Modeling of Large Motions of Floating Bodies in Waves, Proc. 5th Int. Conf. Num. Ship Hydrodynamics, Hiroshima, pp. 257–277.

She K., Greated, C.A. and Easson, W.J. (1992). Development of a Two-Dimensional Numerical Wave Tank, Proc. 2nd Int. Offshore Polar Eng. Conf., San Francisco, ISOPE, Vol. 3, pp. 102–109.

Skourup, J. (1994). Evolution and Kinematics of a Modulated Wave Train by Use of the Boundary Element Method., Proc. Int. Symp.: Phys. and Num. Modeling., University of British Columbia, Canada, pp. 871–880.

Skourup, J. and Schoeffer, H.A. (1997). Wave Generation and Active Absorption in a Numerical Wave Flume, Proc. 7th Int. Offshore Polar Eng. Conf., Honolulu, ISOPE, Vol. 3, pp. 85–91.

Skourup, J. and Schoeffer, H.A. (1998). Simulation with a 3-D Active Absorption Method in a Numerical Wave Tank, Proc. 8th Int. Offshore Polar Eng. Conf., Montreal, ISOPE, Vol. 3, pp. 248–255.

Sung, H.G., Hong, S.Y. and Choi, H.S. (1997). A Rational Approach to Free Surface Flow by Using Higher-Order Boundary Elements, Proc. 7th Int. Offshore Polar Eng. Conf., Honolulu, ISOPE, Vol. 3, pp. 56–62.

Tanaka, Y. and Nakamura, T. (1994). Time Domain Simulation Considering Finite Amplitude Slow Drift Oscillation of a Floating Structure, Proc. Int. Symp. Waves Phys. and Num. Modeling., University of British Columbia, Canada.

Tanizawa, K. (1996). Nonlinear Simulation of Floating Body Motions, Proc. 6th Int. Offshore Polar Eng. Conf., Los Angeles, ISOPE, Vol. 3, pp. 414–420.

Tanizawa, K. and Naito, S. (1997). A Study on Parametric Roll Motions by Fully Nonlinear Numerical Wave Tank, Proc. 7th Int. Offshore Polar Eng. Conf., Honolulu, ISOPE, Vol. 3, pp. 69–75.

Tanizawa, K. and Naito, S. (1998). An Application of Fully Nonlinear Numerical Wave Tank to the Study on Chaotic Roll Motions, Proc. 8th Int. Offshore Polar Eng. Conf., Montreal, Vol. 3, pp. 280–287.

Tanizawa, K. and Yue, D.K.P. (1991). Numerical Computation of Plunging Wave Impact Loads on a Vertical Wall, Proc. 6th Int. Workshop Floating Bodies, Woods Hole, MA, pp. 223–226.

Tanizawa, K. and Yue, D.K.P. (1991). Numerical Computation of Plunging Wave Impact Loads on a Vertical Wall — Part 2. The Air Pocket, Proc. 7th Int. Workshop Floating Bodies, Val de Reuil, pp. 189–192.

Vinje, T. and Brevig, P. (1981a). Numerical Simulation of Breaking Wave, Proc. 3rd Int. Conf. Finite Elements Water Resources, University of Mississippi, Oxford, MS, Vol. 5, pp. 196–210.

Vinje, T. and Brevig, P. (1981b). Numerical Calculations of Forces from Breaking Waves, Proc Int. Symp. on Hydrodynamics Ocean Eng., pp. 547–566.

Vinje, T. and Brevig, P. (1981c). Nonlinear Ship Motions, Proc. 3rd Int. Conf. Num. Ship Hydrodynamics Ocean Eng., pp. 257–268.

Wang, P. Yao, Y. and Tulin, M.T. (1993). Wave Group Evolution, Wave Deformation, and Breaking: Simulations Using LONGTANK, a Numerical Wave Tank, Proc. 3rd Int. Offshore Polar Eng. Conf., Singapore, ISOPE, Vol. 3, pp. 27–34.

Wu, G.X. and Eatock Taylor, R. (1996). Transient Motion of a floating Body in Steep Water Waves, Proc. 11th Int. Workshop Water Waves Floating Bodies, Hamburg.

Xü, H. and Yue, D.K.P. (1992). A Numerical Study of Kinematics of Nonlinear Water Waves in Three Dimensions, Proc. Int. Conf. Civil Eng. Oceans., pp. 81–98.

Yang, C., Liu, Y.Z. and Takagi, N. (1989). Time-Domain Calculation of the Non-Linear Hydrodynamics of Wave-Body Interaction, Proc. 5th Int. Conf. Num. Ship Hydrodynamics, pp. 341–350.

Zou, J. and Kim, C.H. (1996). A Note on Preconditioned GMRES Solver, Proc. 6th Int. Offshore Polar Eng. Conf., Los Angeles, ISOPE, Vol. 3, pp. 44–49.

Zou, J. and Kim, C.H. (2000). Generation of Strongly Asymmetric Wave in Random Seaway, Proc. 11th Int. Offshore Polar Eng. Conf., Vol. 3, pp. 95–102.

Index